Environmental Chemistry for a Sustainable World

Volume 11

Other Publications by the Editors

Books
Environmental Chemistry
http://www.springer.com/978-3-540-22860-8

Organic Contaminants in Riverine and Groundwater Systems
http://www.springer.com/978-3-540-31169-0

Sustainable Agriculture
Volume 1: http://www.springer.com/978-90-481-2665-1
Volume 2: http://www.springer.com/978-94-007-0393-3

Book series
Environmental Chemistry for a Sustainable World
http://www.springer.com/series/11480

Sustainable Agriculture Reviews
http://www.springer.com/series/8380

Journals
Environmental Chemistry Letters
http://www.springer.com/10311

Agronomy for Sustainable Development
http://www.springer.com/13593

More information about this series at http://www.springer.com/series/11480

K M Gothandam • Shivendu Ranjan
Nandita Dasgupta • Chidambaram Ramalingam
Eric Lichtfouse
Editors

Nanotechnology, Food Security and Water Treatment

Springer

Editors
K M Gothandam
School of Bio Sciences and Technology
VIT University
Vellore, Tamil Nadu, India

Nandita Dasgupta
Computational Modelling and
Nanoscale Processing Unit
Indian Institute of Food Processing
Technology
Thanjavur, Tamil Nadu, India

Eric Lichtfouse
CEREGE, Aix-Marseille University
Aix en Provence, France

Shivendu Ranjan
Computational Modelling and
Nanoscale Processing Unit
Indian Institute of Food Processing
Technology
Thanjavur, Tamil Nadu, India

Chidambaram Ramalingam
School of Bio Sciences and Technology
VIT University
Vellore, Tamil Nadu, India

ISSN 2213-7114 ISSN 2213-7122 (electronic)
Environmental Chemistry for a Sustainable World
ISBN 978-3-319-70165-3 ISBN 978-3-319-70166-0 (eBook)
https://doi.org/10.1007/978-3-319-70166-0

Library of Congress Control Number: 2017960815

Printed on acid-free paper

This Springer imprint is published by Springer Nature
The registered company is Springer International Publishing AG
The registered company address is: Gewerbestrasse 11, 6330 Cham, Switzerland

Dedicated to all real sufferers for the lack of a clean environment

Preface

Food security and pollution are global issues that will get bigger due to the increasing population, industrialisation and climate change. One-third of food produced for human consumption is lost or wasted globally, which amounts to about 1.3 billion tons per year, according to the Food and Agriculture Organization. There is therefore a need for advanced technology to save food and clean the environment. This book reviews advanced nanotechnology in food, health, water and agriculture. In food, nanobiosensors display an unprecedented efficiency for the detection of allergens, genetically modified organisms and pathogens, as explained in Chaps. 1, 2 and 3 (Fig. 1). In agriculture, nanofertilisers improve plant nutrition by releasing nutrients slowly and steadily (Chap. 4). Chapter 5 reviews the toxicological impact of carbon nanomaterials on plants, whereas Chap. 10 presents a modelling method to predict the toxicity of pollutants. Classical and advanced methods for water desalinisation are then described in Chap. 6. Bioremediation and nanoremediation of waters and metals are reviewed in Chaps. 7, 8 and 9.

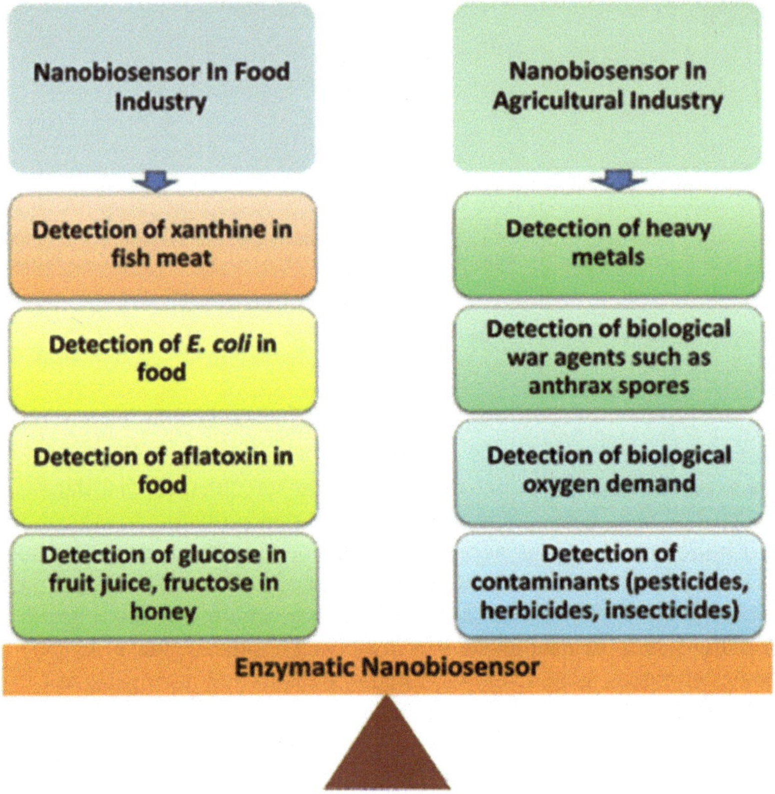

Fig. 1 Nanobiosensor, a unique combination of high-order enzyme specificity and quantum property of nanomaterial, provides many applications in agri-food industry by rapid and ultrasensitive detection of various contaminants (Verma, 2017; Env Chem Lett, doi:10.1007/s10311-017-0640-4)

Vellore, Tamil Nadu, India	K M Gothandam
Thanjavur, Tamil Nadu, India	Shivendu Ranjan
Thanjavur, Tamil Nadu, India	Nandita Dasgupta
Vellore, Tamil Nadu, India	Chidambaram Ramalingam
Aix en Provence, France	Eric Lichtfouse

Contents

About the Editors and Contributors

Editors

Dr. K M Gothandam is currently working as dean and professor at the School of Bio Sciences and Technology, VIT University, Vellore, Tamil Nadu, India. His main area of research is plant biotechnology and environmental biotechnology. He has published many scientific research, reviewed articles in international peer-reviewed journals and also refereed for many journals of high-impact factor.

http://orcid.org/0000-0003-1576-5324

Shivendu Ranjan has extensive expertise in Micro/Nanotechnology and is currently working at Indian Institute of Food Processing Technology, Thanjavur, Tamil Nadu, India. He has founded and drafted the concept for the first edition of the "VIT Bio Summit" in 2012, and the same has been continued till date by the university. He has worked in CSIR-CFTRI, Mysuru, India as well as UP Drugs and Pharmaceutical Co. Ltd., India. His research interests are multidisciplinary which include: Micro/Nanobiotechnology, Nano-toxicology, Environmental Nanotechnology, Nanomedicine, and Nanoemulsions. He is the associate editor of Environmental Chemistry Letters – a Springer journal of 3.59 impact factor – and an editorial board member in Biotechnology and Biotechnological Equipment (Taylor and Francis, USA). He is serving as executive editor of a journal in iMed Press, USA, and also serving as editorial board member and referee for reputed international peer-reviewed journals. He has published six edited books and one authored book in Springer, Switzerland and two with CRC Press, USA. He has recently finished his contract of three volumes of book in Elsevier, four volumes in CRC Press and one with Wiley. He has published many scientific articles in international peer-reviewed journals and has authored many book chapters as well as review articles. He has bagged several awards and recognitions from different national as well as international organizations.

Nandita Dasgupta has vast working experience in Micro/Nanoscience and is currently working at Computational Modelling and Nanoscale Processing Unit, Indian Institute of Food Processing Technology, Thanjavur, Tamil Nadu, India. She has exposure of working at the university, research institutes and industries including VIT University, Vellore, Tamil Nadu, India; CSIR-Central Food Technological Research Institute, Mysore, India; and Uttar Pradesh Drugs and Pharmaceutical Co. Ltd., Lucknow, India. Her areas of interest include Micro/Nanomaterial fabrication and its applications in various fields – medicine, food, environment, agriculture biomedical. She has published six edited books and one authored book in Springer, Switzerland and two with CRC Press, USA. She has finished a contract for three book volumes in Elsevier, one volume with Wiley and two book volumes in CRC Press. She has authored many chapters and also published many scientific articles in international peer-reviewed journals. She has received the Certificate for "Outstanding Contribution" in Reviewing from Elsevier, Netherlands. She has also been nominated for the advisory panel for Elsevier Inc., Netherlands. She is the associate editor of Environmental Chemistry Letters – a Springer journal of 3.59 impact factor – and also serving as editorial board member and referee for reputed international peer-reviewed journals. She has received several awards and recognitions from different national and international organizations.

Chidambaram Ramalingam is currently working as senior professor at the School of Bio Sciences and Technology, VIT University, Vellore, Tamil Nadu, India. He was former dean of the School of Bio Sciences and Technology, VIT University, Vellore, and also was associate dean of academic research at VIT University. Before coming to academia, he has more than 15 years of experience in manufacturing and R&D units of national and multinational food industries.

His areas of research include food process technology and bioremediation. He has co-authored many book chapters. He has published many scientific articles in international peer-reviewed journals.

Eric Lichtfouse is a soil scientist at the French National Institute for Agricultural Research (INRA), CEREGE, Aix-en-Provence, France. He has invented the ^{13}C-dating method allowing to measure the dynamics of soil organic molecules, thus opening the field of molecular-level investigations of soil carbon sequestration. He is chief editor and founder of the journal *Environmental Chemistry Letters* and the book series Sustainable Agriculture Reviews. He is lecturing scientific writing and communication in universities worldwide. His publication assistance service at the INRA has founded the newsletter Publier La Science. He has published the book *Scientific Writing for Impact Factor Journals*. This textbook describes in particular the micro-article, a new tool to identify the novelty of experimental results. Further details are available on SlideShare, LinkedIn, ResearchGate, ResearcherID and ORCID.

Contributors

Neama Abdalla Plant Biotechnology Department, Genetic Engineering Division, National Research Center, Giza, Egypt

Tarek Alshaal Soil and Water Deparment, Faculty of Agriculture, Kafrelsheikh Uni., Kafr El-Sheikh, Egypt

Megahed Amer Soils, Water and Environment Research Institute (SWERI), Agricultural Research Center, Giza, Egypt

Kavita Arora Advanced Instrumentation Research Facility, Jawaharlal Nehru University, New Delhi, India

Balamurali MM Department of Chemistry, School of Advanced Sciences, VIT University, Chennai, Tamil Nadu, India

Yousry Bayoumi Horticulture Department, Faculty of Agriculture, Kafrelsheikh Uni., Kafr El-Sheikh, Egypt

Brindha J Department of Chemistry, School of Advanced Sciences, VIT University, Chennai, Tamil Nadu, India

Kaushik Chanda Department of Chemistry, School of Advanced Sciences, VIT University, Vellore, Tamil Nadu, India

Anuj N. Chaudhari Department of Biotechnology, School of Life Sciences, North Maharashtra University, Jalgaon, Maharashtra, India

Shiao-Shing Chen Institute of Environmental Engineering and Management, National Taipei University of Technology, Taipei, Taiwan

Anamika Das School of Bio Sciences and Technology, VIT University, Vellore, Tamil Nadu, India

Nandita Dasgupta Computational Modelling and Nanoscale Processing Unit, Indian Institute of Food Processing Technology, Thanjavur, Tamil Nadu, India

Éva Domokos-Szabolcsy Plant Biotechnology Department, Debrecen Uni., Debrecen, Hungary

Tamer Elbana Soils and Water Use Department, National Research Centre (NRC), Cairo, Egypt

Ahmed El-Henawy Soil and Water Deparment, Faculty of Agriculture, Kafrelsheikh Uni., Kafr El-Sheikh, Egypt

Mohammed Elmahrouk Horticulture Department, Faculty of Agriculture, Kafrelsheikh Uni., Kafr El-Sheikh, Egypt

Hassan El-Ramady Soil and Water Deparment, Faculty of Agriculture, Kafrelsheikh Uni., Kafr El-Sheikh, Egypt

Miklós Fári Plant Biotechnology Department, Debrecen Uni., Debrecen, Hungary

Prakash M. Gopalakrishnan Nair Department of Applied Bioscience, College of Life and Environmental Sciences, Konkuk University, Seoul, South Korea

K M Gothandam School of Bio Sciences and Technology, VIT University, Vellore, Tamil Nadu, India

Silvia Haneklaus Institute of Crop and Soil Science (JKI), Federal Research Centre for Cultivated Plants, Braunschweig, Germany

Arun G. Ingale Department of Biotechnology, School of Life Sciences, North Maharashtra University, Jalgaon, Maharashtra, India

Piet N. L. Lens UNESCO-IHE Institute for Water Education, Delft, The Netherlands

Eric Lichtfouse CEREGE, Aix-Marseille University, Aix en Provence, France

Yarlagadda V. Nancharaiah UNESCO-IHE Institute for Water Education, Delft, The Netherlands

Nguyen Cong Nguyen Faculty of Environment and Natural Resources, Dalat University, Dalat, Vietnam

Hau-Thi Nguyen Faculty of Environment and Natural Resources, Dalat University, Dalat, Vietnam

Jabez William Osborne School of Bio Sciences and Technology, VIT University, Vellore, Tamil Nadu, India

Marinus Pilon Department of Biology, Colorado State University, Fort Collins, CO, USA

Elizabeth A. H. Pilon-Smits Department of Biology, Colorado State University, Fort Collins, CO, USA

József Prokisch Institute of Bio- and Environmental Enegetics, Debrecen Uni., Debrecen, Hungary

Chidambaram Ramalingam School of Bio Sciences and Technology, VIT University, Vellore, Tamil Nadu, India

Shivendu Ranjan Computational Modelling and Nanoscale Processing Unit, Indian Institute of Food Processing Technology, Thanjavur, Tamil Nadu, India

Saikat Sinha Ray Institute of Environmental Engineering and Management, National Taipei University of Technology, Taipei, Taiwan

Eldon R. Rene UNESCO-IHE Institute for Water Education, Delft, The Netherlands

Dhanaraj Sangeetha Department of Chemistry, Vellore Institute of Technology, Vellore, Tamil Nadu, India

Raghunath Satpathy Department of Biotechnology, MITS Engineering College, Rayagada, Odisha, India

Ewald Schnug Institute of Crop and Soil Science (JKI), Federal Research Centre for Cultivated Plants, Braunschweig, Germany

Dirk Selmar Applied Plant Science Department, Institute for Plant Biology, TU Braunschweig, Braunschweig, Germany

Manisha Shakya UNESCO-IHE Institute for Water Education, Delft, The Netherlands

Tarek Shalaby Horticulture Department, Faculty of Agriculture, Kafrelsheikh Uni., Kafr El-Sheikh, Egypt

College of Agricultural and Food Sciences, King Faisal University, Al-Hassa, Saudi Arabia

Said Shehata Vegetable crops Department, Faculty of Agriculture, Cairo University, Giza, Egypt

Attila Sztrik Institute of Bio- and Environmental Enegetics, Debrecen Uni., Debrecen, Hungary

Mohamed Yousry Soils and Water Use Department, National Research Centre (NRC), Cairo, Egypt

Chapter 1
Advances in Nano Based Biosensors for Food and Agriculture

Kavita Arora

Abstract Nanotechnology is revolutionizing development in almost all technological sectors, with applications in building materials, electronics, cosmetics, pharmaceuticals, food processing, food quality control and medicine. In particular, nano-based sensors use nanomaterials either as sensing material directly or as associated materials to detect specific molecular interactions occurring at the nano scale. Nano biosensors are used for clinical diagnostics, environmental monitoring, food and quality control. Nano biosensors can achieve *on site, in situ* and *online* measurements.

This chapter reviews nanobiosensors and nanosensors, and their applications to food and agriculture. Nanosensors exhibit an unprecedented level of performance and the ability to 'nano-tune' various properties to achieve the desired levels of sensitivity and detection limit. Nanobiosensors are used for the monitoring of food additives, toxins and mycotoxins, microbial contamination, food allergens, nutritional constituents, pesticides, environmental parameters, plant diseases, and genetically modified organisms. Applications include: a nano-diagnostic briefcase kit for *in situ* crop investigation; a dip stick nanosensor kit '4-my-co-sensor' for multi-analyte detection; a barcode assay for genetically modified organisms (GMO) using Surface Enhanced Raman Spectroscopy (SERS); and a mobile barcode enzymatic assay.

Keywords Nanoparticles • Nanobiosensors • Nanosensors • Food • Agriculture • Environmental monitoring • GMOs

Contents

K. Arora (✉)
Advanced Instrumentation Research Facility, Jawaharlal Nehru University, New Delhi, India
e-mail: kavitaa@gmail.com; kavitaarora@mail.jnu.ac.in

K M Gothandam et al. (eds.), *Nanotechnology, Food Security and Water Treatment*, Environmental Chemistry for a Sustainable World 11,
https://doi.org/10.1007/978-3-319-70166-0_1

1.1 Introduction

Nano-based biosensors and nanosensors are sensors designed to sense parameters of interest either by measuring chemical, physical, biological 'signals or interactions' at nano scale or by making use of nanomaterials for measuring desired parameters in specific application range. Applications of sensors and biosensors can be traced all around us, from our bathroom, kitchen, laundry through clinical diagnostics, environmental monitoring, safety alarms to industrial process etc. to almost every technology that involves measurement of some parameter. This becomes very important to understand basics of sensor and biosensor before understanding a nano based biosensors (Dasgupta et al. 2015, 2017; Shukla et al. 2017; Jain et al. 2016; Ranjan et al. 2014).

A typical sensor is a device, which detects or measures a physical property and then responds, records and indicates the measured phenomena into understandable form by observer or an instrument. It consists of three parts viz. sensor, transducer, detector and coupled to output display device as shown in Fig. 1.1. This device responds to electrical or optical or mechanical signal and converts that physical parameter with the help transducer to be detected into a signal output. Physical parameter can be temperature, blood pressure; humidity etc. Simplest example of sensor is thermometer that has mercury that expands when temperature increases, which is measured through visual movement of the mercury at a calibrated scale of 1 atmosphere pressure. In order to be a good sensor, it must have accuracy, specificity, ability to measure in the desired analyte range along with easy calibration, good resolution, reusability and low cost.

A Biosensor is a self-contained analytical device that incorporates a biologically active material in intimate contact with an appropriate transducer to qualitatively or quantitatively sense chemical or biochemical phenomena occurring at sensor surface. It converts a biological recognition response into an electrical signal (Arnold 1985) which is further processed to be represented as output display. The schematic arrangement of a typical biosensor is shown in Fig. 1.2. It consists of three primary components: bio-receptor, transducer and amplifier coupled to display output.

A biosensor may use biomolecule as a bio-receptor component such as tissue, microorganisms, organelles, cell receptors, enzymes, antibodies, nucleic acids, etc. interfaced to a desired transducer component (Chaubey and Malhotra 2002). Signals generated due to biomolecular interaction can be electrical, electrochemical,

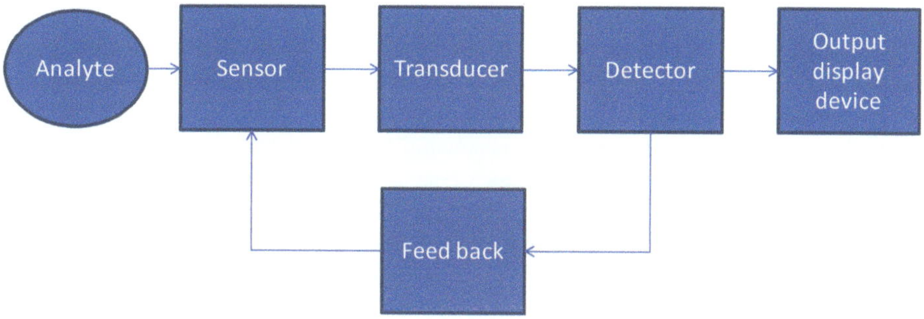

Fig. 1.1 A typical sensor consisting of sensor, transducer and detector connected to output display unit to collectively sense process and display change in parameter/analyte of interest

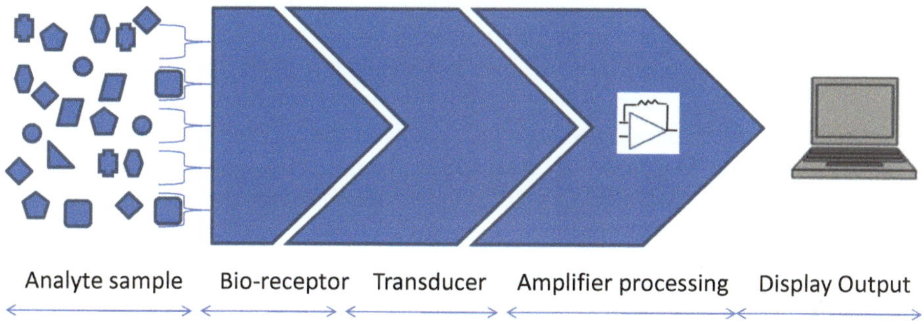

Fig. 1.2 A simple biosensor consisting of biomolecule coupled or linked to substrate/sensor surface in close contact with transducer-amplifier and display unit for signal to be expressed in user-desired scale/ units of measurements

physicochemical, optical, piezoelectric or thermal, which is converted into electrical signal *via* desired transducer that is easily measured, quantified, amplified and processed to associated electronics for display as output in user friendly form or desired units/scale of measurement (Gerard et al. 2002, Arora et al. 2006a, b). A variety of signals can be generated from the different types of biomolecular interactions which can be measured and processed using different types of transducers such as potentiometric, amperometric, voltammetric, surface conductivity, electrolyte conductivity, fluorescence, colorimetric measurements, absorption, reflection, surface plasmon resonance, resonance frequency of peizocrystals, heat of reaction, heat of absorption etc.

Nanosensors are basically chemical sensors, which help in detection of presence of chemical species or monitor various parameters through use of nanomaterials / nanostructures that may or may not lie at nano-scale. These may include electronic nose, miniaturized point of care devices, silicon computer chips, nano robots etc. that are urbanized to operate at nanoscale and give extraordinary sensation aptitude at cellular or molecular lever. Their vocation is by scheming and quantifying ups downs and adapts dislodgment, dislocations, concentration, volume, acceleration, external forces pressure or temperature. Henceforth, nano based biosensors are set of sensing devices that make use chemical or physical or mechanical or biological phenomena to measure change in parameters (biological/nonbiological) of interest at nano-scale and may make use of nanostructures or materials as integral part through use of biological molecules as sensing (recognition) material.

Use of nanotechnology in the area of sensing technology has offered wider opportunities to construct sensors to provide high product competence that has influenced all areas including home, communication, transportation, medicine, agriculture, and industry. Nanomaterials are materials with structure at the nano scale that have unique optical, electronic, physical or mechanical properties that are absent in the bulk form and can be used for various applications. These unique and bracing features of nanomaterials facilitate opportunities to improvise and enhance the performance characteristics for various sensing applications too. Nano materials can exist in single, fused, aggregated or agglomerated forms with various shapes such as spherical, tubular, and irregular shapes. Depending on structure, composition and configuration nanomaterials can be made from carbon, metals or organic or inorganic materials. Common types of nanomaterials may include nanotubes, dendrimers, quantum dots, nanoparticles, nanowires and fullerenes. Diverse spectrum of anisotropic nanomaterials reported in the literature may include nanorods (Pérez-Juste et al. 2005), nanowires (Chen et al. 2007), nanotubes (Hu et al. 1999), triangles (Jin et al. 2001; Millstone et al. 2005), plates and sheets (Wang et al. 2005), ribbons (Swami et al. 2003), and so on.

As per US National Nanotechnology initiative, nanotechnology has moved from first generation- passive nanostructures (2000-dispersed nanostructured metals, polymers, ceramics, composites) to second generation-active nanostructures (2005- bioactive drugs, biodevices, amplifiers, actuators, transistors etc.) to third generation – systems of nanosystems (2010- guided assemblies, 2D networking, robotics, evolutionary structures etc.) to fourth generation- molecular nanosystems (2015 onwards- by designing molecular devices, emerging functions etc.) to molecular manufacturing. Nano based biosensors developed through nano molecular systems can play a far larger and vital role in healthcare and biomedical industry. Although, nano based implications impend future productivity of counting robotics, transportation, construction, energy storage, food management, environmental monitoring, security, surveillance and military (Touhami 2014). Production processes still holds it back for nanosensor development due to challenges imposed through high cost and technical limitations involved in fabrications to design physical nano based biosensors or nanosensors.

This chapter intends to bring in detailed review some important nano based biosensors and nanosensors while explaining role of nanomaterials towards enhancing various working principles and performance characteristics of the intended devices for various applications towards food and agriculture. Attempts have been made to include various arenas in food and agriculture for measurement of food additives, toxins and mycotoxins, microbial contamination, food allergens, nutritional constituents in food, pesticides, environmental parameters, in food and environment, plant diseases, genetically modified organisms/plants (GMOs), pH etc. reported in past 5 years.

1.2 Nano Based Biosensors and Nanosensors for Food and Agriculture

The requisite objective of any sensor especially a nano based biosensor or a nanosensor is to spot any chemical or biophysical or biochemical indication occurring at lone molecular or cellular levels. As explained earlier, use of nanomaterials offers miniaturization of a sensor dimension to achieve enormous resourcefulness for assimilation into multiplexed, mobile, convenient, wearable, *in situ* and even implantable medical devices. This also incorporates application areas to be limited not only to industrial production processes, environmental monitoring and molecular diagnostic purposes in health care but lot more including food and agriculture. Besides, the dominating biomedical applications and need to achieve point-of-care diagnostics, nano based biosensors and nanosensors appear to be the major step and the panorama impact of these nano-molecular systems for *onsite* or *online* testing remains unrivalled.

Nano based biosensors made from various carbon, metal based nanomaterials and screen printed electrodes generally utilize electrochemical mode of measurement and/or microfluidics based system to achieve simple and compact analytical devices for detection of toxins, various applications in food, agriculture and environmental monitoring (Fig. 1.3a, Reverté et al. 2016; Hughes et al. 2016). Although, several reports do exist on various electrochemical/ acoustics nano based biosensors, till date majority of them are based on optical methods due to feasibility of ease of visual detection. Demchenko 2006, had elaborated on advantages and application of fluorescence probes for probing and sensing for proteins, cells and bio membranes. He explained that two band maxima containing two different dyes can be simultaneously used to demonstrate two different phenomena occurring at nanostructure levels (Fig. 1.3b, Demchenko 2006). This phenomenon made use of the principle of coupling of wavelength shifts with two-band ratiometric response in fluorescence intensities. Different intermolecular interactions resulted in a strongly amplified fluorescence signal, where two fluorescence dyes at ground state are denoted as N and T and two excited species as N* and T* in dynamic equilibrium. For each fluorophore change of intermolecular interactions leads to change of energy separation between ground (N or T) and excited (N* or T*) states, expressed through shifts of their "green" and "red" fluorescence bands. These shifts are common and can be used in fluorescence sensing. Some examples of such dyes include 3-Hydroxychromone dyes, 3-hydroxyquinolones etc.

Food and agricultural analysis may involve: quality check for presence of toxins, microbial/fungal/viral contamination, rotting; food production quality control i.e., control of various parameters like, pH, temperature pKa, sugar/glucose content; or monitoring environmental parameters for qualitative/quantitative analysis of soil, water, fertilizers, pesticides/herbicide etc.to achieve desired level of food and agricultural production. Next sections are categorized to facilitate nano based biosensors reportedly available to achieve for aforesaid objectives.

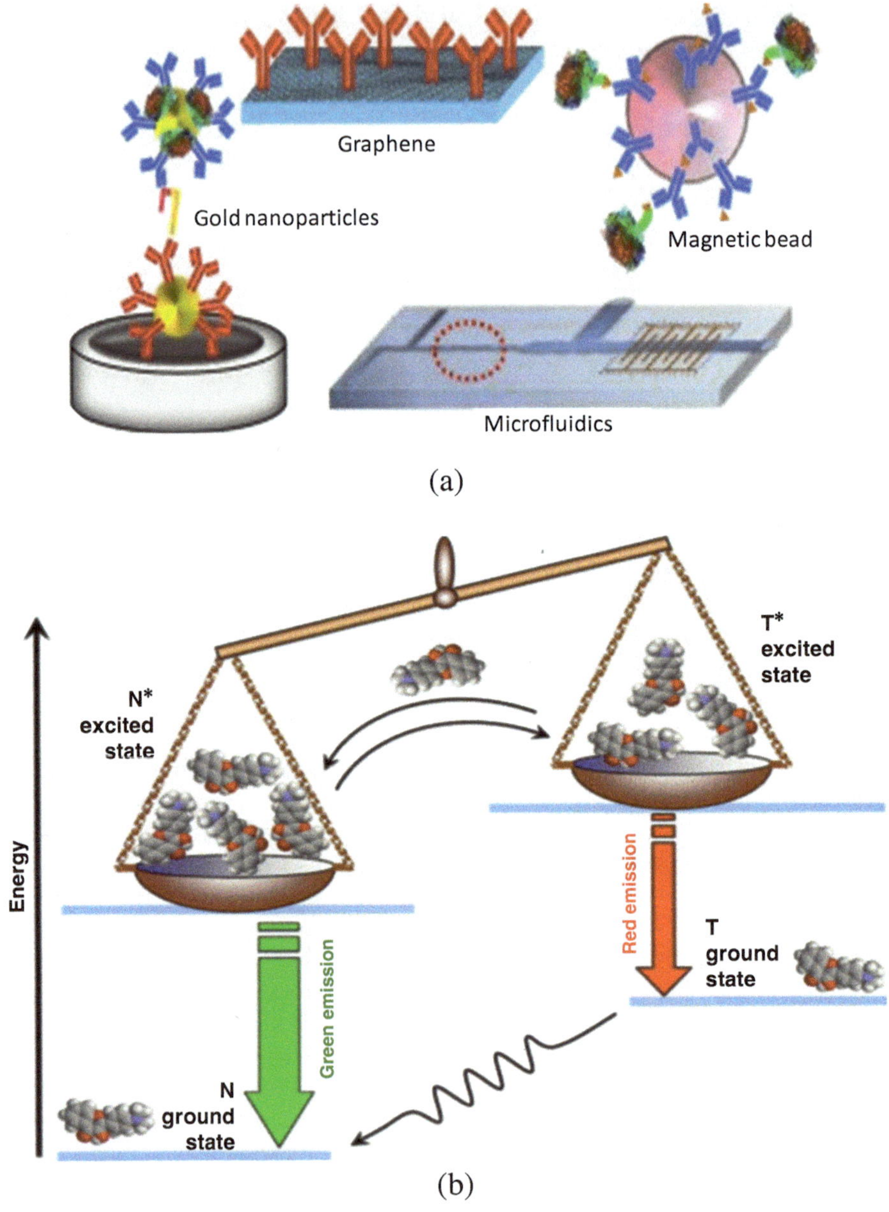

Fig. 1.3 (**a**) Electrochemical nano based biosensors that measure biorecognition event through change in electrochemical properties at receptor/sensors surface (Reprinted from Reverté et al. 2016 with © permission from Elsevier Publishing company), (**b**) Optical nano based biosensors that use fluorescence response of two different fluorophores (N and T) giving two different fluorescence signals (N* to N, T* to T) with change in intermolecular interaction occurring at receptor/sensor surface (Reprinted from Demchenko 2006 with © permission from John Wiley and Sons Publishing Company)

1.2.1 Food Additives

Present day food industry is governed by changing customer interests that has drifted the attention of producers towards the attractive looks, colour, flavor and taste rather than the nutritional values. Intentional and unintentional additives in food have led to significant health problems which points towards the need for food analysis. Food additives may include artificial colours, flavours, texturants, antibiotics, pesticides etc.

Sivasankaran et al. reported a fluorometric nanosensor for detection of blue food colorant Brilliant blue FCF in food samples like sports drink and candies, demonstrating its potential in food analysis (Sivasankaran et al. 2016). They had developed a L-cysteine capped cadmium sulphide quantum dots based nanosensor in a fluorometric quenching assay (Fig. 1.4a) for discriminative detection and determination up to 3.50×10^{-7} M and a linear range of 4.00×10^{-5}–4.50×10^{-6} M Brilliant Blue FCF.

Melamine is an additive, which is often added in dry milk powder, dried egg and protein powders as a food adulterant to increase protein content, which has been shown to have toxic effects for humans. Chondroitin sulfate-reduced gold nanoparticles (using green synthesis) based nanosensor was used to detect melamine by measuring absorbance (surface plasmon resonance band) ratio (A620/A530). This nano based biosensor was reported to have melamine linear range 0.1–10 μM and was used to quantify melamine spiked in real infant formula at concentrations as low as 12.6 ppb (Noh et al. 2013). Wu et al. 2015 have reported combination of upconversion nanoparticles and gold nanoparticles composite based nanosensor for detection of melamine (Fig. 1.4b). As it can be seen that up conversion nanoparticles were prepared from sodium Yttrium fluoride doped with rare earth metals lanthanides (Ytterbium-Yb and Erbium-Er) i.e., $NaYF_4:Yb^{3+},Er^{3+}$ (explained in Sect. 1.2.2.). $NaYF_4:Yb^{3+}$, Er^{3+} possess unique fluorescence properties, that get quenched by associate gold nanoparticles under normal conditions. When melamine is added, gold nanoparticles get released from the surface of up conversion nanoparticles since melamine could cause gold nanoparticles to aggregates by N-Au interaction, resulting fluorescence of up conversion nanoparticles. This easily operatable nanosensor showed linear response to 32.0–500 nM melamine with a detection limit of 18.0 nM at pH (7.0) with 12 min incubation time and sensitivity of 0.968 in raw milk samples.

Formalin/formaldehyde is constituent of many fruits and vegetables at low concentrations, which is known to cause cancer at high dose. This is a commonly used additive to various foods like fish, milk and fruits to facilitate and sustain their shelf life. Nano emraldene-polyaniline based nanosensor was described to detect low concentrations of formaldehyde ranging from 0.0003 to 0.9 ppm in a dose dependent manner (Omara et al. 2016).

Urea is one of the metabolic products of protein metabolism and has a strategic function in the marine nitrogen cycle as a source of excreted nitrogen by invertebrates and fish. Likewise, the bacterial decomposition of nitrogenous materials and terrestrial drainage are influenced by urea. That is why, estimation of urea is very crucial in clinical diagnostics, food science and environmental-monitoring

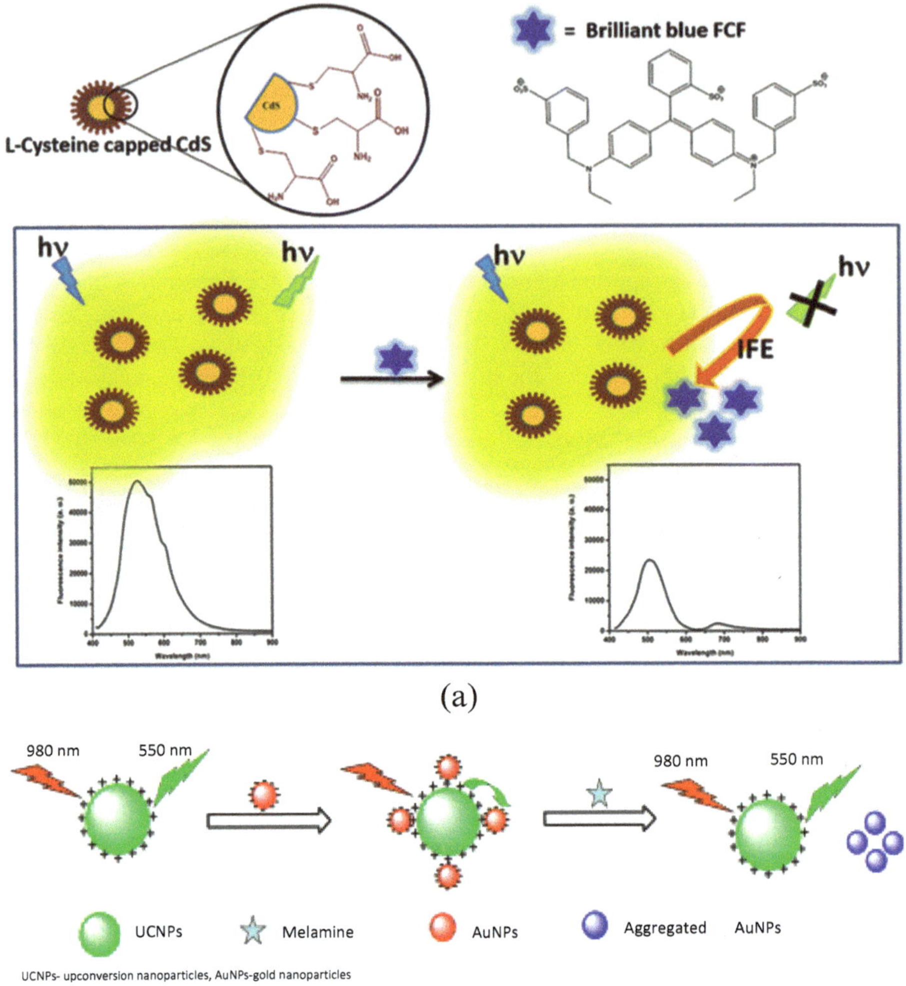

Fig. 1.4 Detection of (**a**) Brilliant Blue FCF using L-cysteine capped Cadmium sulfide (CdS) quantum dots based nanosensor that shows quenching in fluoresce signal upon addition of analyte (Reprinted from Sivasankaran et al. 2016 with © permission from Springer Publishing company) and (**b**) Melamine using up conversion nanoparticles (UCNPs) and gold nanoparticles (AuNPs) via fluorescence resonance energy transfer (FRET) phenomena based fluorescence 'turn on' assay (Reprinted from Wu et al. 2015 with © permission from Elsevier Publishing company)

(Saeedfar et al. 2013). Urea is used as fertilizer too and annual worldwide production of urea exceeds 100 million metric tons where overuse of nitrogen fertilizer application can lead to decrease in soil pH and pest problems (increasing birth rate, longevity, and overall fitness of certain pests etc.). Urease (from *Arthrobacter creatinolyticus*) immobilized membrane (PAN-[poly(acrylonitrile-methylmethacrylate-sodium vinylsulfonate)] membrane) was employed in analysis of urea spiked milk samples that showed detection range of urea concentration from

1 to 100 mM (Ramesh et al. 2015).The immobilized urease had good storage stability for a period of 70 days at 4 °C and could be effectively reused for 13 cycles.

Intentional addition of various antibiotics in food and its products is a usual practice to increase its shelf life throughout the world. Although, repercussions of excessive use of antibiotics has been realized and despite the fact that now there are known adverse affects to human health, very few countries could impose regulations of their uses. Tendency of these compounds to get accumulated, warrants need of easy *onsite/in situ* sensing devices for suspected antibiotics in various food matrices. Danofloxacin is one broad spectrum antibacterial fluoroquinolone compound used for treatment of respiratory diseases in human and veterinary diseases. At higher concentrations, i.e., after accumulation, this may have adverse reactions and can detrimentally affect muscle, central nerve system, peripheral nerve system, liver, and skin. Therefore, prescreening and determination of the level of danofloxacin in foods or food products becomes very important. An surface plasmon resonance based nanosensor was reported that used RNA (ribonucleic acid) aptamers for danofloxacin (Han et al. 2014). The selected specific RNA aptamer were shown to have potential for specific detection of danofloxacin that could be uploaded on sensor systems and was found to be useful as a rapid, selective, and sensitive monitoring/ diagnostic/ detection of ligand for danofloxacin in food animals. In a similar row, a chemiluminescence biosensor based on aptamer functionalized gold nanoparticles for detection of p53, a tumor suppressor protein up to 10 pg/ml and showed 10-fold improvement in p53 detection gold nanoparticles based colorimetric assay (Shwetha et al. 2013). Counting on similar kinds of reports mentioned in this chapter, the potential of aptamers as specific biorecognition elements could substantially enhance the performance of nanobiosensors.

Tetracyclin, is a widely overused antibiotic whose exact and rapid quantification in an aqueous buffer solutions and complex biological samples such as milk is of high importance. An ultra long zinc oxide (ZnO) nano walls based nanobiosensor was developed and demonstrated for real-time electrical measurement of dynamic molecular interactions via monitoring phenomena of binding of the tetracycline repressor (TetR) to its operator DNA (deoxyribonucleic acid) and its inducible release by the addition of tetracycline (Menzel et al. 2013). This exciting method allows ultra-sensitive measurements of tetracycline concentrations as shown in Fig. 1.5a. When tetracycline is added, the induced switching and release causes a down bending of the surface energy bands (E_V – valence band and E_C – conduction bands, E_F – Fermi energy level) due to the reduction of negatively charged molecules. The process is reversed when TetR molecules are attached to the surface again.

Tobramycin, a aminoglycoside is water soluble antibiotic which is utilized to treat the infections caused by aerobic Gram-negative and some Gram-positive microorganisms) and excessive use of this drug may result in ototoxicity and nephrotoxicity. Tobramycin imprinted poly(2-hydroxyethyl methacrylate–methacryloyl amidoglutamic acid) [p(HEMA–MAGA)] molecular imprinted

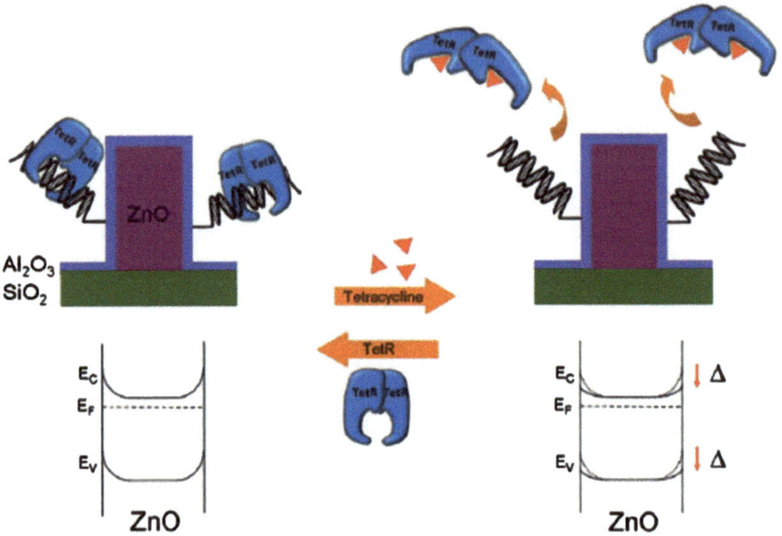

(a)

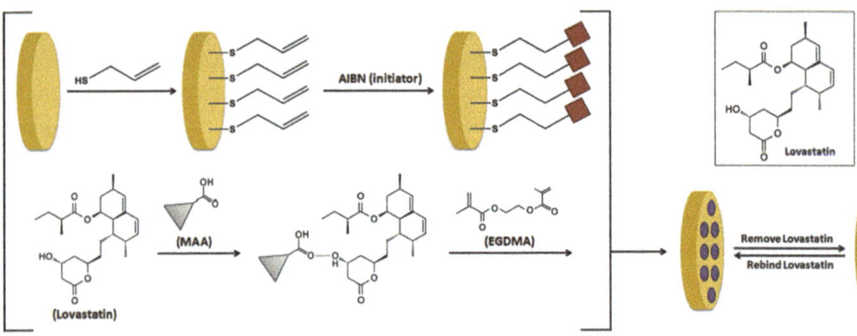

(b)

Fig. 1.5 Detection of (**a**) tetracyclin using zinc oxide/aluminum oxide (ZnO/Al_2O_3) nanowall nanobiosensor (cross section) that uses affinity of tetracyclin with its repressor /operator DNA where binding of tetracyclin results in down bending of surface energy bands (*where TetR-tetracyclin repressor, zinc oxide/aluminum oxide -ZnO/Al_2O_3, SiO_2 – silicon oxide, E_V – valence band and E_C – conduction bands, E_F – Fermi energy level))* (Reprinted from Menzel et al. 2013 with © 2013 permission from Royal Society of Chemistry); (**b**) lovastatin using molecular imprinted polymer (pMAA) – gold – quartz crystal based nanosensor where binding of analyte shall be indicated by directly proportional change in vibrational frequency of quartz crystal *(where EDGMA-ethylene glycol dimethacrylate, AIBN- N,N′-azobis-iso-butyro-nitrile, pMAA- poly2-hydroxy ethyl methacrylate–methacryloyl amido aspartic acid* (Reprinted from Eren et al. 2015 with © 2015 permission from Elsevier Publishing company) and (**c**) Small drug molecule using a plasmonic nanosensor in a sandwich structure through anchored capture antibodies onto substrate and gold nanocluster labeled antibodies where presence of analyte shall facilitate formation of sandwich structure and will favour formation of gold nanoparticles *(where gold nano clusters-AuNCs, gold nanoparticles- GNPs, $HAuCl_4$- auric chloride and H_2O_2- hydrogen peroxide*) (Reprinted from Zhao et al. 2016 with © 2016 permission from ACS publications)

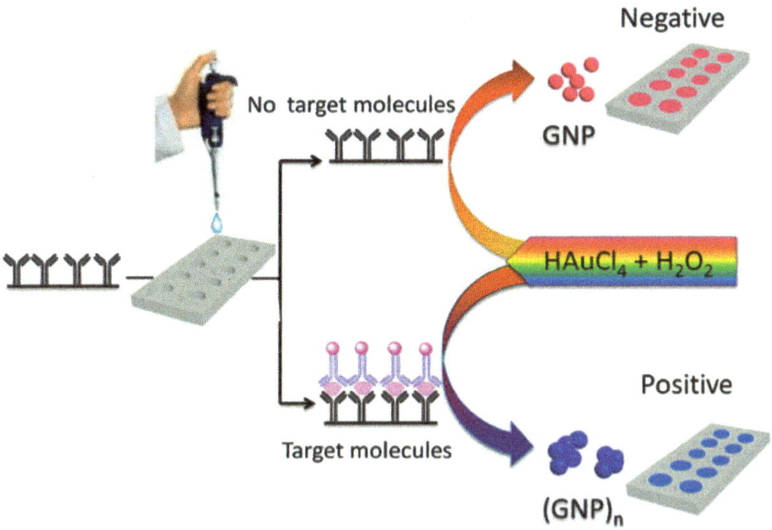

(c)

Fig. 1.5 (continued)

polymer film was generated on the gold surface to prepare a nanosensor for tobramycin (Yola et al. 2014). This nanosensor was described to give linearity range and detection limit of 1.7×10^{-11}–1.5×10^{-10} M and 5.7×10^{-12} M, respectively for pharmaceuticals, and food samples like chicken egg white and milk extract.

Lovastatin is a member of the class of statins, which are produced through fermentation process and are used to lower the cholesterol content in hypercholesterolemia. Red yeast rice is a dietary supplement in south Asia and this, being fermentation product grown on rice, contain lovastatin drug residue. Increased use of this food supplement is causing cardiovascular diseases and posing serious risk of the over release of lovastatin drug residue to the environment that may cause increased incidences of coronary artery disease, muscle and liver damage. Therefore, a simple, sensitive and quick molecular imprinted gold quartz crystal microbalance chip based nanosensor (Fig. 1.5b) was developed to detect lovastatin in natural samples (Eren et al. 2015). Lovastatin imprinted poly(2-hydroxyethyl methacrylate–methacryloyl amido aspartic acid) [p(HEMA–MAAsp)] nano film was attached on the mercapto propane based self assembled monolayer deposited gold surface of quartz crystal microbalance chip. The fabricated specific nanosensor gave linear performance for lovastatin at 0.10–1.25 nM and detection limit of 0.030 nM in red yeast rice.

A plasmonic nanosensor using gold nanoclusters was fabricated to enable visually quantitative determination of ultra-trace target molecules like synthetic small molecules or drugs. This method combines enzyme-mimetic gold nanoclusters assisted visual color change exhibited by gold nanoparticles in visible range in presence of desired analyte (Fig. 1.5c) (Zhao et al. 2016). In this sensor, a target analyte can be captured by its antibody anchored on a solid surface and further covered by a layer of same antibody tagged with enzyme mimetic gold nano-clusters. Now, the formation of sandwich structure shall favor the formation of gold nanoparticles when immersed in into a solution of $HAuCl_4$ and H_2O_2, thereby leading to visual colour change. This system was demonstrated for protein avidin, cancer antigen 15-3 (a breast cancer biomarker shortened as CA15-3), 3,5,3′ -L –tri-iodo thyronine thyroid hormone (T3), and even synthetic small molecular drug such as methamphetamine. This systme possess potential to be utilized for its applications in analytical requirements of food and agriculture.

Toxic metal content in food, pharmaceutical industry and clinical diagnostics is one of the area of concern, therefore, monitoring trace levels is desired for various applications (Maddinedi et al. 2015, 2017; Tammina et al. 2017; Siripireddy et al. 2017; Sannapaneni et al. 2016). Cu^{2+} ions are among frequently monitored species, especially where strict purity guidelines are implemented e.g., medical industry, pharmaceutical applications, dialysis water, microelectronics and manufacturing of integrated circuit semiconductor chips etc. Kacmaz et al. 2015 reported a nanosensor based on fluoroionophore DMK7 or 2-{[(2-aminophenyl)imino]methyl}-4,6-di-tert-butylphenol doped nano-fibrous (polymeric ethyl cellulose) films to detect ultra-low concentrations of Cu ions giving detection limit of 3.3×10^{-13} M and detection range of 5.0×10^{-12}–5.0×10^{-5} M (Fig. 1.6a). Additionally, this extremely specific nanosensor exhibited high selectivity over convenient cations like Na^+, K^+, Ca^{2+}, Mg^{2+}, NH_4^+ and Ag^+, Al^{3+}, Ba^{2+}, Co^{2+}, Cr^{3+}, Fe^{3+}, Fe^{2+}, Hg^{2+}, Li^+, Mn^{2+}, Ni^{2+}, Pb^{2+},Sn^{2+} and Zn^{2+}.

Selenium is known as an essential nutrient responsible for immunity and anti-oxidant activity. Its deficiency and excess intake causes both have been reported to be unsafe for human health, therefore, the accurate detection of trace amounts of Se has great significance on environmental, medical and nutritional sciences. A ratiometric fluorescent nanosensor for accurate and *on-site* sensing of SeO_3^{2-} by linking the recognition molecule 3,3′-diaminobenzidine onto the surface of carboxyl group modified cadmium telluride embedded silica nanospheres or quantum dots that were was explained to have single fluorescence peak at 655 nm (Chen et al. 2016a, b). Addition of SeO_3^{2-}onto nanosensor results in two emissions peaks (530 and 655 nm) of Se-diaminobenzidine and Se-cadmium telluride embedded silica nanosphere quantum dots under a single excitation wavelength as shown in Fig. 1.6b. This nanosensor presented detection range 0–2.5 μM and detection limit of 6.68 nM (0.53 ppb) of selenium ions. No interference to the performance of nanosensor was observed for other common anion ions and some amino acids, such as NO^{2-}, CO_3^{2-}, SO_3^{2-}, SO_4^{2-}, S^{2-}, HS^-, HSO_3^-, ClO^-,HPO_4^{2-}, $H_2PO_4^-$ Br^-, NO_3^-, H_2O_2, GHS and Cys under the same experimental conditions. Nano sensor was tested on real water samples spiked with different amounts of Se(IV) and in food samples like rice, lettuce and radish.

Mercury is the most toxic water soluble elements known in ecosystems which is non-biodegradable and can only get absorbed through plants and water resources to be subsequently accumulated in food chain. Monitoring Hg^{2+} level in environmental, food and biological samples is an important issue to understand its distribution and potential pollution. A dual emission fluorescent probe nanosensor for Hg^{2+} detection was developed by Tan et al. 2015, that used lanthanide combination of green emitting terbium (Tb^{3+}) embedded and red emitting europium (Eu^{3+}) covalently tagged SiO_2 nanoparticles. In dual-emission fluorescent probe, one fluorophore functions as reference unit and another as response moiety to ensure naked eye distinction and accuracy in quantification. As shown in Fig. 1.6c, two lanthanide (Tb^{3+} and Eu^{3+}) chelates were synthesized by the chemical coordination dipicolinic acid (2.6-pyridinedicarboxylic acid) denoted as Tb- dipicolinic acid chelate and Eu- dipicolinic acid chelate) and the surface of SiO_2 nanoparticle doped with Tb- dipicolinic acid chelate was functionalized by diethylene tri-amine penta acetic acid to immobilize Eu- dipicolinic acid chelate at periphery. Diethylene tri-amine penta acetic acid as functional ligand offers its carboxyl groups to coordinate with Eu- dipicolinic acid chelate to form 'Diethylene tri-amine penta acetic acid-Eu-dipicolinic acid' ternary complex on surface of the SiO_2 nanoparticle and also assist dipicolinic acid to offer selective response to Hg^{2+}. Since Hg^{2+} has higher binding constant ($K = 10^{26.4}$) compared to Eu^{3+} ($K = 10^{22.39}$) the binding of Hg^{2+} is favored to enhance its detection. Upon addition of Hg^{2+} onto nanosensor, the fluorescence of Eu^{3+} chelates gets selectively quenched, while the fluorescence of Tb^{3+} chelates remained unchanged (Fig. 1.6c) and this nanosensor gave excellent selectivity and high sensitivity up to 7.07 nM detection limit in drinking water and milk samples.

Bisphenol analogs or popularly known as BPAs are compounds, which are ubiquitously involved in our daily commodities and for this reason this has become a part of our food ingredients due to unintentional leaching from all around. BPA is known as ubiquitous endocrine disrupter and considering its serious adverse human health risks; its use has been banned in many countries. Since, tyrosinase being ortho-hydroxylation oxidase can oxidize BPA to corresponding o -diphenols and o -quinones (Ragavan et al. 2013), it has been used to fabricate metal – organic frameworks and chitosan based tyrosinase nanosensor (Lu et al. 2016). This nanosensor consists of Cu- metal organic frame works i.e., metal nodes connected/linked to organic chains or network to lead to a nano-porous materials. In this work, two organic ligands, chitosan and tyrosinase were used to sense bisphenol analogs (BPAs). The Cu- metal organic frame works based nanobiosensor showed a wide linear range for BPE from 5.0×10^{-8} to 3.0×10^{-6} mol L^{-1} with sensitivity as 5.51 A M^{-1} cm^{-2}, and the limit of detection as 15 nmol L^{-1} ($S/N = 3$). This nanosensor showed sensitive response to bisphenol A, bisphenol F, bisphenol E, bisphenol B, and bisphenol Z in order of sensitivity as BPE > BPF > BPA > BPB > BPZ ranging from 5.51 to 1.13 A M^{-1} cm^{-2} and anti-interference ability to anti-interference ability to heavy metals like Hg^{2+}, Pd^{2+}, Cu^{2+}, Fe^{2+}, Co^{2+}, Ba^{2+}, Zn^{2+}, Cd^{2+}, and Ni^{2+}. Authors also illustrated the advantage of using Cu metal organic frame works, as BPA tends to preconcentrate on the

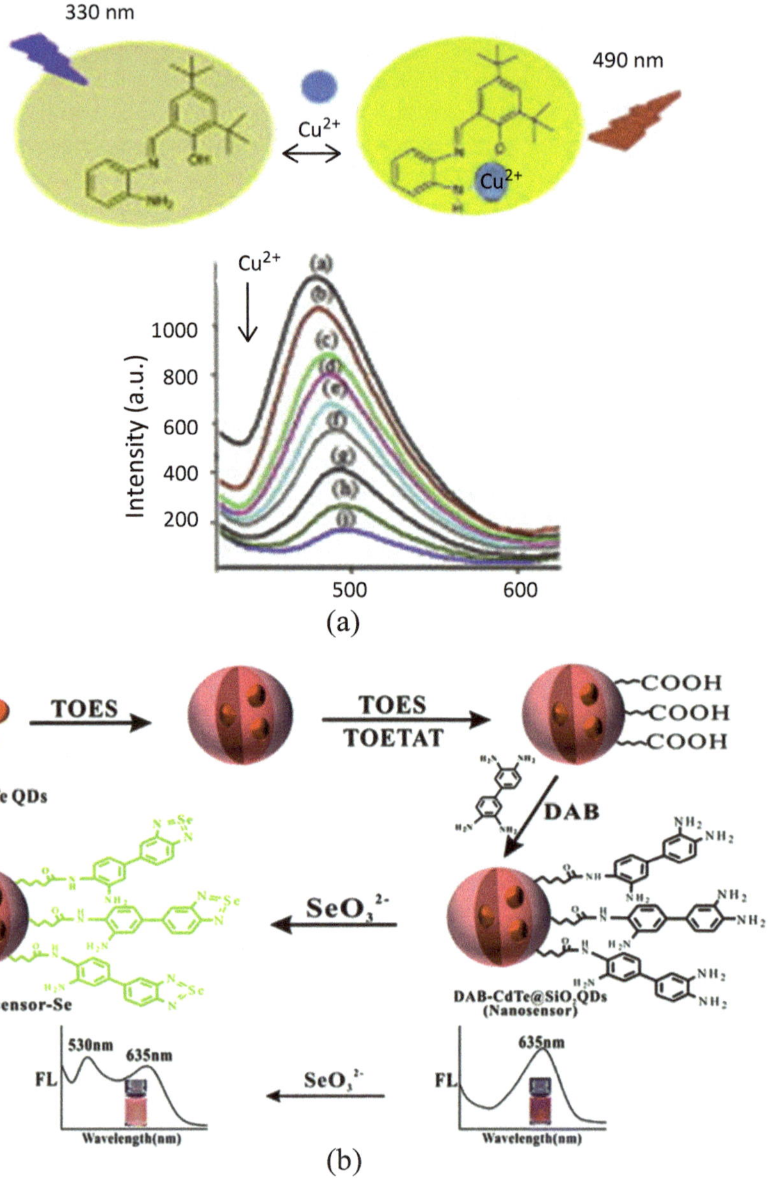

Fig. 1.6 Detection of (**a**) Copper ions (Cu^{2+}) using nano-scale fluorescent chemo-nanosensor where selective binding of analyte resulted in fluorescence quenching (Reprinted from Kacmaz et al. 2015 with © 2015 permission from Elsevier Publishing company); (**b**) Selenium ions (SeO_3^{2-}) using diamino benzidine (DAB) – cadmium telluride coated silicon oxide ($CdTe@SiO_2$) quantum dot (QuD) nanosensor where presence of analyte causes an additional emission peak at 530 nm (TOETAT-N-((trimethyloxy)silylpropyl) ethylene diamine triacetic acid trisodium salt, TEOS- tetra ethyl orthosilicate) (Reprinted from Chen et al. 2016b with © permission from Royal Society of Chemistry) and (**c**) mercury ions (Hg^{2+}) using dual-emission fluorescent probe Tb-DPA@SiO_2-Eu-DPA nanosensor where presence of analyte favors quenching of fluorescent surface via selective replacement of Eu^{3+} ions from nanosensor surface (*where Tb- terbium and Eu- europium and dopants to DPA-dipicolinic acid; DTPA- diethylene tri-amine penta acetic acid; SiO_2- silicon oxide*) (Reprinted from Tan et al. 2015 with © 2015 permission from Elsevier Publishing company)

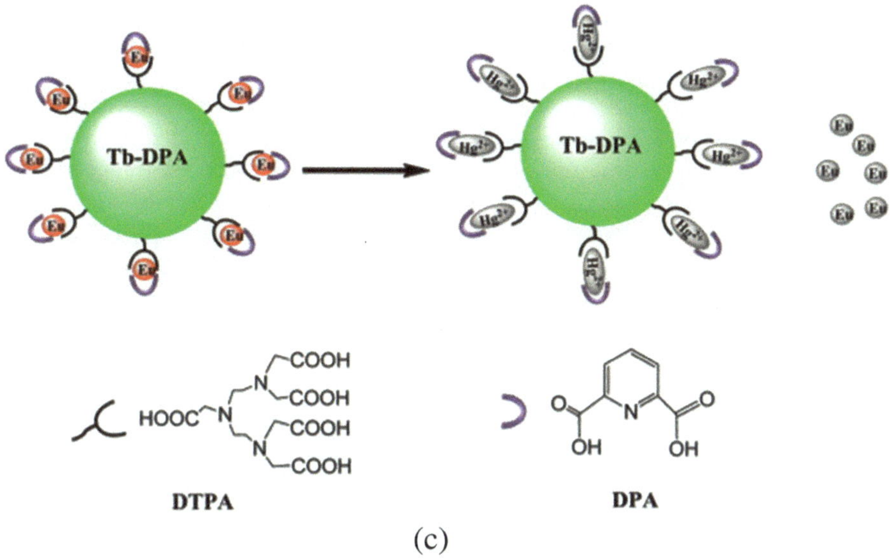

Fig. 1.6 (continued)

biosensor surface through a $\pi - \pi$ stacking interaction between the aromatic rings of BPA and the organic ligands of metal organic frame works coupled with favorable immobilization of tyrosinase in a biologically stable environment.

A super paramagnetic nanoparticle and tannic acid hybrid nanosensor was shown to detect polyphenol (dihydroxybenzene derivatives and their polymers) content in blueberries by using square wave voltammetry (Magro et al. 2016). This unique core–shell hybrid nanomaterial was formulated due to ability of metal organic frame works for stable colloidal suspensions without organic or inorganic coating i.e., no aggregations and at the same time to be able to bind to specific to organic molecules to form composites and associating properties of tannic acid (P-penta-O-galloyl-d-glucose) to form easy complexes with Fe^{3+} ions imparting low solubility in water and corrosion inhibition (Iglesias et al. 2001). This nanosensor exhibited square wave voltammetric based studies to sense tannic acid in linear range of 25–500 μM with sensitivity 312.81 nCμ M^{-1} cm^{-2} and detection limit of 8.57 μM.

1.2.2 Toxins and Mycotoxins

Mycotoxins are secondary metabolites that are produced by fungal/microbial contamination of crops and foods. These are highly resistive in nature and cause severe toxic effects leading to teratogenic, carcinogenic, and nephrotoxic situations in humans. Conventionally mycotoxins are detected by diode arrays, multi-chromatographic and enzyme linked immunosorbent assay (popularly known as ELISA) based immunological techniques that require sample pretreatment, laborious synthetic procedures and expensive instrumentations.

A nanostructured cerium oxide film-based immunosensor was also developed for the detection of food-borne mycotoxins ochratoxin-A (Kaushik et al. 2009). Then, a nanobiosensor using aflatoxin B1 antibodies linked cysteamine capped gold nanoparticles attached onto a 4-mercaptobenzoic acid self assembled monolayer coated gold electrode was used to detect aflatoxin B1 in the range of 10–100 ng L^{-1} (Sharma et al. 2010). Subsequently, a sol–gel derived nano-zinc oxide based immunosensor was developed for ochratoxin A (Ansari et al. 2010). This group and many other groups made attempts to develop and review nano based biosensors for mycotoxins (Maragos 2016; Ruscito et al. 2016; Lin and Guo 2016; Chauhan et al. 2016; Turner et al. 2015; McPartlin et al. 2016) for detection mycotoxins such as aflatoxins, ochratoxin B, citrinin, patulin, ergot alkaloids, fumonisins, trichothecenes, zearalenone etc. and multi-mycotoxin detection nanobiosensor (Mak et al. 2010). Around the same time, a new signal transduction by ion nano-gating sensors for the ultrasensitive detection of mycotoxins was described, with a detection limit up to 100 fg mL^{-1} (Actis et al. 2010; Lattanzio et al. 2012).

A nanodiagnostic kit was developed as 'lab in a box' system having sophisticated measuring devices, reagents, power supply and other features packed in a briefcase like box that can be implemented to field for *in situ* crop investigations to prevent disease epidemics (Goluch et al. 2006; Pimentel 2009). Recently, a dip stick multi parameter detecting nanosensor kit '4-my-co-sensor' based on competitive antibody assay for the real-time detection of mycotoxins such as zearalenone, trichothecene (T-2/HT-2), deoxynivalenol and fumonisin (B1/B2) for corn, wheat, oat and barley samples was reported (Lattanzio et al. 2012). This proposed immunoassay protocol was fast, cheap, easy-to-use and suitable for the purpose of quick screening of mycotoxins in cereals.

Immunoglobulin (anti-mycotoxin viz. anti-aflatoxin B1 and anti-deoxynivalenol) coupled rare earth-doped up conversion nanoparticles i.e., trivalent ions (ytterbium-Yb^{3+}, holmium-Ho^{3+}/thulium-Tm^{3+} and gadolinium-Gd^{3+}) doped sodium-yttrium-fluoride ($NaYF_4$) nanoparticles were used to simultaneously detect mycotoxins (aflatoxin B1 and deoxynivalenol) linked to SiO_2 magnetic nanoparticles having sensing range of 0.001–0.1 ng ml^{-1} with the limit of detection of 0.001 ng ml^{-1} in adulterated peanut oil (Chen et al. 2016a, b). Antigen-modified magnetic nanoparticles were employed as biosensing probes and antibody-functionalized improved up conversion nanoparticles were used as signal probes. As shown in Fig. 1.7, this method involved magnet-assisted separation of antigen-antibody complex and subsequent discriminative (aflatoxin B1 and deoxynivalenol) fluorescence bioassay facilitated by Ho^{3+} and Tm^{3+} doped up conversion nanoparticles coupled to anti-AFB1 and anti-DON, respectively. Discriminative fluorescence/ luminescence properties were introduced via doping rare earth metals (Ho^{3+}/Tm^{3+}) to $NaYF_4$ nanoparticles that can efficiently convert a long wavelength radiation (e.g. near-infrared light) into a sharp and short wavelength luminescence emission (e.g. visible light) in narrow bandwidth giving large anti-Stokes shifts and

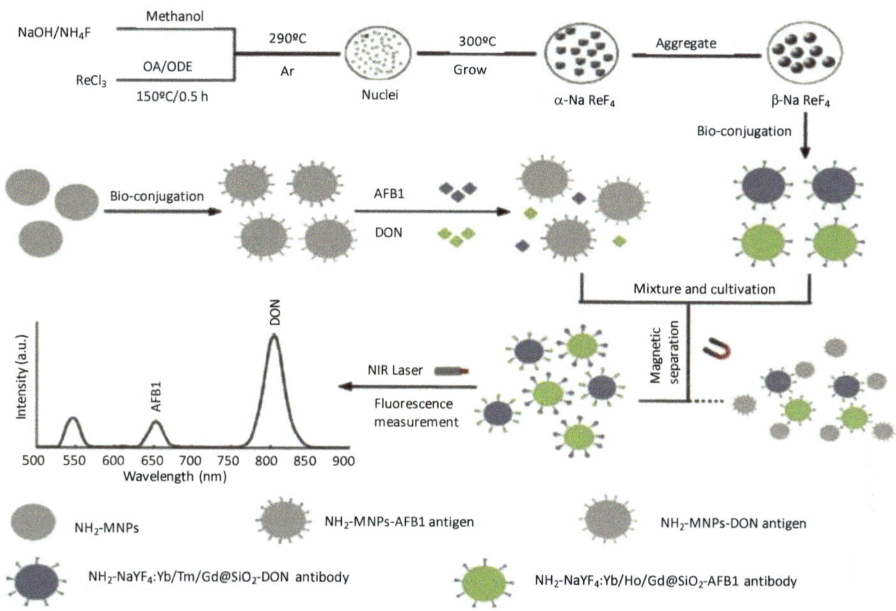

Fig. 1.7 A photo-luminescence nanobiosensor for simultaneous detection of mycotoxins: aflatoxin B1 (AFB1) and deoxynivalenol (DON) is shown to be made of antigen (AFB1 or DON)-modified magnetic nanoparticles (MNPs) and antibody-functionalized upconversion nanoparticles (UCNPs) signal probes. Selective detection of both mycotoxins: AFB1 and DON in single reaction is facilitated by selective/ separate doping of rare earth metals Ho^{3+} and Tm^{3+}, respectively (where $NaReF_4$: Sodium-Rare earth Fluoride with Re = ytterbium-Yb^{3+}, holmium-Ho^{3+}/thulium-Tm^{3+} and gadolinium-Gd^{3+}, SiO_2 silicon oxide nanoparticles) (Reprinted from Chen et al. 2016a, b with © 2016 permission from Elsevier Publishing company)

improved signal to noise ratio. Flexible chemical features and low toxicities for *in vitro* and *in vivo* systems also make them suitable for biological applications.

Detection of aflatoxin B1 was achieved via aptamer-gold nanoparticles based nanosensor in a colourimetric (red to purple) analysis showing linear range aflatoxin B1 concentrations from 80 to 270 nM and the detection limit of 7 nM (Hosseini et al. 2015).

Phylotoxins are some potent marine toxins found in temperate waters. These are known worldwide for their extreme toxicity and ability to contaminate seafood thereby causing intoxications and/or fatalities. Zamolo et al. 2012 developed a chemiluminescence based nanobiosensor that was able to produce a concentration-dependent light signal, allowing phylotoxins quantification in mussels, with a limit of quantification (LOQ = 2.2 μg kg^{-1} of mussel) more than 2 orders of magnitude more sensitive than that of the commonly used detection techniques, such as liquid chromatography-mass spectrometry/mass spectrometry (popularly known as LC-MS/MS). This method used anti-PITX linked to multiwalled carbon nanotubes bound on polysuccinimidyl acrylate-indium tin oxide substrate and Ruthenium

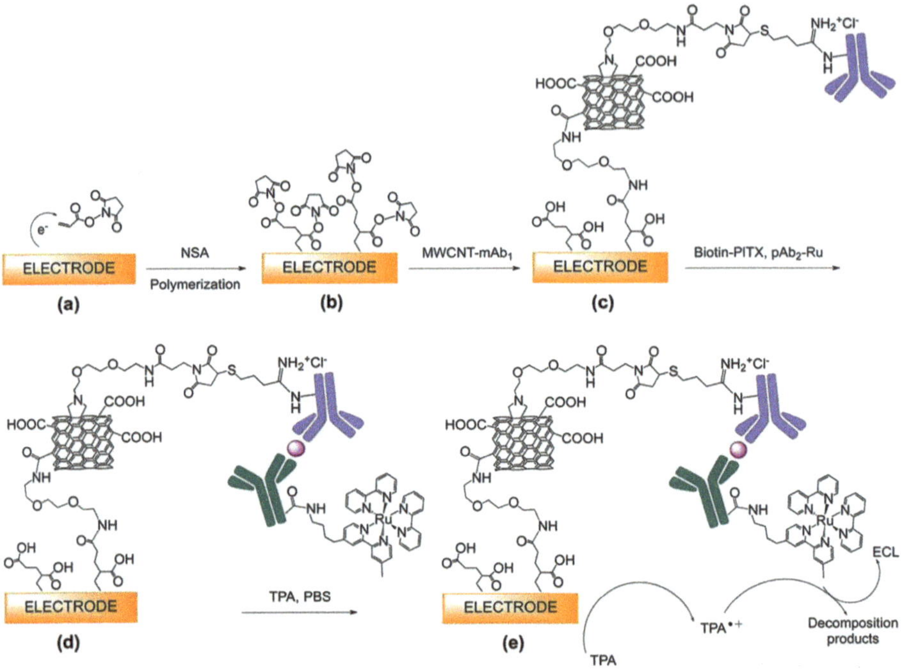

Fig. 1.8 A nanobiosensor for phylotoxin (purple sphere, Biotin-PITX) showing (**a**) Electrografting of indium tin oxide (ITO) with N-succinimidyl acrylate (NSA), (**b**) Functionalization of multiwalled carbon nanotube with antibodies against PITX (MWCNT-mAb1); (**c**) Addition of biotin-PITX (purple sphere) followed by addition of Ruthenium complex labelled antibodies (pAb2 –Ru) to PITX and (**d**) Addition of the tripropyl amine (TPA) co-reactant for electrochemiluminescence (ECL) generation and (**e**) ECL measurement which resulted in concentration dependent light generation (Reprinted from Zamolo et al. 2012 with © 2012 permission from ACS publications)

complex linked anti-phylotoxins with tripropyl amine co-reactant for detection of phylotoxin as shown in Fig. 1.8.

1.2.3 Microbial Contamination

Microbial contamination in food and water is known to cause major food borne outbreaks that has major impact on human health. *E. coli (O157:H7), Salmonella, Campylobacter, Staphylococcus, Shigella, Clostridium, L. monocytogenes, Bacillus cereus* are most common microbes known to cause food borne outbreaks (Arora et al. 2006a, b). Most food pathogens are easily transmitted through untreated water supply, undercooked or raw meat, milk, fruits, vegetables, food. Use of common facilities make easy contamination and provide higher probabilities of causing outbreaks. This is important to know that sometimes presence of 1 cfu *E.coli* O157: H7 in 25 g of food is considered at its dangerous level!

A simple approach was developed for rapid determination of *Escherichia coli* using a flow-injection system where microbial metabolism induced $K_3Fe(CN)_6$, reduction was electrochemically measured, and used as direct evidence of microbial metabolism (Hashimoto et al. 2008). This method allowed the quantitative determination of bacteria / fungi in 20 min. This new biosensor system gave opportunity for rapid diagnosis of soil-borne diseases which consisted of two biosensors made up of equal quantities of two different microbes, each individually immobilized on an electrode (Hashimoto et al. 2008).

Raman spectroscopy has been a routine practice for label free analysis chemical and biological components of a sample at micrometer scale. Surface Enhanced Raman Scattering (popularly named as SERS) is one of the available technique that is increasingly being used to detect changes occurring at the surface through antigen-antibody based specific binding (Chae et al. 2013). This convenient and reliable nanobiosensing technique was demonstrated for detection of bacteria *E. coli* using antibody (against *E. coli*) bound to gold nanoparticles deposited Indium Tin Oxide substrate chip via studying concentration dependent SERS peak intensity Raman shift in raw milk sample. Likewise, use of nanoparticles have potential to enhance Raman signal in the order of 10^4–10^6 using surface enhanced Raman spectroscopy via surface plasmon resonance (SPR) phenomena giving extended applications in detection, imaging and bacterial discrimination. Due to higher negative charge availability onto to surface of gram positive bacteria compared to gram negative bacteria, significantly distinguishable SERS spectra can be obtained through use of nanoparticles over wide range of wavelengths. A magnetic–plasmonic Fe_3O_4–Au core–shell nanoparticle synthesis was used to concentrate, detect and identify different bacterial cells by applying an external point magnetic field through SERS (Zhang et al. 2012). A silver nanoparticle coating was used to design a nanobiosensor for the detection of live bacteria in drinking water (Zhou et al. 2014) as well as anthrax spores on nanosphere substrates (Zhang et al. 2005) through simple mixing process. This enabled external spectra of the bacterial cells which are very much similar for two categories of bacterial (Gram positive and Gram negative). In this is row, an interesting label-free near infrared SERS based nanosensor using silver nanoparticles was reported. This method discriminated wide range bacteria in water by analysing inner side of the cell wall through synthesis of silver nanoparticles within bacterial culture (*E. coli, P. aeruginosa, Listeria monocytogenes, L. innocua* and Methicillin-resistant *Staphylococcus aureus*) in presence of cell membrane disruption agent (Triton X 100) in less than 5 min. This could enable to achieve distinguishable SERS spectra of inner side of bacterial walls avoiding an additional sample preparation step (i.e., isolating bacterial plasma) (Chen et al. 2015).

A label-free ultrasensitive nanosensor based on Surface Enhanced Raman Scattering (SERS) for detection of bacteria is recently reported via one step assembly

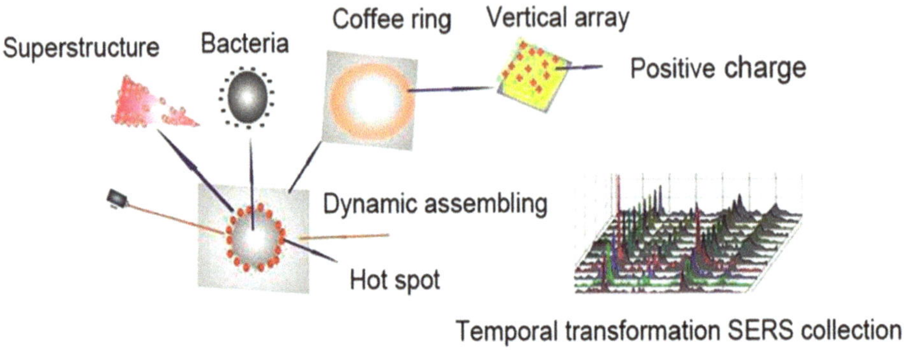

Fig. 1.9 One-step assembling and Surface Enhanced Raman Spectrometry (SERS) based detection of bacteria by adding plasmonic superstructure on bacteria – Au@Ag nanorod columnar array resulting in formation of coffee ring (distinguishable SERS spectra) (Reprinted from Qiu et al. 2016 with © 2016 permission from ACS Publications)

phenomena guided by electrostatic attraction of negatively charged bacteria with positively charged plasmonic nanoparticles (silver @ gold core shell (Ag@Au) nanorods) and two-dimensional bifacial nanoparticle liquid crystalline superstructure (SH-polyethylene glycol-NH_2 coated triangular gold nanoplates – gold nanospheres based bifacial plasmonic assembly) (Qiu et al. 2016). In this method, a 'bifacial superstructure-bacteria-columnar array' assembles when nanoparticle liquid crystalline superstructure is added onto bacterial sample placed onto a 'columnar array of Au@Ag nanorods' as shown in Fig. 1.9. Presence of food borne Gram positive bacteria resulted in formation of dynamic optical hotspots leading to a hybridized nano-assembly under wet – dry critical state, thereby amplifying efficiently the weak vibrational modes. A SERS spectrum was measured at 730 cm^{-1} for detection of bacteria on a nanorod columnar array using bifacial triangular gold nanoplates – gold nanospheres superstructure. This method represents an attractive detection approach that can detect presence of bacteria in various samples/matrices. Moreover, in this report this method limits its application to be not able to distinguish likewise bacteria detected (*S. xylosus, L. monocytogenes, and E. faecium*).

An *E. coli* (O157:H7) specific 22mer oligonucleotide functionalized SiO_2 nanostructure (70 nm sized) coated shear horizontal surface acoustic wave YX $LiNbO_3$ substrate was fabricated to detect high performance nanobiosensor for detection of *E. coli* showing sensitivity of 0.6439 nM/0.1 kHz and detection limit of 1.8 femtomolar (1.8×10^{-15} M) (Ten et al. 2016).

Recently, a co-polymer brushes based functional coating was used to exhibit high fouling resistance and biorecognition capabilities for variety of food matrices (like milk, spinach, cucumber, hamburger, and lettuce) for detection of bacterial contamination using a surface plasmon resonance shift as function of binding with specific antibodies against bacteria like *E.coliO157:H7*, *E.coliO145:H2*, and

Salmonella. Detection parameters were found to be within concentrations ranging from 1.5× 10^{-2} to 1.5 × 10^{-7} CFU mL^{-1} (colony forming unit per millilitre), 1.5 × 10^{-2} to 1.5 × 10^{-7} CFU mL^{-1}, and 2.5 × 10^{-2} to 2.5 × 10^{-7} CFU mL^{-1}, respectively (Lísalová et al. 2016). *Bacillus cereus* spore-forming, gram-positive bacilli (found in diverse environmental conditions, including soil and food such as dairy products, rice and vegetables) was electrochemically detected using electrochemical gold nanoparticles and DNA (single stranded DNA of nheA gene) based nanobiosensor in milk. The nano based biosensor showed up to 10 colony forming units per milliliter (CFU mL^{-1}) with a detection limit of 9.4 × 10^{-12} mol L^{-1}. The infected milk sample was pre-treated and extracted for the specific target DNA prior to detection using nanobiosensor (Izadi et al. 2016).

Some of recent reviews describe variety of available nanobiosensors for detection of waterborne bacteria (Deshmukh et al. 2016); use of nanotechnology for microbial biosensors (Lim et al. 2015) and detection of pathogenic microbes (Yoo and Lee 2016) to demonstrate the potential of nanobiosensors and nanosensors for detection of microbes in wide range of matrices.

1.2.4 Food Allergens

Food borne allergies share a major food safety and public health concern globally and impose huge cost to patients and sometimes death. Since there exist no cure for any kind of allergies, the only significant way is to avoid intake of allergen containing food (Alves et al. 2016).Till date food allergens are being tested by immunological and DNA based methods that involves use of ELISA based kits and use of polymerase chain reaction (known as PCR) based methods (such as PCR-ELISA, real-time PCR, PCR-peptide nucleic acid-high performance liquid chromatography, duplex PCR and multiplex real-time PCR etc.). Recently, Alves et al., have reported available biosensors both immunosensors and DNA biosensors for detection of allergens in various food matrices along with sample preparation methods (Table 1.1).This review contains few nano based biosensors that make use of optical and electrochemical modes of signal transduction mechanisms explained as follows.

Beta-lactoglobulin, a milk based allergen was detected via sensitive label-free graphene modified screen printed based voltammetric immunosensor (Eissa et al. 2012). Cyclic and differential pulse voltammetry (DPV) based measurement were taken to study the response of fabricated nano based biosensor as a function of b–lactoglobulin concentration at pg mL^{-1} level (Table 1.1) in different samples from cake, cheese snacks and biscuits.

Mascini and group reported large number of publications on food allergens/ contaminants. Hazelnut allergens were reportedly detected by voltammetric genosensor made up of screen printed eight sample-DNA-array for PCR amplified samples from various food items at nanomolar range. This method provided favorable poor non-specific signal and high sensor stability (Bettazzi et al. 2008;

Table 1.1 Biosensors for detection of food allergens

	Biosensors		Matrix	Allergen	Limit of Detection	References
	Colorimetric	Genosensor	Cereal bar Chocolate Biscuit Fried mud . . .	Lectin, Ara h 3, Ana o 3, mitochondrion DNA (beef, chicken), 16S rRNA (fish, shrimp)	Absolute: 0.5 pg of cashew DNA Practical:0.001% (w/w)	Wang et al. (2011)
Optical	Surface Plasmon resonance (Spr)	Immunosensor	Chocolate	Peanut allergens	0.7 μg/mL of extract; 7 ppm peanut sample	Mohammed et al. (2001)
			Olive oil	Hazelnut protein	0.08 μg/g of hazelnut proteins in olive oil	Bremer et al. (2009)
			Chocolate pasta Bread Surimi . . .	Ovomucoid, ovotransfemin, β-lactoglobulin, tropomyosin, proteins from hazelnut, peanut, and sesame	<1–12.5 μg/g in samples	Yman et al. (2006)
			Processed sea food products	Parvalbumin	3.55 μg/L	Lu et al. (2004)
			Chocolate bars	Ara h 1	0.09 μg/mL	Pollet et al. (2011)
		Aptasensor	–	Ara h1	–	Tran et al. (2013)
	Localized Spr	Immunosensor	Milk	Casein	10 ng/mL	Hiep et al. (2007)
	Imaging Spr	Immunosensor	Cookies and dark chocolate products	Peanut, hazelnut, milk, soy, lupine, egg, pine nut, and tree nuts allergens	Cookies: 0.2–3.2 mg/kg Dark chocolates: 0.4–5 mg/kg	Raz et al. (2010)
	Resonance Enhanced	Immunosensor	Egg-containing food products	Ovalbumin Ovomucoid	1 ng/mL	Maier et al. (2008)
	Absorption		Processed milk matrices	β-lactoglobulin	–	Hohensinner et al. (2007)

Electrochemical	Voltammetry	Genosensor	Chocolates	Hazelnut allergens (Cor a 1.03 and Cor a 1.04)	Cor a 1.03: 0.3 mol/L	Bettazzi et al. (2008)
			Snacks		Cor a 1.04: 0.1 mol/L	
			Biscuits			
			Soy milk, peanut			
	Voltammetry	Immunosensor	Cake, cheese snacks, biscuit	β-lactoglobulin	0.85 pg/mL	Eissa et al. (2012)
	Impedance	Immunosensor	–	Ara h 1	0.04 µg/mL	Singh et al. (2010a, b)
			–	Ara h 1	<0.3 nM	Huang et al. (2008)
Piezoelectric/ electrochemical		Immunosensor	Shrimp	Pen a 1	0.333 µg/mL	Xiulan et al. (2010)

Reprinted from Alves et al. 2016 with © permission from Taylor and Francis

Farabullini et al. 2007) and allows simultaneous analysis of eight samples. This makes use of PCR derived biotinylated hybrids binding to streptavidin–alkaline phosphatase conjugate in naphthyl phosphate solution (that acts as electro-active indicator) giving current proportional to the target analyte.

To detect ovalbumin or ovomucoid allergens, a resonance enhanced absorption based colorimetric solid-phase immunoassay on a planar chip using biofunctionalized gold nanoparticles as signal transducers in a highly sensitive distance-dependent interferometric setup (Maier et al. 2008). Resonance enhanced absorption involves use of labeled detection reagents and when noble metal nanoclusters are deposited at nanometric distances from the highly reflective mirror of an interferometric setup in optical near field and change in absorption is measured as a function of binding. Binding of gold nanoparticles – conjugated IgGs (immunoglobulin-G, i.e., polyclonal rabbit antisera against ovalbumin and ovomucoid) was successfully demonstrated for food allergens or antigen (ovalbumin or ovomucoid) immobilized on the surface of the optically transparent distance layer (poly(styrene-methyl methacrylate)copolymer) of a Aluminium foil chip. In a similar type of setup, a sandwich assay was described for the detection of lactoglobulin (milk allergen) in processed milk matrices using antibody (purified polyclonal rabbit anti-bovine-lactoglobulin, IgG) pre-coated matrix to capture the antigen (Hohensinner et al. 2007). And, detection of antigen was accomplished by second gold nanoparticles-labeled readout antibody, which within a certain resonance distance generated a visually detectable colorimetric signal (strong blue color) on the chip that could be photometrically read for a semi-quantitative measurement.

A peanut protein 'Ara h1', known to be responsible for peanut allergy was detected using gold-coated nanoporous polycarbonate based impedance immunosensor by measuring the change in the pore conductivity (Singh et al. 2010a, b). These authors reported to study binding of Ara h 1 to antibody bound within nanopore as a function of the membrane pore diameter (15, 30 and 50 nm) and the protein concentration. Interestingly, highest sensitivity was achieved with the smallest pore diameter membrane with improved limit of detection of 0.04 m g/mL compared to SPR based immune assay for Ara h1 detection (0.09 m g/mL, as mentioned in Table 1.1).

Recently, aptamer/quantum dots-functionalized grapheme oxide biosensor is reported for food allergen (peanut, Arah1) detection on a microfluidic platform. It utilizes fluorescence quenching and recovering properties of graphene oxide through the adsorption and desorption of Quantum Dots conjugated aptamers (Weng and Neethirajan 2016). This microfluidic platform introduced features like decreased sample/reagent consumption and rapid fluorescent signal detection on a miniature size optical detection while avoiding probe immobilization procedures as shown in Fig. 1.10. This microfluidic system is governed by powerless sampling that can be generated by hexagons capillary pump, which introduce capillary force, and favours liquid sucking into the microfluidic channel. Capillary-driven retarding inlet valve (Mohammed and Desmulliez 2013) help avoiding air capture in

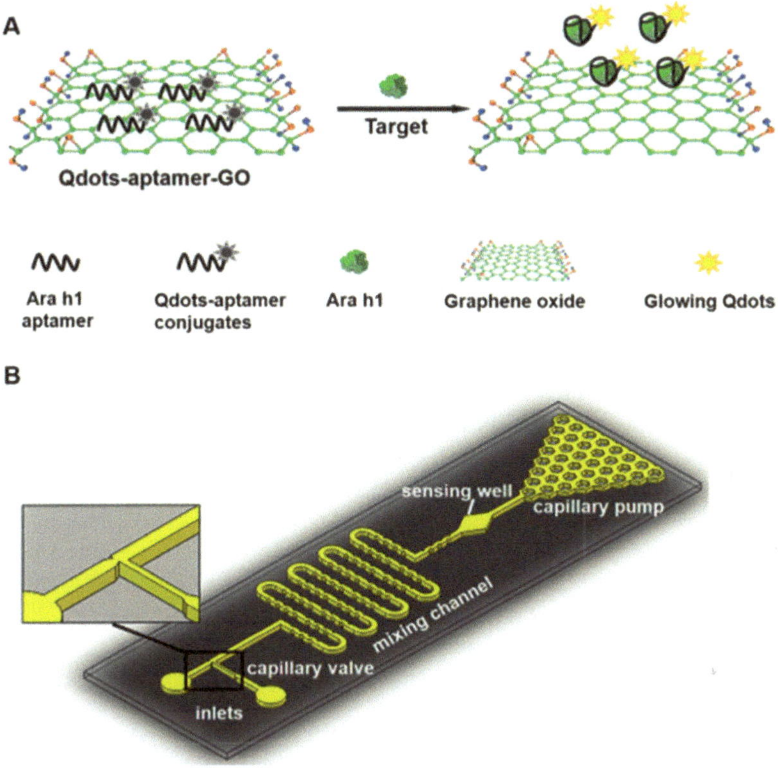

Fig. 1.10 Schematic showing (**a**) sensing mechanism of Quantum dots (Qdots)-aptamer-Graphene oxide (GO) quenching system. (**b**) microfluidic chip driven capillary forces consisting of two inlets for loading the Quantum dots-aptamer-Graphene oxide probe mixture and the Ara h 1 sample, mixing/incubation channel (zig zag), diamond shaped sensing well aligned to sensing Si photodiode window and a capillary pump at the end (Reprinted from Weng and Neethirajan 2016 with © 2016 permission from Elsevier Publishing company)

the microchannel while dispensing the Qdots-aptamer-GO probe mixture and the Ara h 1 sample into the inlets. Ara h1 was demonstrated to be detected in linear response region between 200 ng mL^{-1} and 2000 ng mL^{-1} with detection limits of 56 ng mL^{-1} within 5 min of quenching and fluorescence recovery time (Weng and Neethirajan 2016). These interesting findings warrants realization of integration of existing nano- and bio- sensor technology with microfluidics. This shall facilitate achieving enhanced features of performance through throughput processing, desired transport for controlling the flow conditions, increase the mixing rate of different reagents, smaller sample / reagents volume (down to nanoliter) leading to increased sensitivity of detection, and utilize the same platform for both sample preparation and detection (Luka et al. 2015).

There are various reports available that show selection of various specific aptamers that are being isolated and have been isolated for potential harmful bacteria (Lee et al. 2012; Han and Lee 2013), suggests that aptamer based nanosensors possess immense potential for monitoring bacterial contamination in various food and agricultural matrices.

1.2.5 *Nutritional Constituents in Food*

Vitamins are important constituents of living things, which are not only measured and monitored, as important ingredients of food items but are also equally important for health care diagnostics. Vitamins are complex group of compounds that are known to play important role in various biochemical pathways and have been found to have chemical structures that give rise to electrochemical properties. Vitamins like vitamin C (L-ascorbic acid), compounds in the B vitamin group: vitamin B_2 (riboflavin); vitamin B_6 (pyridoxine); vitamin B_7 (biotin); vitamin B_9 (folic acid); and vitamin B_{12} (cyan) are either naturally electroactive or electroactive under modified conditions (Ho et al. 2010). Hence, their properties continue to be exploited using electrochemical techniques. Nanobiosensors for vitamin C (linear range 50.00–400.00 μM), polyphenols, vanilla flavours, and isoflavones were reported by Crevillén et al. 2007, 2008 using a multi-walled carbon nanotube screen printed carbon electrode coupled with a capillary electrophoresis micro-chip device through amperometric detection. A gold nanoparticle enhanced biotin (vit B_7) detection was carried out in a competitive electrochemical immunosensor assay by Ho et al. 2010.

Choline is an essential food ingredient (mainly milk) that has nutritional value for all age groups as it is required for essential roles in brain development, metabolic functions, signalling in central nervous system, memory functions etc. Although, it is synthesized in body, its dietary intake is also required to sustain appropriate development especially in kids/infants. Any disbalance in choline metabolism has been found to lead to Alzheimer's, Parkinson's and prostate cancer (Richman et al. 2012; Zeisel and Blusztajn 1994). Recommended levels of choline intake ranges from 125 mg day^{-1} in infants to 550 mg day^{-1} in males over age 14 years and in breast-feeding women (Zeisel and Blusztajn 1994). An chemiluminescence bi-enzyme (choline oxidase and horseradish peroxidase) linked Zinc oxide nanorods film based nanobiosensor was developed for detection of choline in range of 0.0005–2 mM as shown in Fig. 1.11 (Pal et al. 2014). In this chemiluminescence assay enzymatic action on analyte choline resulted in production of hydrogen peroxide which was used for quantification through HRP enzyme in presence of chemiluminescence indicator luminal to generated photons directly proportional to the analyte. This nanobiosensor presented a promising example a stable assembly that retained 78% enzyme activity till 28 days.

Glucose is an important food ingredient and an central parameter which needs to be monitored due to various clinical and food production reasons. Magnetite (Fe_2O_3) – Prussian blue nano-composites were described to sense glucose exhibiting fast response time 3–4 s, lower detection limit of 0.5 μM, wide linear

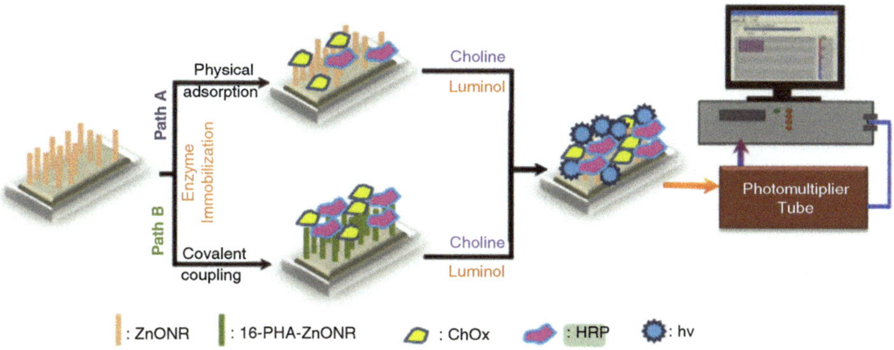

Fig. 1.11 Zinc oxide (ZnO) nanorods (NR) based choline nanobiosensor using bi-enzyme-cholesterol oxidase/horseradish peroxidase (ChOx/HRP) through physical adsorption (Path A) and covalent coupling (Path B), showing detection of choline using chemiluminescence assay where production of hydrogen peroxide by ChOx is quantified by second enzyme HRP in presence of luminol which intern generates photon directly proportional to analyte choline (Reprinted from Pal et al. 2014 with © 2014 permission from Elsevier Publishing company)

range from 5 μM to 1.2 mM, sensitivity of 32 μA mM^{-1} cm^{-2} and good long-term stability (Jomma and Ding 2016).

Lactate is another important analyte for clinical analysis, sports medicine and food industry. Besides, its clinical importance, lactate in food is an import analyte that is monitored in food industry. It can indicate microbial contamination leading to lactate fermentation (Gyawali and Ibrahim 2012; Muyanja et al. 2012). An amperometric lactate nano-biosensor based on lactate dehydrogenase functionalized graphene was reported to have sensitively from 0.08 mM to 20 mM, with a fast steady- state measuring time of 2 s y measuring formation of hydrogen peroxide as a result of enzymatic reaction (Labroo and Cui 2013).

Amino acid phenylalanine and proteins like lysozyme (Lys) and bovine serum albumin was detected through a gold nanorods based nanobiosensor via use Surface Enhanced Raman Spectra (SERS) based detection mechanism (Fazio et al. 2016). In this novel approach radiation pressure was utilized to locally push gold nanorods and induce their aggregation in buffered solutions of biomolecules, achieving biomolecular SERS detection at almost neutral pH reaching detection limits in the μg mL^{-1} range and achieve single molecule sensitivity. The addition of nanoparticles aggregates to protein solutions paved the way to quantitative estimation. SERS exploits electromagnetic enhancement of localized surface plasmon resonance of metal nanoparticles to tailor molecular sensitivity, by creating SERS-active clusters i.e., nanoparticle embedded molecules leading to 'hotspots' that enhance Raman Scattering for magnified detection. In a similar way, SERS based detection of uric acid in human serum was achieved with limits of detection (LOD) ~240 μ M (equivalent to 40 μ g mL^{-1}) (Zakel et al. 2011).

Polyphenolics are a broad class of compounds present in many fruits, vegetables, and their products, including grapes and wines. Wines, particularly red wines are known contain biologically active poly-phenols which provide antioxidant properties and contribute substantially to the quality by color, flavor, stability, and aging behavior. Tyrosinase-immobilized nanobiosensors based on poly(acrylic acid)-grafted multi-wall carbon nanotube and poly(maleic anhydride)-grafted multi-wall carbon nanotube as substrate materials were fabricated (Kim et al. 2010). This amperometric nanobiosensor could show sensing range of 0.2–0.9 mM and 0.1–0.5 mM for phenol in phosphate buffer solution, respectively for both nanobiosensors. Laccase (p-diphenol oxidase containing copper ions) was co-immobilized with Tyrosinase in a sol–gel matrix of diglycery silane) to detect wide range of phenolic compounds present in wine (Montereali et al. 2010). Yang et al. 2009 used a tyrosinase nanobiosensor based on polyglucosyl 4-vinyl phenyl boronate-multiwall carbon nanotube using cyclic voltammetry and suggested that the amounts of phenolic compounds in commercial red wines range between 68.50 and 655.0 mg L^{-1} for Lindemans wine to Duchessa Lia wine. About ten times higher content of these phenolic compounds is responsible for typical bitter taste of the Duchessa red wine.

Heterocyclic amines are constituents of cooked food, which are formed as a result of incomplete combustion process of proteins at high temperature. As per International Agency for Research on Cancer (IARC 1993) some of the heterocyclic aromatic amines as possible human carcinogens e.g., Class 2B: (2-amino-3,4-dimethyl-imidazo [4,5-f]-quinoline, 2-amino-3,8-dimethyl-imidazo [4,5-f] quinoxaline and 2-amino-3,4,8- trimethylimidazo [4,5-f]-quinoxaline, 2-amino-3,7,8-trimethyl-imi-dazo [4,5-f]-quinoxaline and Class 2A: 2-amino-3-methylimidazo [4,5-f]-quinoline (Puangsombat et al. 2012, John and Beedanagari 2014, IARC 1993). A carbon dots based nanosensor was prepared from lactose using microwave process and was used as such without any further functionalization for detection of four different heterocyclic aromatic amines. Specific binding of heterocyclic aromatic amines, reportedly quenched the fluorescence of carbon dots at 455 nM and facilitated detection of heterocyclic aromatic amines in exponential manner in concentration range of 0.35–0.45 mg L^{-1} (Lopez et al. 2015).

1.2.6 Monitoring Environmental Parameters for Food and Agricultural Applications

Phosphate levels in aquatic environments are very important tool in understanding the quality of water to facilitate production of fishes and aquatic plants as well as to sustain balance in ecosystem for various purposes. Phosphorus is usually present in the natural water as phosphates (orthophosphates, polyphosphates, and organically-bound phosphates) (Nollet and Gelder 2013; Spellman 2013) at a very low concentration of 0.025–0.1 mg L^{-1} (Fadiran et al. 2008a, b). Use of excessive fertilizers, industrial effluents, laundry, human and animal waste has caused increase in phosphate level of water bodies. This over-fertilized situation of aquatic plants results in "eutrophication" (Smith et al. 1999) i.e., explosive growth of the plants and algae due to oversupply of the nutrients. Thereby, causing hypoxia condition in water for fishes due to consumption of all O_2 (Upadhyay and Verma 2015). Therefore, a stable amperometric phosphate nanobiosensor was developed to quntitate phosphate levels in water samples using pyruvate oxidase and its cofactors, thiamine pyrophosphate and flavin adenine dinucleotide closely integrated with a highly ordered gold nanowires array (Fig. 1.12a). This nanobiosensor gave detection limit of 0.1 mM, a linear concentration range of 12.5–1000 μM, and a sensitivity of 140.3 $\mu A\ mM^{-1}\ cm^2$ (Ogabiela et al. 2015).

Cyanide is acutely toxic to mammals by all routes of administration. These are produced by plants biologically and also by anthropogenic activities which can be used a potential bio war agent. Therefore, detection of cyanide contamination in food and water is extremely important. A nanosensor for cyanide detection was developed using a pair of two luminescent hetero-trinuclear complexes $[Pt_2Ag(\mu\text{-dpppy})_2(CuCC_6H_4R\text{-}4)_4](ClO_4)$(R = H, 1; R = CH_3, 2) as self-assembling building blocks of $[Pt(CuCC_6H_4R\text{-}4)_4]^{2-}$ and $[Ag_2(\mu\text{-dpppy})_3]^{2+}$ to achieve Pt_2Ag (platinum silver) acetylides. These were further associated with mono dispersed silica nanoparticles to form new kind of luminescent nanoparticles called '$Pt_2Ag@SiO_2$ nanoparticles' (Lin et al. 2014a, b). This interesting 'platinum-silver @ silicon oxide' nano assembly had specific fluorescence quenching response phenomena for cyanide anions and had improved water solubility, stability and luminescence signal enhancement that could be observed through naked eye (Fig. 1.12b). Nanosensor was reported to have ratio of the luminescence intensity (I_0/I) vary in a linear relationship with concentration of cyanide anions in the range of 0.1–10.0 μM ($R^2 = 0.9984$) having detection limit of 0.08 μM at S/N = 3. This nanosensor could sense in the range of acceptable cyanide ion limits (<1.9 μM) prescribed for drinking water by World Health Organization (WHO). A new two-photo excitation nanosensor using graphene quantum dots @gold nanoparticle conjugate for sensing and imaging endogenous biological CN^-ions was reported (Wang et al. 2015). This nanosensor could detect CN^- up to 0.52 μ M and at deeper penetration depth (about 400 μ m) in sample matrices, to realize *in situ* sensing and imaging of CN^- ions in different types of plant tissues and food samples using fluorescent sensing and imaging (Wang et al. 2015). In this unique hybrid nanosensor a peptide-mediated graphene quantum dots /gold nanoparticles hybrid assembly was achieved through $\pi - \pi$ stacking of graphene quantum dots and peptide to form the nanosensor. The nanosensor underwent disassembly upon addition of CN^-, gold nanoparticles were etched to lead to fluorescence of graphene quantum dots to be released to be able to sense and image presence of CN^- ions. Whereas in the absence of CN^-, the fluorescence of graphene quantum dots would remain quenched by gold nanoparticles through the fluorescence resonance energy transfer (commonly known as FRET) (Fig. 1.12c).

Green fluorescent protein (commonly known as GFP)-tagged sensor proteins, ArsR-GFP and CadC-GFP, were used as nano-biosensors for simple and low-cost quantification of As(III) or Cd(II) in drinking water (Siddiki et al. 2012). The sensor

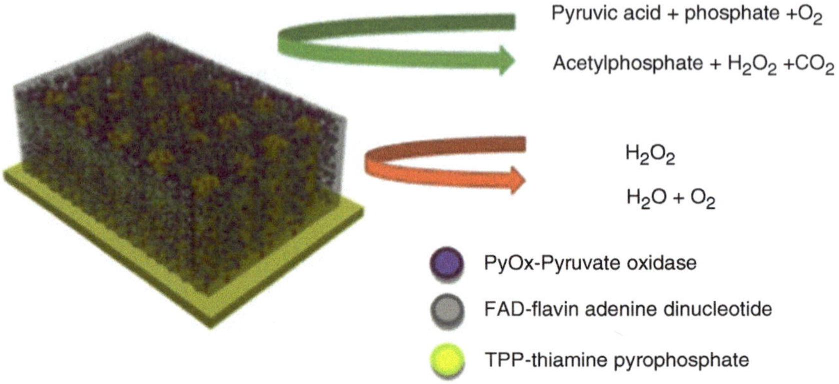

(a)

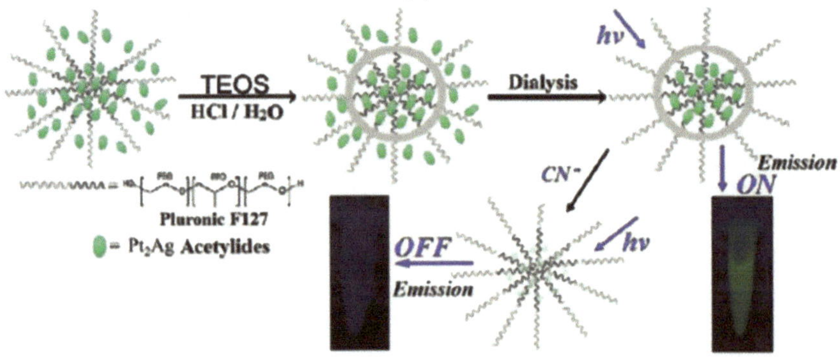

(b)

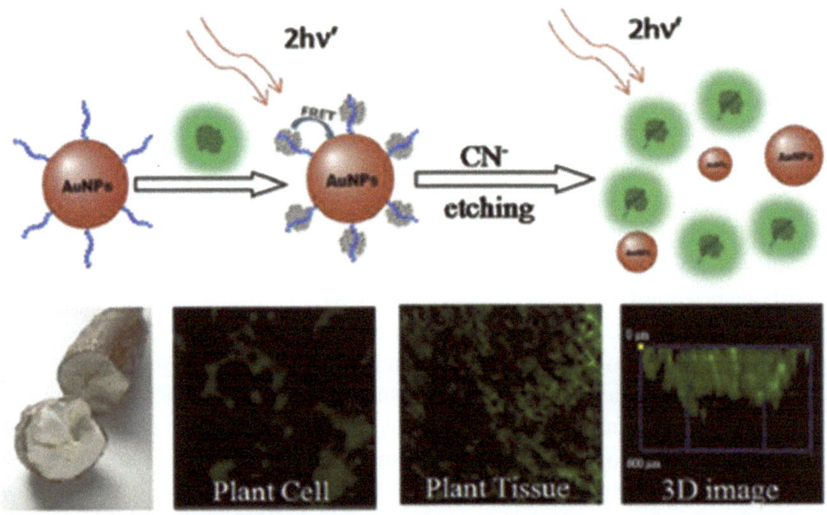

(c)

Fig. 1.12 Detection of (**a**) phosphate ions using gold nanowires array–thiamine pyrophosphate (TPP)–pyruvate oxidase (PyOx) nanobiosensor that amperometrically measures enzymatic degradation of analyte (Reprinted from Ogabiela et al. 2015 with © 2015 permission from Elsevier

protein-promoter DNA complexes bound to surfaces of magnetic particles of different sizes so that they can be separated by magnets, and can release different amounts of GFP-tagged protein, as per metal concentrations within 5 min leading to increases in fluorescence. A detection limit of 1 μg/L for As(III) and Cd(II) in purified water was obtained only with the nanoparticles exhibiting enough magnetization after heat treatment for 1 min.

1.2.7 Pesticides in Food and Environment

Use of pesticides is an indispensable component of modern crop management practices as they are believed to improve the nutritional value of food and minimize the loss in agricultural productivity caused by insects and pests. Pesticides like methyl parathion, organophosphorus compounds, ethyl parathion, malathion, 2,4-dichlorophenoxyacetic acid (2,4-D), atrazine, dichlorodiphenyltrichloroethane (DDT) and hexachlorocyclohexane (HCH) are some examples of chemicals that had been used in agriculture for controlling weeds, insects and rodents to increase productivity. On the contrary, use of pesticides, herbicides and chemical fertilizers have resulted in various harmful effects to environment and disbalance in ecosystem besides bringing in revolution in the production quantity. Hazardous health effects and toxicological interventions due to use of pesticides has led to wide range of diseases throughout the world, that monitoring pesticide levels in various matrices like water, food, soil, ground water, and food at extremely low concentration i.e., pg levels has become need of hour to comply with environmental, product specification and government regulatory norms.

As per Nature Asia 2008, A nucleic acid and intrinsically conducting polymer based biosensor was reported, that could detect minute amounts of toxic organophosphate insecticides at nanomolar levels for screening of environmental samples such as drinking water, waste water and industrial effluents (Nature Asia, doi: https://doi.org/10.1038/nindia.2008.213 Published online 29 May 2008). The sensor could track chlorpyrifos (0.5–200 ppb) and malathion (0.005–10 ppm) a conducting polymer (polyaniline-polyvinyl sulfonate) and nucleic acid as a biorecognition element within 30 s exposure time having stability of about 6 months (Prabhakar et al. 2008). In a similar series of studies, double stranded calf thymus

Fig. 1.12 (continued) Publishing company); (**b**) cyanide anions using 'platinum-silver @ silicon oxide' ($Pt_2Ag@SiO_2$) nano-assembly based nanosensor where fluorescence quenching occurs in presence of analyte, which is shown to have affinity for especially designed nano-assembly (TEOS-tetraethyl orthosilicate) (Reprinted from Lin et al. 2014a, b with © 2014 permission from Royal Society of Chemistry) and (**c**) CN^- ions using gold nanoparticle (AuNP)-Peptide @ graphene quantum dot (GQD) nanosensor assembly where presence of analyte would dissociate the nano assembly to stop fluorescence resonance energy transfer (FRET) between AuNP and GQD facilitating *in situ* sensing and imaging in plant tissues (Reprinted from Wang et al. 2015 with © 2015 permission from ACS publications)

deoxyribonucleic acid entrapped polypyrrole-polyvinyl sulphonate films fabricated onto indium-tin-oxide coated glass plates were used to detect organophosphates such as chlorpyrifos (0.0016–0.025 ppm), malathion (0.17–5.0 ppm), 2-aminoantharcene (0.01–20 ppm) and o-chlorophenol (0.1–30 ppm) (Prabhakar et al. 2007, Arora et al. 2006a, b).

A nanosensor based silver nanoparticles was shown to be sensitive to herbicide in a solution that induced variation in colour of the nanoparticles from yellow to orange red and finally to purple in concentration depended manner (Dubertret et al. 2001). This method represents and interesting useful approach for detection contaminants, such as organic pollutants and microbial pathogens in water bodies and in the environment. In a similar way intrinsically conducting polymers like polyaniline, polythiophene and polypyrrole are being used to fabricate fast nanosensors that can detect molecular signals with very low intensity of chemicals/toxins spoilage and food-borne pathogens (Sekhon 2010).

Deltamethrin (a replacement of organochlorines and organophosphorus insecticides) was detected using cadmium telluride embedded molecularly imprinted polymers of fluorescent silica (SiO_2) quantum dots or nano spheres in concentration range of 0.5–35.0 g mL^{-1}, and corresponding detection limit of 0.16 g mL^{-1} in fruit and vegetable samples using fluorescence measurements (Ge et al. 2011, Fig. 1.13a). This nanosensor was reported to be highly specific to deltamethrin due to existence of a quenching mechanism (that facilitates electron transfer from the cadmium telluride-silica quantum dots to the deltamethrin species through the strong binding to the template molecule) and was shown to give no signal for trichlorfon, carbofuran, lambda-cyhalothrin, cypermethrin, permethrin and various ions (NH_4^+,NO_3^-, Na^+, Cl^+, CO_3^{2-}, Fe^{3+}).

Cartap, (a nereistoxin derivative and a pesticides used in agriculture) was detected using a novel up conversion nanoparticles [$NaYF_4$:Yb,Ho/Au nanocomposite or lanthanide (ytterbium-Yb and holmium-Ho) doped sodium yttrium fluoride-$NaYH_4$ nanocomposite on gold nanoparticles] through chemo-nano-sensor up to 10 ppb via luminescence resonance energy transfer (LRET) (Wang et al. 2013). LRET is reported to occur between upconversion nanocrystals and the gold nanoparticles and upon specific hydrogen binding of cartap to mercaptopropionic acid bound on gold nanoparticles that lie in close association with upconversion nanoparticles ($NaYF_4$:Yb,Ho nanocrystal) LRET is facilitated as shown in Fig. 1.13b.

Same group of researchers worked on detection of organophosphorus pesticides and reported a novel nanosensor based on FRET between upconversion nanoparticles ($NaYF_4$:Yb,Er nanocomposite or lanthanide (Ytterbium-Yb and Erbium-Er) doped sodium yttrium fluoride-$NaYH_4$ nanocomposite) and gold nanoparticles as shown in Fig. 1.13c (Long et al. 2015). The detection mechanism is based on the fact that gold nanoparticles quench the fluorescence of upconversion nanoparticles and organophosphorus pesticides inhibit the activity of acetylcholinesterase which catalyzes the hydrolysis of acetylthiocholine into thiocholine. Acetylthiocholine is an analog of acetylcholine, a substrate of acetylcholinesterase, and it can be easily hydrolyzed to generate thiocholine. The electrostatic

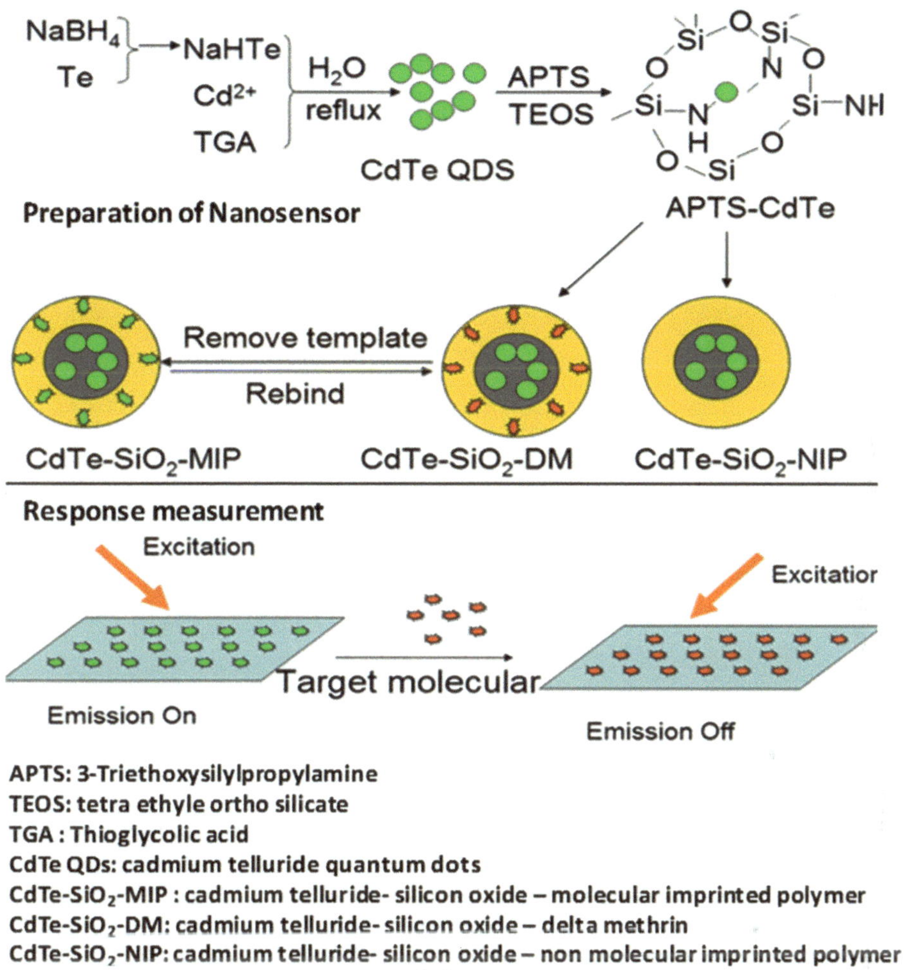

(a)

Fig. 1.13 Detection of: (**a**) deltamethrin using cadmium telluride embedded molecularly imprinted polymers of SiO_2 based nanosensor where binding of target analyte results in quenching of fluorescence (Reprinted from Ge et al. 2011 with © 2011 permission from Elsevier Publishing company); (**b**) cartap using up conversion nanoparticles (UCNPs) and gold nano particles based LRET nano sensor, where a- Hydrogen-bonding between cartap and MPA, b-LRET between UCNPs and GNPs is favoured in presence of analyte (Reprinted from Wang et al. 2013 with © 2013 permission from Elsevier Publishing company); (**c**) organophosphate (OPs) pesticide using upconversion nanoparticles-gold nanoparticles (UCNPs-AuNPs) based nanosensor where presence of analyte deactivates acetylcholinesterase enzyme (AChE) and formation of thiocholine (ATC) goes down thereby leading to disintegration of nanosensor assembly to stop FRET between UCNPs and gold nanoparticles or no/less response signal and vice versa (Reprinted from Long et al. 2015 with © 2015 permission from Elsevier Publishing company); (**d**) Acetamiprid using aptamer-upconversion nanoparticles (UCNPs)–gold nanoparticles (GNPs) based nanosensor where presence of analyte leads to fluorescence signal due to no FRET between UCNPs and GNPs (Reprinted from Hu et al. 2016 with © 2016 permission from Elsevier Publishing company) and (**e**) organophosphorus (OPs) and carbamate Pesticides using colorimetric Sensor Array where presence of analyte results in decrease in colourimetric signal due to inhibition of Acetylcholinesterase (AchE) activity (ChOx-choline oxidase, S-ACh- acetyl thiol choline, ACh-acetyl choline) (Reprinted from Qian and Lin 2015 with © 2015 permission from ACS publications)

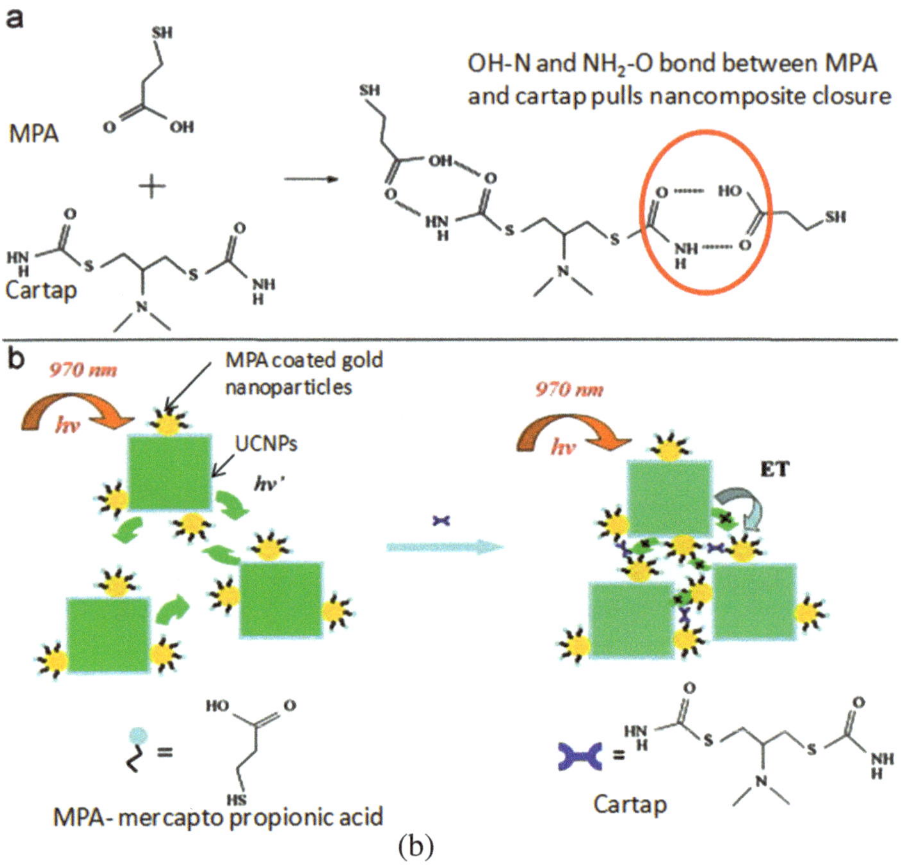

Fig. 1.13 (continued)

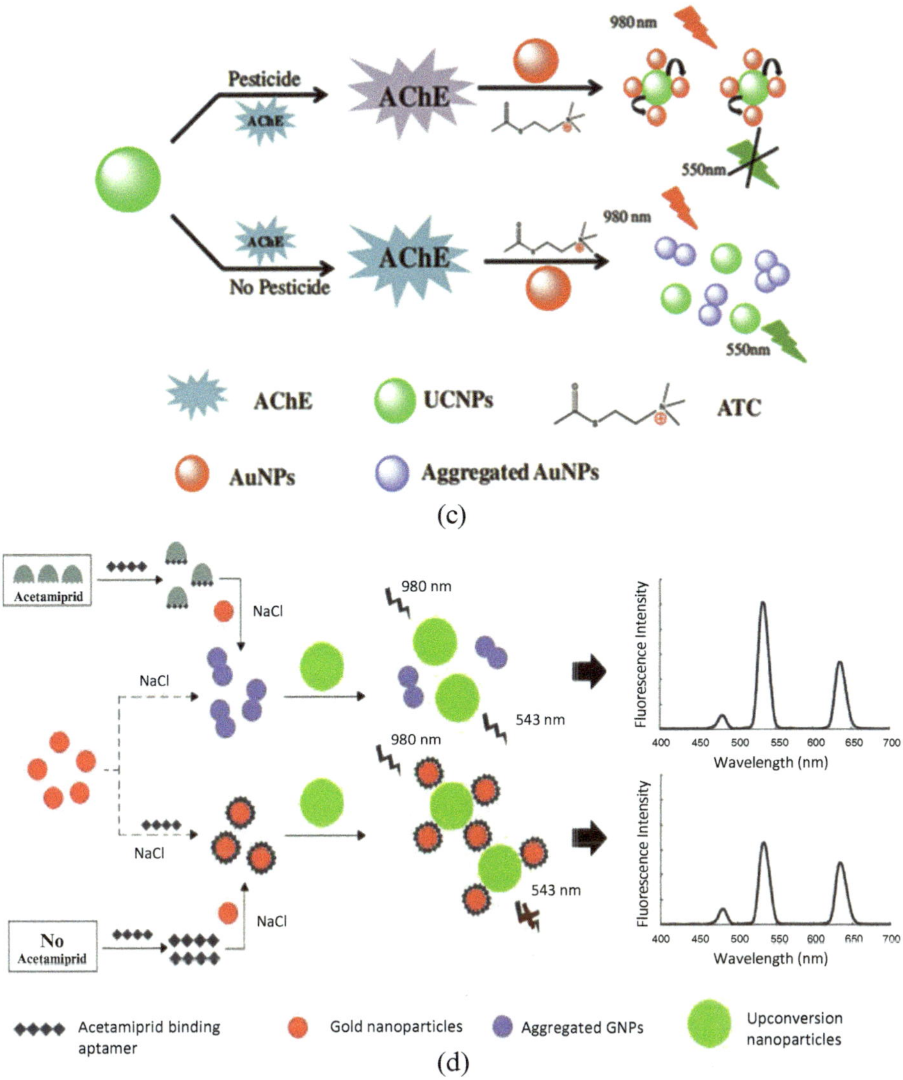

Fig. 1.13 (continued)

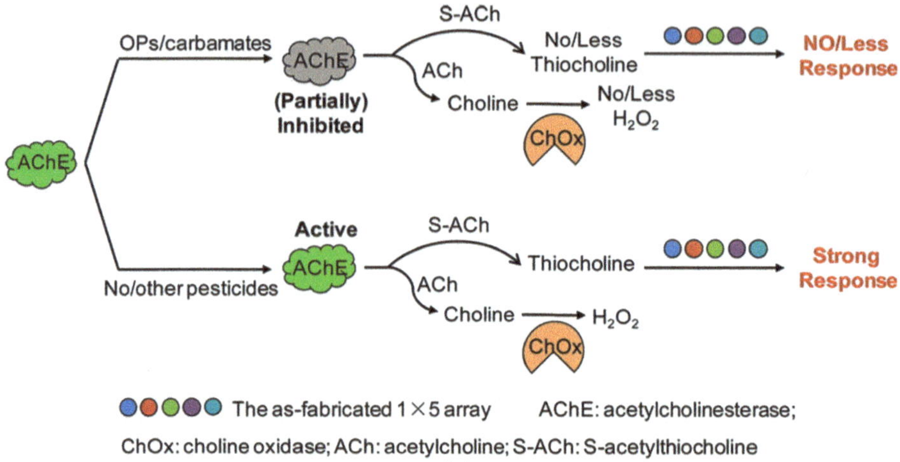

(e)

Fig. 1.13 (continued)

interactions and gold–thiols interaction between thiocholine and gold nanoparticles resulted the disintegration of the gold nano particles/upconversion nanoparticles assembly and the aggregation of gold nanoparticles. In presence of pesticides, activity of acetylcholinesterase is inhibited by pesticides that prevented the generation of thiocholine and facilitated the FRET system resulting in quenching of fluorescence of upconversion nanoparticles. The logarithm of the pesticides concentration was proportional to the inhibition efficiency offering detection limits of parathion-methyl, monocrotophos and dimethoate reached to be 0.67,23, and 67 ng L^{-1}.

In a similar strategy, a upconversion nanoparticles and gold nanoparticles based nanosensor was developed via use of specific aptamer for detection of acetamiprid as shown in Fig. 1.13d (Hu et al. 2016). Acetamiprid is a chloropyridinyl neonicotinoid, is widely used in agriculture and garden markets for its low toxicity and high insecticidal activity. As it is mentioned earlier, for most of the pesticides uncontrolled use of acetamiprid can also lead to dangerous levels of residues in food and its exposure of non-target organisms leading to various harmful effects. Hu et al. 2016 reported of linear detection range of acetamiprid to be from 50 nM to 1000 nM and detection limit of 3.2 nM in adulterated tea sample.

A colorimetric sensor array comprising five inexpensive and commercially available thiocholine and H_2O_2 sensitive indicators for the simultaneous detection and identification of organophosphates and carbamates was reported (Qian and Lin 2015) as shown in Fig. 1.13e. This system makes use of irreversible inhibition acetylcholinesterase activity in presence of organophosphates and carbamates thereby, preventing production of thiocholine and H_2O_2 from S-acetylthiocholine and acetylcholine, resulting in decreased or no color reactions by indicator arrays.

Anti-atrazine based immunosensor for detection of atrazine was reported using directly deposited gold nanostructures onto ITO glass slides for environmental monitoring (Singh et al. 2013a). Fabricated nanobiosensor could sense atrazine through Square Wave Voltammetry and was found to have dynamic linear range from 50 aM to 1 nM (10.78 fg mL^{-1}–215 pg mL^{-1}) in 60 s antigen exposure time. This nanobiosensors was also shown to retain substantial stability till 12 weeks upon storage at 4 °C in desiccated condition and showed no binding with non-specific antigens like malathion, parathion, 2-amino anthracene, albendazole etc.

Graphene oxide-magnetic (Fe_3O_4) nanocomposites based 'on-chip' enzymatic microreactor was developed for ultrasensitive organophosphorus pesticide, dimethoate. This novel on-chip enzymatic (acetylcholine esterase) microreactor exhibited a linear relationship between the inhibition rates of acetylcholine esterase as a function of dimethoate concentration from 1 to 20 μg L^{-1} with a detection limit of 0.18 μg L^{-1} (S/N = 3) (Liang et al. 2013).

Methyl parathion an organophosphorus pesticide which is toxic to both vertebrates and invertebrates (that act by inhibiting acetylcholinesterase enzyme in nerve tissue), is needed to be detected in wide range of matrices due to its extensive use in agriculture and fishes. In the reported smartphone-readable barcode assay, yellow color barcode formation is inhibited due to absence of the activity of acetylcholinesterase (enzyme) (Fig. 1.14).AchE hydrolyses acetyl-thiocholine iodide (substrate) to intermediate (thiocholine), which reacts with DTNB(dithio-bis-nitrobenzoate, a chromogenic reagent) to generate TNB (thionitrobenzoate), giving strong absorbance peak centered at 412 nm with a high extinction coefficient (14,150 M^{-1} cm^{-1}) in dilute buffer solutions (Guo et al. 2015). It can be seen that presence of pesticides(+) leads to enzyme inactivation and hence absence of yellow barcode. While, absence of pesticides(−) results in yellow bar code which is indicated as error code 39 in mobile application.

1.2.8 Plant Diseases

Increased food production is requirement throughout the globe to cope up with the increasing demands of exponentially rising population. Crop infections occurring due to pathogens like bacteria, viruses and fungi are major cause of agricultural losses past many centuries. As per projections an additional 70% of food production is required by 2050 throughout the world (Godfrey et al. 2010) to cope up with the daily nutritional needs especially for lower economies and developing countries. Although, decrease in agricultural productivity can be attributed to a variety of reasons, damage caused by pests and pathogens plays a significant role in crop losses throughout the world. This becomes highly important to ensure agricultural sustainability through addressing specific reasons of the losses and taking preventive measures. Identifying plant pathogens via conventional techniques may take several days and therefore researchers need rapid detection tools that can provide

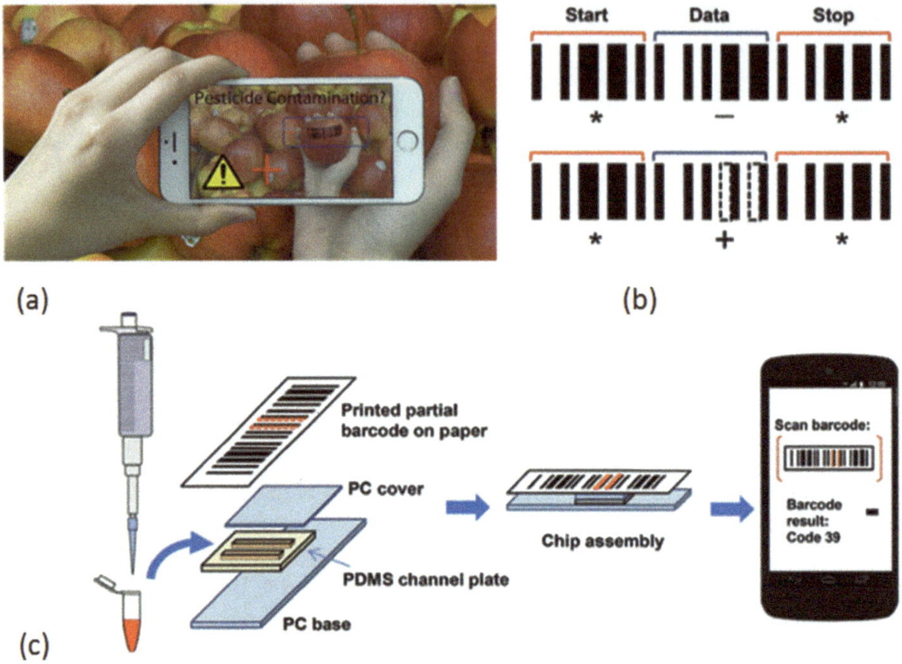

Fig. 1.14 (**a**, **b** and **c**) A smartphone-readable barcode assay for detection. Yellow bar code is formed due to active acetylcholinesterase. In presence of analyte yellow barcodes is not formed and quantitation of pesticide residues colorimetrically read as yellow bar code 39 in absence of pesticide (−) and absence of barcode 39 indicates presence of pesticide (+) (Reprinted from Guo et al. 2015 with © 2015 permission from Royal society of Chemistry)

results within a few hours. Conventionally, direct methods like, PCR: polymerase chain reaction; FISH: fluorescence *in-situ* hybridization; ELISA: enzyme-linked immunosorbent assay; IF: immunofluorescence; FCM: flow cytometry etc.(with detection up to 10^3–10^6 CFU (colony forming unit) mL^{-1}) (Fang and Ramasamy 2015) and indirect methods like, thermography, fluorescence imaging, hyperspectral techniques, gas chromatography are used for detection of various plant diseases. Potential nanotechnology applications in plant pathology can not only facilitate in detection of plant pathogens but also in plant disease control (Khiyami et al. 2014).

Wide range of nanomaterials based biosensors and sensors have been reported for detecting plant diseases by utilizing DNA, antibody and enzymes as biorecognition element making use of variety of measurable signals/changes occurring onto infection containing plants, besides various other parameters that possess potential to be used for food and agricultural applications. As per a recent review from Fang and Ramasamy 2015, some examples of optical nanobiosensors include, antibody assisted fluorescent silica nanoparticles based nanobiosensor for *Xanthomonas axonopodis* that causes bacterial spot disease in *Solanaceae* plant (Yao et al. 2009a, b); gold nanoparticle-based optical immunosensors for karnal bunt disease in wheat using surface plasmon resonance (SPR) (Singh et al. 2010a, b); DNA hybridization fluorescent oligo probes based single probe sensors/ nano-chips based microarrays using for bacteria and viruses (Lopez et al. 2009); quantum dot-fluorescence resonance energy transfer based nanosensor for witches' broom disease of lime caused by *Candidatus* Phytoplasma aurantifolia (*Ca. P. aurantifolia*) using immunosensing at a detection limit of 5 *ca.* P. aurantifolia/ μL (Rad et al. 2012); fluorescent silica nanoparticles based immunosensor for *Xanthomonas axonopodis* pv. *vesicatoria* for bacterial spot disease in tomatoes and peppers (Yao et al. 2009a, b); Rhizomania (a most destructive disease) in sugar beet caused by *beet necrotic yellow vein virus* (BNYVV) carried by vector *Polymyxa betae* (Keskin) (Safarpour et al. 2012); copper oxide (CuO) nanoparticles and nanostructural layer immunosensors for detecting the *A. niger* fungi (Etefagh et al. 2013) and metal oxide nanoparticles (such as Au, SnO_2 and TiO_2) based nanobiosensor for *p*-ethylguaiacol (a volatile organic compounds (VOC)known to be released in various plant diseases) by infected strawberry (Fang et al. 2014a, b) at nanomolar concentration etc.

A recent report suggests detection of viral infection using novel combination of quantum dot incorporated Bacillus spore as nanosensor (Zhang et al. 2015). A spore-based mono disperse microparticles were used to form nanocomposites of spore-based mono disperse microparticles loaded with Cadmium telluride quantum dots. As shown in Fig. 1.15a cadmium telluride quantum dots were multicolor-coded microspheres that have quantum dots of different emission spectra and the capture antigen porcine parvovirus. This was coated on the microparticles loaded cadmium telluride quantum dots surface. The surface reactivity of the microparticles loaded cadmium telluride quantum dots was tested for immunoassay of porcine parvovirus antibody in swine sera using microparticles loaded cadmium telluride quantum dots as suspension beads in a heterogeneous assay system and reporter (labeled with Alexa fluor 647) secondary antibody by monitoring luminescence color by fluorescence spectroscopy, flow cytometry and isothermal titration microcalorimetry.(Fig. 1.15b). This method possesses excellent potential to be implemented for viral infections in the field of food and agriculture.

Amperometric immunosensors for plant pathogens viz. bacteria, viruses and fungi include *Cowpea mosaic virus*, *Tobacco mosaic virus*, *Lettuce mosaic virus*, *Fusarium culmorum*, *Puccinia striiformis*, *Phytophthora infestans*, orchid viruses, chlorotic mottle virus and *Aspergillus niger* (Fang and Ramasamy 2015). Some examples of nano-immunosensors include gold nanorods functionalized antibodies for *Cymbidium mosaic virus* (CymMV) or *Odontoglossum ringspot virus* (ORSV) infections with limits of detection 48 and 42 pg mL^{-1}, respectively, using surface plasmon resonance, 1 ng using quartz crystal microbalance technique in leaf saps (Lin et al. 2014a, b) and polypyrrole nanoribbon modified chemiresistive sensors for *Cucumber mosaic virus* (CMV) up to 10 ng mL^{-1} using amperometric technique (James 2013).

Nucleic acid based nanobiosensors use unique complementary nucleic acid sequences specific to bacterial/viral/fungal pathogens through DNA-DNA, DNA-RNA hybridization as biorecognition event (Arora et al. 2008, Singh et al.

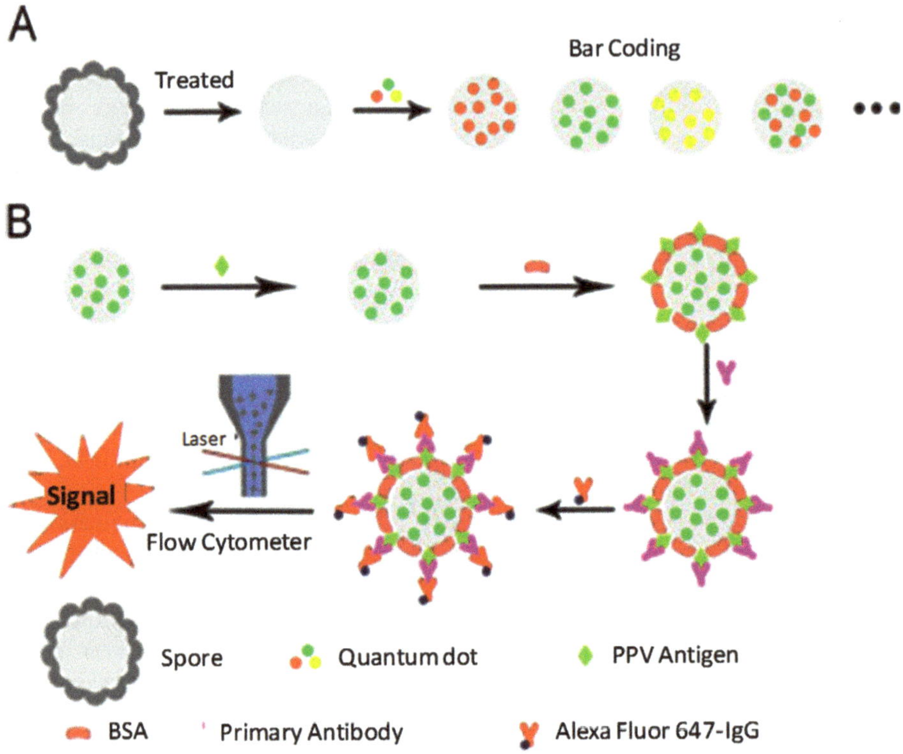

Fig. 1.15 Quantum dot-encoded Bacillus spores (**a**) and microparticles loaded cadmium telluride quantum dots for immunoassay (**b**) where PPV- porcine parvovirus, BSA- bovine serum albumin. Presence of analyte virus is captured by immune labeled microparticles loaded cadmium telluride quantum dots, which is further detected via fluorescent labeled secondary antibody (Reprinted from Zhang et al. 2015 with © 2015 permission from Elsevier Publishing company)

2013b, 2015a, b). Examples of DNA nanobiosensors include use of molecular beacons and quantum dots for two orchid viruses—*Cymbidium mosaic virus* (CymMV) and *Odontoglossum ringspot virus* (ORSV) to detect viral RNA of both orchid viruses up to 0.5 ng of viral RNA in 100 mg orchid leaves using fluorescent probe (Eun and Wong 2000) and presence of target RNA up to 1 ng and 10 ng in the crude sap using quartz crystal microbalance based detection (Eun et al. 2002a, b). A 5′ end fluorescent and 3′ end – gold nanoparticle labelled DNA oligonucleotide was used as a nano transducer to diagnose flavescence dorée phytoplasma of grapevine. Fluorescent signal was measured consequent to hybridization event occurring with complementary target (Firrao et al. 2005).

Enzymatic biosensors for plant pathogen detection usually utilize detection of volatile organic compound which are released in the infected plants only. Studies have shown that various redox enzymes catabolize several of phytohormones and these plant chemicals can be detected using enzyme-based nanobiosensors. Some examples for detection of plant infection are: methyl salicylate with a bi-enzymatic system where analyte determination involves conversion of methyl salicylate to methanol and salicylic acid and then oxidation of methanol (Fang et al., 2014), alcohols and aldehydes such as *cis*-3-hexen-1-ol and *trans*-2-hexanal by alcohol dehydrogenase enzymes (Jansen et al. 2009), common phytohormones such as auxin, cytokinins and gibberellins by oxidases (e.g., Gibberellin by GA-2-oxidases for plant disease prediction)(Thomas et al. 1999; Kulagina et al. 1999).

Advancements in plant science research to analyse plant genomics, gene function, crop improvement and pathogen detection have taken edge with the help of nanotechnological tools and techniques. A recent example of such application suggests use of nanopore technology (Oxford Nanopore Technologies) and "DNA transistor" technology (companies like IBM and Roche are working on this) that can be used for DNA sequencing in minutes instead of hours and days (Niedringhaus et al. 2011, Ozsolak 2012, Zhang et al. 2011). Portable genome sequencer (MinION) is already available to sequence 10 kb of a single sense and anti-sense DNA strand to enable next-generation sequencing (NGS). A protein nanopore and enzyme were designed to control a single strand of DNA, and as the DNA goes through the nanopore a direct electronic analysis is conducted (Clark et al. 2009). The protein nanopore is inserted in a polymer bilayer membrane across the top of a microwell. Each microwell has a sensor chip that measures the ionic current as the single molecule passes through the nanopore. However, the speed at which the DNA strand travels through the nanopore is too fast for accurate identification.

Present day analytical nanosensors associated with biomolecular recognitions coupled with latest tools-technologies, possess capacity to detect and quantify minute amounts of contaminants such as viruses bacteria, fungi, toxins and other bio-hazardous substances in the agriculture and food systems. These nano-sensors can be linked to a global positioning system (GPS) for real-time monitoring of disease and distributed throughout the field to monitor soil conditions, water quality/quantity, ecological changes and crop health. In fact, Khiyami et al. had suggested that, 'nanosensors will allow us to identify plant diseases before visible symptoms appear and thus will facilitate their control and also that recision farming will allow improved agriculture production by providing precise data, helping growers to make better decisions' (Khiyami et al. 2014).

1.2.9 Genetically Modified Organisms (GMOs)

GMOs such as rice (e.g., stress/saline resistant or golden rice to added to its nutritional value), mustard, resistant cotton, tomatoes, fruits and many more had been the need of hour to bring in green revolution in agricultural production while avoiding food spoilage/losses and facilitating nutritional food available to poorest people throughout the world. Lot of GM crops/foods are available throughout the world nowadays, however, there is ongoing debate over use of GMOs and various government based regulations have been applied to put a control on use of GMO crops. European continent has put complete ban over use of GMOs based food materials, however, US continent has been less stringent and liberal enough to allow declaration of use GMO on product packaging. Lot of GM crops and food products are in markets of US, Canada and Asian subcontinents. Presently, there are lot of methods available in the market which are based on polymerase chain reaction (PCR, multiplex PCR, qPCR etc.), microarrays, southern blotting, ELISA, western blots, strip tests including biosensors etc. are available for detection of GM crops or organisms (Singh et al. 2011).

Surface Enhanced Raman Spectroscopy (SERS)-barcoded nanosensor was reported that sensed *Bacillus thuringiensis* (Bt) gene transformed rice, expressing insecticidal proteins. This method used specific oligonucleotide conjugated silica encapsulated gold nanoparticles as 'SERS-barcoded nanoparticles spectroscopic tags' (Chen et al. 2012). The Bt genes usually used in rice are cry1Ab, cry1Ac, therefore, transition between the cry1A(b) and cry1A(c) fusion gene sequence was used to construct a specific SERS-based detection method and sucrose phosphate synthase (a rice gene) was used as interior reference gene of rice that gave detection limit of 0.1 pg mL^{-1} as shown in Fig. 1.16. The SERS-barcode nanosensor had sensitivity and accuracy comparable with real-time PCR. The SERS-barcoded analytical method provided precise detection of transgenic rice varieties but also informative supplement to avoid false positive outcomes.

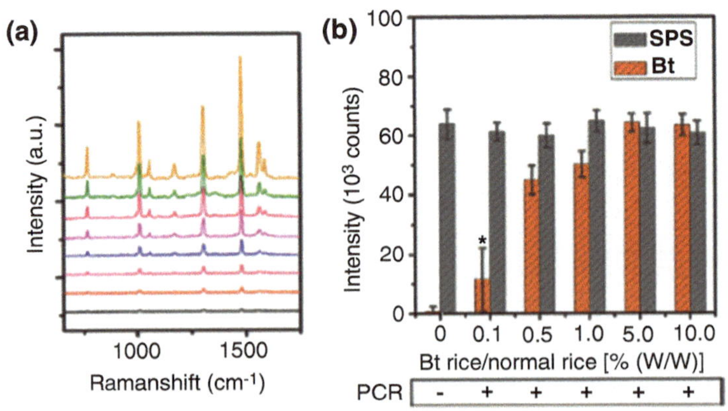

Fig. 1.16 (**a**) SERS spectra of the nanosensor for Bt (from bottom to top: 0.0 pg mL^{-1}, 0.05 pg mL^{-1}, 0.1 pg mL^{-1}, 1.0 pg mL^{-1}, 10 pg mL^{-1}, 100 pg mL^{-1}, 1.0 ng mL^{-1}, 10 ng mL^{-1}, black line: blank control by using sucrose phosphate synthase). (**b**) Normalized Raman intensity and real-time PCR results in the presence of different amount of Bt rice relative to normal rice. Error bars were calculated based on the standard deviation of eight measurements. *Comparison of obtained result with real-time PCR)Reprinted from Chen et al. 2012 with © 2012 permission from Elsevier Publishing company)

1.2.10 Measurement of pH

Food production requires large scale fermentation and bioreactors to achieve desired food product. In this context measurement of pH is one of important parameter not only at industrial scale but also at intra and inter cellular levels for various food and agricultural applications. Various pH control devices have been reported for such applications that involve use of nano based biosensors and/or nanosensors.

A triple fluorescent pH sensor was reported as a new tool for pH measurement that can measure pH 3.9–7.3 by simultaneously incorporating two complemental pH-sensitive fluorophores in a same nanoparticle in a cellular compartment by making use of octaarginine which mimics human immunodeficiency virus-1, Tat protein (a cell penetrating moiety) (Ke et al. 2016). Ratiometric pH nanosensors with tunable pK(a) were prepared by entrapping combinations of two pH-sensitive fluorophores (fluorescein isothiocyanate dextran and Oregon Green(®) dextran) and a reference fluorophore (5-(and-6)-carboxytetramethylrhodamine dextran), in a biocompatible polymer matrix (Chauhan et al. 2011). Dual-fluorophore pH nanosensors permit the measurement of an extended dynamic range, from pH 4.0 to 7.5. A polyacrylamide-based nanosensor with two pH-sensitive fluorophores, fluorescein and Oregon Green was reported to sense (pH 3.1–7.0) having pH-insensitive fluorophore rhodamine as a reference fluorophore. These nanosensors are spontaneously taken up via endocytosis and directed to the lysosomes where dynamic changes in pH can be measured with live-cell confocal microscopy (Sondergaard et al. 2014). Henceforth, it can be stated that nanosensors are available as exciting tools for determining *on line / in situ* pH in the micro and nano environments of living cells as well as in food production bioreactors, thereby allowing measurements of absolute values of pH at places that have so far been restricted by the limited sensitivity range of nanosensors, calibration challenges and the complexity of image analysis.

1.3 Future Prospects of Nano Based Biosensors

Nano based biosensors and nanosensors have demonstrated exceptional amount of developments and has combated various challenges of contemporary as well as competitive methods of detection of various parameters/analytes of interest to achieve unprecedented levels performance i.e. to sense ultra trace amounts with unsurpassed sensitivity. Having been through with entire set of available research in the literature, potential of existing various tools and technologies has been realized. Various techniques like molecular imprinted polymers, microfluidics, plasmonic nanosensors, Surface Enhanced Raman Scattering (SERS)/ fluorescence/chemiluminescence/ quartz crystal microbalance /advanced electrochemical measurements coupled with additional features i.e., ability to 'nano-tune' various properties of fabricated 'nano-bio-molecular assembly systems' as per custom requirements offers limitless possibilities. Merging of chemical and biological components into a single platform can allow new opportunities for future nano sensing/ nano based biosensing applications with additional features of portability, disposability, real-time detection, unprecedented accuracies, and allowing simultaneous analysis of different analytes in a single device. However, it needs lot more to integrate and enable current methodologies to reach to desired level of performance characteristics and open doors to reach to realize *on site, in-situ, on-line* measurements. Moreover, this is also true that, there exists lot of unexplored potential in nano based biosensors, which have not been utilized for various food and agricultural applications till date. Achievements made so far suggest that nano based biosensors are the pioneers for the future diagnostic devices that offer unlimited opportunities to be tapped.

1.4 Conclusions

Nano based biosensors and nanosensors have witnessed successful demonstration of their potential to provide unsurpassed levels of detection limits and sensitivity utilizing various unique properties, features and affinities of biological as well as nanomaterials for various food and agricultural applications. Most of these available reports demonstrate their applications for measurement of food additives, toxins and mycotoxins, microbial contamination, food allergens, nutritional constituents in food, pesticides, environmental parameters, in food quality control, environment, plant diseases, genetically modified organisms/plants/crops (GMOs) etc. However, it can be established that most of these works continue to develop at their primitive stages and exists only till laboratory or researcher level. There had been very few techniques that are successfully translated to real world applications and lot more attention/efforts are needed to smoothen up and remove the barriers to bring these new developments to the market and serve masses throughout the world.

Acknowledgements DBT sponsored projects IYBA-2008 (BT/B1/12/045/2008) and RGYI-2009 (BT/PR13127/GBD/27/195/2009), JNU capacity build up funds (UPE-I) and UPE-II(158, 161) and DST PURSE–I&II are sincerely acknowledged.

References

Actis P, Jejelowo O, Pourmand N (2010) Ultrasensitive mycotoxin detection by STING sensors. Biosens Bioelectron 26(2):333–337

Alves RC, Barroso MF, González-García MB, Beatriz M, Oliveira PP, Matos CD (2016) New trends in food allergens detection: toward biosensing strategies. Crit Rev Food Sci Nutr 56 (14):2304–2319

Ansari AA, Kaushik A, Solanki PR, Malhotra BD (2010) Nanostructured zinc oxide platform for mycotoxin detection. Bioelectrochemistry 77(2):75–81
Arnold MA (1985) Enzyme-based fiber optic sensor. Anal Chem 57:565–566
Arora K, Chand S, Malhotra BD (2006a) Recent developments in bio-molecular electronics detection techniques for food pathogens. Anal Chim Acta 568:259–272
Arora K, Chaubey A, Singhal R, Singh RP, Pandey MK, Samanta SB, Malhotra BD, Chand S (2006b) Application of electrochemically prepared polypyrrole–polyvinyl sulphonate films to DNA biosensor. Biosens Bioelectron 21:1777–1783
Arora K, Prabhakar N, Chand S, Malhotra BD (2007) Conducting polyaniline based reusable electrochemical genosensor for detection of *E coli*. Anal Chem 79:6152–6158; Err. 2008, 80:1833
Bettazzi F, Lucarelli F, Palchetti I, Berti F, Marrazza G, Mascini M (2008) Disposable electrochemical DNA array for PCR amplified detection of hazel nut allergens in food stuffs. Anal Chim Acta 614(1):93–102
Bremer MGEG, Smits NGE, Haasnoot W (2009) Biosensor immunoassay for traces of hazelnut protein in olive oil. Anal Bioanal Chem 395:119–126
Chae EJ, Lee JH, Oh BK, Choi JW (2013) Label-free Nanobiosensor to detect infectious Bacterica based on SERS. J Biomed Nanotechnol 9(4):659–663
Chaubey A, Malhotra BD (2002) Mediated biosensors. Biosens Bioelectron 17(6–7):441–456
Chauhan VM, Burnett GR, Aylott JW (2011) Dual-fluorophore ratiometric pH nanosensor with tuneable pKa and extended dynamic range. Analyst 136(9):1799–1801
Chauhan R, Singh J, Sachdev T, Basu T, Malhotra BD (2016) Recent advances in mycotoxins detection. Biosens Bioelectron 81:532–545
Chen M, Feng YG, Wang X, Li TC, Zhang JY, Qian DJ (2007) Silver nanoparticles capped by oleylamine: formation, growth, and self-organization. Langmuir 23:5296–5304
Chen K, Han H, Luo Z, Wang Y, Wang X (2012) A practicable detection system for genetically modified rice by SERS-barcoded nanosensors. Biosens Bioelectron 34:118–124
Chen L, Mungroo N, Daikuara L, Neethirajan S (2015) Label-free NIR-SERS discrimination and detection of foodborne bacteria by in situ synthesis of Ag colloids. J Nanobiotechnol 45-54 (1–9):13
Chen Q, Hu W, Sun C, Li H, Ouyang Q (2016a) Synthesis of improved upconversion nanoparticles as ultrasensitive fluorescence probe for mycotoxins. Anal Chim Acta 938:137–145
Chen L, Tian X, Zhao Y, Li Y, Yang C, Zhou Z, Liu X (2016b) A ratiometric fluorescence nanosensor for highly selective and sensitive detection of selenite. Analyst 141:4685–4693
Clarke J, HC W, Jayasinghe L, Patel A, Reid S, Bayley H (2009) Continuous base identification for single-molecule nanopore DNA sequencing. Nat Nanotechnol 4(4):265–270
Crevillén AG, Avila M, Pumera M, González MC, Escarpa A (2007) Food analysis on microfluidic devices using ultrasensitive carbon nanotube detectors. Anal Chem 79:7408–7415
Crevillén AG, Pumera M, González MC, Escarpa A (2008) Carbon nanotube disposable detectors in microchipcapillary electrophoresis for water-soluble vitamin determination: analytical possibilities in pharmaceutical quality control. Electrophoresis 29:2997–3004
Dasgupta N, Ranjan S, Mundekkad D, Ramalingam C, Shanker R, Kumar A (2015) Nanotechnology in agro-food: from field to plate. Food Res Int 69:381–400
Dasgupta N, Ranjan S, Ramalingam C (2017) Applications of nanotechnology in agriculture and water quality management. Environ Chem Lett 15(4):591–605
Demchenko AP (2006) Visualization and sensing of intermolecular interactions with two-color fluorescent probes. FEBS Lett 580:2951–2957
Deshmukh RA, Joshi K, Bhand S, Roy U (2016) Recent developments in detection and enumeration of waterborne bacteria: a retrospective minireview. Microbiology 5(6):901–922
Dubertret B, Calame M, Libchaber AJ (2001) Single-mismatch detection using gold-quenched fluorescent oligonucleotides. Nat Biotechnol 19(4):365–370
Eissa S, Chaker T, L'Hocine L, Zourob M (2012) Electrochemical immunosensor for the milk allergen b-lactoglobulin based on electrografting of organic film on graphene modified screen-printed carbon electrodes. Biosen Bioelectron 38:308–313

Eren T, Atar N, Yola ML, Karimi-Maleh H (2015) A sensitive molecularly imprinted polymer based quartz crystal microbalance nanosensor for selective determination of lovastatin red yeast rice. Food Chem 185:430–436

Etefagh R, Azhir E, Shahtahmasebi N (2013) Synthesis of CuO nanoparticles and fabrication of nanostructural layer biosensors for detecting *Aspergillus niger* fungi. Sci Iranian 20 (3):1055–1058

Eun AJC, Wong SM (2000) Molecular beacons: a new approach to plant virus detection. Phytopathology 90:269–275

Eun AJC, Huang L, Chew FT, Li SFY, Wong S-M (2002a) Detection of two orchid viruses using quartz crystal microbalance-based DNA biosensors. Phytopathology 92:654–658

Eun AJC, Huang L, Chew FT, Li SFY, Wong SM (2002b) Detection of two orchid viruses using quartz crystal microbalance (QCM) immunosensors. J Virol Methods 99:71–79

Fadiran AO, Dlamini SC, Mavuso AA (2008a) Comparative study of the phosphate levels in some surface and ground water bodies of Swaziland. Bull Chem Soc Ethiop 22(2):197–206

Fadiran AO, Dlamini SC, Mavuso A (2008b) A comparative study of the phosphate levels in some surface and ground water bodies of Swaziland. Bull Chem Soc Ethiop 22(2):197–206

Fang Y, Ramasamy RP (2015) Current and prospective methods for plant disease detection. Biosensors (Basel) 5(3):537–561

Fang Y, Umasankar Y, Ramasamy RP (2014a) Electrochemical detection of p-ethylguaiacol, a fungi infected fruit volatile using metal oxide nanoparticles. Analyst 139(15):3804–3810

Fang Y, Umasankar Y, Ramasamy RP (2014b) The electrochemical society; Pennington, NJ, USA: 2014. Plant volatile sensor: enzymatic transducer for selective and sensitive determination of methyl salicyalte. ECS meetings Abstract MA 01:414

Farabullini F, Lucarelli F, Palchetti I, Marrazza M, Mascini M (2007) Disposable electrochemical genosensor for simultaneous analysis of bacterial food contaminants. Biosens Bioelectron 22 (7):1544–1549

Fazio B, D'Andrea C, Foti A, Messina E, Irrera A, Donato MG, Villari V, Micali N, Maragò OM, Gucciardi PG (2016) SERS detection of biomolecules at physiological pH via aggregation of gold Nanorods mediated by optical forces and Plasmonic heating. Nat Sci Rep 6(1–13):26952

Firrao G, Moretti M, Ruiz-Rosquete M, Gobbi E, Locci R (2005) Nanobiotransducer for detecting flavescence dorée phytoplasma. J Plant Pathol 87(2):101–107

Ge S, Lu J, Ge L, Yan M, Yu J (2011) Development of a novel deltamethrin sensor based on molecularly imprinted silica nanospheres embedded CdTe quantum dots. Spectrochim Acta A 79:1704–1709

Gerard M, Chaubey A, Malhotra BD (2002) Application of conducting polymers to biosensors. Biosens Bioelectron 17(5):345–359

Godfray HC, Beddington JR, Crute IR, Haddad L, Lawrence D, Muir JF, Pretty J, Robinson S, Thomas SM, Toulmin C (2010) Food security: the challenge of feeding 9 billion people. Science 327(5967):812–818

Goluch ED, Nam JM, Georganopoulou DG, Chiesl TN, Shaikh KA, Ryu KS, Barron AE, Mirkin CA, Liu CA (2006) Bio-barcode assay for on-chip attomolar-sensitivity protein detection. Lab Chip 6(10):1293–1299

Guo J, Wong JXH, Cui C, Li X, Yu HZA (2015) Smartphone-readable barcode assay for the detection and quantitation of pesticide residues. Analyst 140(16):5363–5790

Gyawali R, Ibrahim SA (2012) Synergistic effect of copper and lactic acid against Salmonella and Escherichia coli O157:H7: a review. Emirates J Food Agric 24:1–11

Han SR, Lee SW (2013) *In Vitro* selection of RNA aptamer specific to Salmonella Typhimurium. J Microbiol Biotechnol 23(6):878–884

Han SR, Yu J, Lee SW (2014) *In vitro* selection of RNA aptamers that selectively bind danofloxacin. Biochem Biophys Res Commun 448(4):397–402

Hashimoto Y, Nakamura H, Asaga K, Karube I (2008) A new diagnostic method for soil-borne disease using a microbial biosensor. Microbes Environ 23(1):35–39

Hiep HM, Endo T, Kerman K, Chikae M, Kim D-K, Yamamura S, Takamura Y, Tamiya E (2007) A localized surface plasmon resonance based immunosensor for the detection of casein in milk. Sci Technol Adv Mat 8:331–338

Ho JA, Hsu WL, Liao WC, Chiu JK, Chen ML, Chang HC, Li CC (2010) Ultrasensitive electrochemical detection of biotin using electrically addressable site-oriented antibody immobilization approach via aminophenyl boronic acid. Biosens Bioelectron 26:1021–1027

Hohensinner V, Maier I, Pittne F (2007) A 'gold cluster-linked immunosorbentassay': optical near-field biosensor chip for the detection of allergenic-lactoglobulin in processed milk matrices. J Biotechnol 130:385–388

Hosseini M, Khabbaz H, Dadmehr M, Ganjali MR, Mohamadnejad J (2015) Aptamer-based Colorimetric and Chemiluminescence detection of Aflatoxin B1 in foods samples. Acta Chim Slov 62(3):721–728

Hu XY, Xiao Y, Chen HY (1999) Adsorption characteristics of Fe $(CN)_6^{3-/4-}$ on Au colloids as monolayer films on cysteamine-modified gold electrode. J Electroanal Chem 466:26–30

Hu W, Chen Q, Li H, Ouyang Q, Zhao J (2016) Fabricating a novel label-free aptasensor for acetamiprid by fluorescence resonance energy transfer between NH2-NaYF4: Yb, Ho@SiO2 and Au nanoparticles. Biosens Bioelectron 80:398–404

Huang Y, Bell MC, Suni II (2008) Impedance biosensor for peanut protein Ara h 1. Anal Chem 80:9157–9161

Hughes G, Westmacott K, Honeychurch KC, Crew A, Pemberton RM, Hart JP (2016) Recent advances in the fabrication and application of screen-printed electrochemical (bio)sensors based on carbon materials for biomedical, Agri-food and environmental analyses. Biosensors 6(4):50

IARC (1993) Monographs on the Evaluation of Carcinogenic Risks to Humans. In: Some naturally occurring substances: food items and constituents, heterocyclic aromatic amines and mycotoxins. IARC, Geneva, p 56

Iglesias J, Saldana EG, Jaen JA (2001) On the Tannic acid interaction with metallic iron. Hyperfine Interact 134:109–114

Izadi Z, Sheikh-Zeinoddin M, Ensafi AA, Soleimanian-Zad S (2016) Fabrication of an electrochemical DNA-based biosensor for Bacillus cereus detection in milk and infant formula. Biosens Bioelectron 80:582–589

Jain A, Ranjan S, Dasgupta N, Ramalingam C (2016) Nanomaterials in food and agriculture: an overview on their safety concerns and regulatory issues. Crit Rev Food Sci Nutr 15:1–21

James C (2013) Polypyrrole nanoribbon based chemiresistive immunosensors for viral plant pathogen detection. Anal Methods 5:3497–3502

Jansen RM, Hofstee JW, Wildt J, Verstappen FW, Bouwmeester H, van Henten EJ (2009) Induced plant volatiles allow sensitive monitoring of plant health status in greenhouses. Plant Signal Behav 4:824–829

Jin R, Cao Y, Mirkin CA, Kelly K, Schatz GC, Zheng J (2001) Photoinduced conversion of silver nanospheres to nanoprisms. Science 294:1901–1903

John K, Beedanagari S (2014) Heterocyclic Aromatic Amines. Encyclopaedia of Toxicology 855–863 (thirded.)

Jomma EY, Ding SN (2016) One-pot hydrothermal synthesis of magnetite Prussian blue nanocomposites and their application to fabricate glucose biosensor. Sensors (Basel). 16(2):243

Kacmaz S, Ertekin K, Mercan D, Oter O, Cetinkaya E, Celik E (2015) An ultra sensitive fluorescent nanosensor for detection of ionic copper. Spectrochim Acta A Mol Biomol Spectrosc 135:551–559

Kaushik A, Solanki PR, Ansari AA, Ahmad S, Malhotra BD (2009) A nanostructured cerium oxide film-based immunosensor for mycotoxin detection. Nanotechnology 20(5):055105

Ke P, Sun H, Liu M, Su Z, Hu K, Zhua H, Guo H, Sun H, Andresen TL, Olsen LF (2016) Polymeric pH nanosensor with extended measurement range bearing octaarginine as cell penetrating peptide. IET Nanobiotechnol 10(1):8–12

Khiyami MA, Almoammar H, Awad YM, Alghuthaymi MA, Abd-Elsalam KA (2014) Plant pathogen nanodiagnostic techniques: forthcoming changes? Biotechnol Biotechnol Equip 28 (5):775–785

Kim KI, Lee JC, Robards K, Choi SH (2010) Immobilization of tyrosinase in carboxylic and carbonyl group-modified MWNT electrode and its application for sensing phenolics in red wines. J Nanosci Nanotechnol 10(6):3790–3798

Kulagina NV, Shankar L, Michael AC (1999) Monitoring glutamate and ascorbate in the extracellular space of brain tissue with electrochemical microsensors. Anal Chem 71:5093–5100

Labroo P, Cui Y (2013) Flexible grapheme bio-nanosensor for lactate. Biosens Bioelectron 41:852–856

Lattanzio VMT, Nivarlet N, Lippolis V, Gatta SD, Huet AC, Delahaut P, Granier B, Visconti A (2012) Multiplex dipstick immunoassay for semi-quantitative determination of Fusarium mycotoxins in cereals. Anal Chim Acta 718:99–108

Lee YJ, Han SR, Maeng JS, Cho YJ, Lee SW (2012) In vitro selection of Escherichia coli O157: H7-specific RNA aptamer. Biochem Biophys Res Commun 417:414–420

Liang RP, Wang XN, Liu CM, Meng XY, Qiu JD (2013) Construction of graphene oxide magnetic nanocomposites-based on-chip enzymatic microreactor for ultrasensitive pesticide detection. J Chromatogr A 1315:28–35

Lim JW, Ha D, Lee J, Lee SK, Kim T (2015 May 11) Review of micro/nanotechnologies for microbial biosensors. Front Bioeng Biotechnol 3:61

Lin X, Guo X (2016) Advances in biosensors, Chemosensors and assays for the determination of Fusarium Mycotoxins. Toxins (Basel) 8(6):173

Lin HY, Huang CH, SH L, Kuo IT, Chau LK (2014a) Direct detection of orchid viruses using nanorod-based fiber optic particle plasmon resonance immunosensor. Biosens Bioelectron 51:371–378

Lin XX, Li ZH, Huang XY, Jiang Y, Wei QH, Chena GN (2014b) Pt_2Ag acetylide-doped silica nanoparticles: enabling luminescence of Pt_2Ag complexes in water and sensors for highly sensitive detection of cyanide anions. Dalton Trans 43:8861–8868

Lísalová HV, Surman F, Víšová I, Milan Vala M, Špringer T, Ermini ML, Šípová H, Šedivák P, Houska M, Riedel T, Georgievski OP, Brynda E, Homola J (2016) Copolymer brush-based ultralow-fouling biorecognition surface platform for food safety. Anal Chem. in press

Long Q, Li H, Zhang Y, Yao S (2015) Upconversion nanoparticle-based fluorescence resonance energy transfer assay for organophosphorus pesticides. Biosens Bioelectron 68:168–174

López MM, Llop P, Olmos A, Marco-Noales E, Cambra M, Bertolini E (2009) Are molecular tools solving the challenges posed by detection of plant pathogenic bacteria and viruses? Curr Issues Mol Biol 11:13–46

López C, Zougagh M, Algarra M, Rodríguez-Castellón E, Campos BB, Esteves da Silva JCG, Jiménez-Jiménez J, Ríos A (2015) Microwave-assisted synthesis of carbon dots and its potential as analysis of four heterocyclic aromatic amines. Talanta 132:845–850

Lu Y, Oshima T, Ushio H (2004) Rapid detection of fish major allergen parvalbumin by surface plasmon resonance biosensor. Food Chem Toxicol 69:C652–C657

Lu X, Wang X, Wu L, Wu L, DL F, Gao Y, Chen J (2016) Response characteristics of bisphenols on a metal–organic framework-based Tyrosinase Nanosensor. ACS Appl Mater Interf 8:16533–16539

Luka G, Ahmadi A, Najjaran H, Alocilja E, DeRosa M, Wolthers K, Malki A, Aziz H, Althani A, Hoorfar M (2015) Microfluidics integrated biosensors: a leading technology towards lab-on-a-Chip and Sensing applications. Sensors (Basel). 15(12):30011–30031

Magro M, Bonaiuto E, Baratella D, Roger JA, Jakubec P, Corraducci V, Tucˇek J, Malina O, Zborˇil R, Vianello F (2016) Electrocatalytic nanostructured ferric tannates: characterization and application of a polyphenol nanosensor. Chem Phys Chem 17:1–9. (An article in press)

Maier I, Morgan MR, Lindner W, Pittner F (2008) Optical resonance-enhanced absorption-based near-field immunochip biosensor for allergen detection. Anal Chem 80:2694–2703

Mak AC, Osterfeld SJ, Yu H, Wang SX, Davis RW, Jejelowo OA, Pourmand N (2010) Sensitive giant magneto resistive-based immunoassay for multiplex mycotoxin detection. Biosens Bioelectron 25(7):1635–1639

Maragos CM (2016) Multiplexed biosensors for Mycotoxins. J AOAC Int 99(4):849–860

McPartlin DA, Lochhead MJ, Connell LB, Doucette GJ, O'Kennedy RJ (2016) Use of biosensors for the detection of marine toxins. Essays Biochem 60(1):49–58

Maddinedi SB, Mandal BK, Ranjan S, Dasgupta N (2015) Diastase assisted green synthesis of size-controllable gold nanoparticles. RSC Adv 5(34):26727–26733

Maddinedi SB, Mandal BK, Patil SH, Andhalkar VV, Ranjan S, Dasgupta N (2017) Diastase induced green synthesis of bilayered reduced graphene oxide and its decoration with gold nanoparticles. J Photochem Photobiol B Biol 166:252–258

Menzel A, Gubeli RJ, Gu¨der F, Weberb W, Zachariasac M (2013) Detection of real-time dynamics of drug–target interactions by ultralong nanowalls. Lab Chip 13:4173–4180

Millstone JE, Park S, Shuford KL, Qin L, Schatz GC, Mirkin CA (2005) Observation of a quadrupole plasmon mode for a colloidal solution of gold nanoprisms. J Am Chem Soc 127:5312–5313

Mohammed MI, Desmulliez MPY (2013) The manufacturing of packaged capillary action microfluidic systems by means of CO_2 laser processing. Microsyst Technol 19(6):809–818

Mohammed I, Mullett W, Lai E, Yeung J (2001) Is biosensor a viable method for food allergen detection? Anal Chim Acta 444:97–102

Montereali MR, Seta LD, Vastarella W, Pilloton R (2010) A disposable Laccase–Tyrosinase based biosensor for amperometric detection of phenolic compounds in must and wine. J Mol Catal B Enzym 64:189–194

Muyanja CMBK, Narvhus JA, Langsrud T (2012) Organic acids and volatile organic compounds produced during traditional and starter culture fermentation of *Bushera*, a Ugandan fermented cereal beverage. Food Biotechnol 26:1–28

Niedringhaus TP, Milanova D, Kerby MB, Snyder MP, Barron AE (2011) Landscape of next-generation sequencing technologies. Anal Chem 83(12):4327–4341

Noh HJ, Kim HS, Cho S, Park Y (2013) Melamine nanosensing with chondroitin sulfate-reduced gold nanoparticles. J Nanosci Nanotechnol 13(12):8229–8238

Nollet L, De Gelder L (2013) Handbook of water analysis. CRC Press, Boca Raton

Ogabiela E, Adeloju SB, Cui J, Wu Y, Chen WA (2015) Novel ultrasensitive phosphate amperometric nanobiosensor based on the integration of pyruvate oxidase with highly ordered gold nanowires array. Biosens Bioelectron 71:278–285

Omara W, Amin R, Elhaes H, Ibrahim M, Elfeky SA (2016) Preparation and characterization of novel polyaniline nanosensor for sensitive detection of formaldehyde. Recent Pat Nanotechnol 9(3):195–203

Ozsolak F (2012) Third-generation sequencing techniques and applications to drug discovery. Expert Opin Drug Discov 7(3):231–243

Pal S, Sharma MK, Danielsson B, Willander M, Chatterjee R, Bhand S (2014) A miniaturized nanobiosensor for choline analysis. Biosens Bioelectron 54:558–564

Pérez-Juste J, Pastoriza-Santos I, Liz-Marzán LM, Mulvaney P (2005) Gold nanorods: synthesis, characterization and applications. Coord Chem Rev 249:1870–1901

Pimentel D (2009) Invasive plants: their role in species extinctions and economic losses to agriculture in the USA. In: Inderjit (ed) Management of invasive weeds, invading nature Springer Series in invasion ecology. Springer, Dordrecht pp 1–7

Pollet J, Delport F, Janssen KP, Tran D, Wouters J, Verbiest T, Lammertyn J (2011) Fast and accurate peanut allergen detection with nanobead enhanced optical fiber SPR biosensor. Talanta 83:1436–1441

Prabhakar N, Arora K, Singh SP, Pandey MK, Singh H, Malhotra BD (2007) Polypyrrole-polyvinyl sulphonate film based disposable nucleic acid biosensor. Anal Chim Acta 589:6–13

Prabhakar N, Sumanaa G, Arora K, Singh H, Malhotra BD (2008) Improved electrochemical nucleic acid biosensor based on polyaniline-polyvinyl sulphonate. Electrochim Acta 53:4344–4350

Puangsombat K, Gadgil P, Houser TA, Hunt MC, Smith JS (2012) Occurrence of heterocyclic amines in cooked meat products. Meat Sci 90:739–746

Qian S, Lin H (2015) Colorimetric sensor array for detection and identification of organophosphorus and carbamate pesticides. Anal Chem 87(10):5395–5400

Qiu L, Wang WQ, Zhang AW, Zhang NN, Lemma T, Ge HH, Toppari JJ, Hytönen VP, Wang J (2016) Core−Shell Nanorod columnar array combined with gold nanoplate−nanosphere assemblies enable powerful in situ SERS detection of bacteria. ACS Appl Mater Interfaces 8:24394–24403

Rad F, Mohsenifar A, Tabatabaei M, Safarnejad MR, Shahryari F, Safarpour H, Foroutan A, Mardi M, Davoudi D, Fotokian M (2012) Detection of Candidatus Phytoplasma aurantifolia with a quantum dots fret-based biosensor. J Plant Pathol 94:525–534

Ragavan KV, Rastogi NK, Thakur MS (2013) Sensors and biosensors for analysis of Bisphenol-A. TrAC Trends Anal Chem 52:248–260

Ramesh R, Puhazhendi P, Kumar J, Gowthaman MK, D'Souza SF, Kamini NR (2015) Potentiometric biosensor for determination of urea in milk using immobilized Arthrobacter creatinolyticus urease. Mater Sci Eng C Mater Biol Appl 49:786–792

Ranjan S, Dasgupta N, Chakraborty AR, Melvin Samuel S, Ramalingam C, Shanker R, Kumar A (2014) Nanoscience and nanotechnologies in food industries: opportunities and research trends. J Nanopart Res 16(6):1–23

Raz SR, Liu H, Norde W, Bremer MG (2010) Food allergens profiling with an imaging surface plasmon resonance-based biosensor. Anal Chem 82:8485–8491

Reverté L, Prieto-Simón B, Campàs M (2016) New advances in electrochemical biosensors for the detection of toxins: nanomaterials, magnetic beads and microfluidics systems. A review. Anal Chim Acta 908:8–21

Richman EL, Kenfield SA, Stampfer MJ, Giovannucci EL, Zeisel SH, Willett WC, Chan JM (2012) Choline intake and risk of lethal prostate cancer: incidence and survival. Am J Clin Nutr 96(4):855–863

Ruscito A, Smith M, Goudreau DN, DeRosa MC (2016) Current status and future prospects for aptamer-based Mycotoxin detection. J AOAC Int 99(4):865–877

Saeedfar K, Heng LY, Ling TL, Rezayi M (2013) Potentiometric urea biosensor based on an immobilised fullerene-urease bio-conjugate. Sensors (Basel). 13(12):16851–16866

Safarpour H, Safarnejad MR, Tabatabaei M, Mohsenifar A, Rad F, Basirat M, Shahryari F, Hasanzadeh F (2012) Development of a quantum dots FRET-based biosensor for efficient detection of Polymyxa betae. Can J Plant Pathol 34:507–515

Sekhon BS (2010) Food nanotechnology – an overview. Nanotechnol Sci Appl 3:1–15

Sharma A, Matharu Z, Sumana G, Solanki PR, Kim CG, Malhotra BD (2010) Antibody immobilized cysteamine functionalized-gold nanoparticles for aflatoxin detection. Thin Solid Films 519(3):1213–1218

Shukla A, Dasgupta N, Ranjan S, Singh S, Chidambram R (2017) Nanotechnology towards prevention of anaemia and osteoporosis: from concept to market. Biotechnol Biotechnol Equip 31(5):863–879

Shwetha N, Selvakumar LS, Thakur MS (2013) Aptamer–nanoparticle-based chemiluminescence for p53 protein. Anal Biochem 441:73–79

Siddiki MSR, Shimoaoki S, Ueda S, Maeda I (2012) Thermoresponsive magnetic nano-biosensors for rapid measurements of Inorganic Arsenic and Cadmium. Sensors (Basel) 12 (10):14041–14052

Singh R, Sharma PP, Baltus RE, Suni II (2010a) Nanopore immunosensor for peanut protein Ara h1. Sens Actuators B Chem 145:98–103

Singh S, Singh M, Agrawal VV, Kumar A (2010b) An attempt to develop surface plasmon resonance based immunosensor for Karnal bunt (*Tilletia indica*) diagnosis based on the experience of nano-gold based lateral flow immuno-dipstick test. Thin Solid Films 519:1156–1159

Singh A, Singh MP, Sharma, V, Verma HN, Arora K (2011) Chapter 13: molecular techniques in edited book on chemical analysis of food: techniques & applications. Elsevier Publishing. http://www.elsevier.com/wps/find/bookdescription.Cws_home/726991/description#description. ISBN: 978-0-12-384862-8

Singh A, Joshi Y, Singh MP, Arora K (2013a) Anti-atrazine functionalized gold nano structures for environmental monitoring. Biosens J 3(1):105

Singh A, Sinsinbar G, Choudhary M, Kumar V, Pasricha R, Singh SP, Verma HN, Arora K (2013b) Graphene oxide-chitosan nanocomposite based electrochemical DNA biosensor for detection of typhoid. Sens Actuators: B 185:675–684

Singh A, Chaudhary M, Singh MP, Verma HN, Singh SP, Arora K (2015a) Direct electrochemical deposition of gold Nano-aggregates and their application towards electrochemical DNA biosensor for detection of *Salmonella typhi*. Bioelectrochemistry 105:7–15

Singh A, Verma HN, Arora K (2015b) Surface Plasmon resonance based label-free detection of salmonella. Appl Biochem Biotechnol 175:1330–1343

Siripireddy B, Mandal BK, Ranjan S, Dasgupta N, Ramalingam C (2017) Nano-zirconia – evaluation of its antioxidant and anticancer activity. J Photochem Photobiol B Biol 170:125–133. https://doi.org/10.1016/j.jphotobiol.2017.04.004

Sivasankaran U, Cyriac S T, Menon S, Kumar KG (2016) Fluorescence turn off sensor for brilliant BlueFCF- an approach based on inner filter effect. J Fluoresc (article in press)

Smith VH, Tilman GD, Nekola JC (1999) Eutrophication: impacts of excess nutrient inputs on freshwater, marine, and terrestrial ecosystems. Environ Pollut 100:179–196

Søndergaard RV, Henriksen JR, Andresen TL (2014) Design, calibration and application of broad-range optical nanosensors for determining intracellular pH. Nat Protoc 9(12):2841–2858

Spellman F (2013) Basic water chemistry. Handbook of water and wastewater treatment plant operations, 3rd edn. CRCPress, Boca Raton, pp 453–470

Swami A, Kumar A, Selvakannan P, Mandal S, Pasricha R, Sastry M (2003) Highly oriented gold nanoribbons by the reduction of aqueous chloroaurate ions by hexadecylaniline Langmuir monolayers. Chem Mater 15:17–19

Tammina SK, Mandal BK, Ranjan S, Dasgupta N (2017) Cytotoxicity study of Piper nigrum seed mediated synthesized SnO2 nanoparticles towards colorectal (HCT116) and lung cancer (A549) cell lines. J Photochem Photobiol B Biol 166:158–168

Tan HL, Li Q, Ma C, Song Y, Xu F, Chen S, Wang L (2015) Lanthanide based dual-emission fluorescent probe for detection of mercury(II) in milk. Biosens Bioelectron 63:566–571

Ten ST, Hashim U, Gopinath SCB, Liu WW, Foo KL, Sam ST, SFA R, Voon CH, Nordin AN (2016) Highly sensitive Escherichia coli shear horizontal surface acoustic wave biosensor with silicon dioxide nanostructures. Biosens Bioelectron. in press

Thomas SG, Phillips AL, Hedden P (1999) Molecular cloning and functional expression of gibberellin 2-oxidases, multifunctional enzymes involved in gibberellin deactivation. Proc Natl Acad Sci U S A 96:4698–4703

Touhami A (2014) Chapter 15: biosensors and nanobiosensors: design and applications. In: Seifalian A, de Mel A, Kalaskar DM (eds) Nanomedicine. Onc Central Press, pp 374–403

Tran DT, Knez K, Janssen KP, Pollet J, Spasic D, Lammertyn J (2013) Selection of aptamers against Ara h 1 protein for FO-SPR biosensing of peanut allergens in food matrices. Biosens Bioelectron 15:245–251

Turner NW, Bramhmbhatt H, Szabo-Vezse M, Poma A, Coker R, Piletsky SA (2015) Analytical methods for determination of mycotoxins: an update (2009–2014). Anal Chim Acta 901:12–33

Upadhyay LSB, Verma N (2015) Recent advances in phosphate biosensors. Biotechnol Lett 37:1335–1345

Wang L, Chen X, Zhan J, Chai Y, Yang C, Xu L et al (2005) Synthesis of gold nano-and microplates in hexagonal liquid crystals. J Phys Chem B 109:3189–3194

Wang W, Han J, Wu Y, Yuan F, Chen Y, Ge Y (2011) Simultaneous detection of eight food allergens using optical thin-film biosensor chips. J Agric Food Chem 59:6889–6894

Wang Z, Wu L, Shen B, Jiang Z (2013) Highly sensitive and selective cartap nanosensor based on luminescence resonance energy transfer between NaYF4:Yb,Ho nanocrystals and gold nanoparticles. Talanta 114:124–130

Wang L, Zheng J, Yang S, Wu C, Liu C, Xiao Y, Li Y, Qing Z, Yang R (2015) Two-photon sensing and imaging of endogenous biological cyanidein plant tissues using Graphene Quantum Dot/Gold nanoparticle conjugate. ACS Appl Mater Interfaces 7:19509–19515

Weng X, Neethirajan S (2016) A microfluidic biosensor using graphene oxide and aptamer-functionalized quantum dots for peanut allergen detection. Biosens Bioelectron 85:649–656

Wu Q, Long Q, Li H, Zhang Y, Yao S (2015) An up conversion fluorescence resonance energy transfer nanosensor for one step detection of melamine in raw milk. Talanta 136:47–53

Xiulan S, Yinzhi Z, Jingdong S, Liyan S, He Q, Weijuan Z (2010) A quartz crystal microbalance-based immunosensor for shrimp allergen determination in food. Eur Food Res Technol 231:563–570

Yang JH, Lee JC, Choi SH (2009) Tyrosinase-immobilized biosensor based on the functionalized Hydroxyl group-MWNT and detection of phenolic compounds in red wines. J Sens 2009:1–9

Yao KS, Li SJ, Tzeng KC, Cheng TC, Chang CY, Chiu CY, Liao CY, Hsu JJ, Lin ZP (2009a) Fluorescence silica nanoprobe as a biomarker for rapid detection of plant pathogens. Multi-Funct Mater Struct II Parts 1 and 2 79–82:513–516

Yao KS, Li SJ, Tzeng KC, Cheng TC, Chang CY, Chiu CY, Liao CY, Hsu JJ, Lin ZP (2009b) Fluorescence silica nanoprobe as a biomarker for rapid detection of plant pathogens. Adv Mater Res 79:513–516

Yman I, Eriksson A, Johansson M, Hellenäs K (2006) Food allergen detection with biosensor immunoassays. J AOAC Int 89:856–861

Yola ML, Uzun L, Özaltın N, Denizli A (2014) Development of molecular imprinted nanosensor for determination of tobramycin in pharmaceuticals and foods. Talanta 120:318–324

Yoo SM, Lee SY (2016) Optical biosensors for the detection of pathogenic microorganisms. Trends Biotechnol 34(1):7–25

Zakel S, Zakel S, Rienitz O, Güttler B, Rainer Stosch R (2011) Double isotope dilution surface-enhanced Raman scattering as a reference procedure for the quantification of biomarkers in human serum. Analyst 136:3956–3961

Zamolo VA, Valenti G, Venturelli E, Chaloin O, Marcaccio M, Boscolo S, Castagnola V, Sosa S, Berti F, Fontanive G, Poli M, Tubaro A, Bianco A, Paolucci F, Prato M (2012) Highly sensitive electrochemiluminescent nanobiosensor for the detection of palytoxin. ACS Nano 6 (9):7989–7997

Zeisel SH, Blusztajn JK (1994) Annual review of nutrition choline and human. Nutrition 14:269–296

Zhang X, Young MA, Lyandres O, Van Duyne RP (2005) Rapid detection of an anthrax biomarker by surface-enhanced Raman spectroscopy. J Am Chem Soc 127:4484–4489

Zhang J, Chiodini R, Badr A, Zhang G (2011) The impact of next-generation sequencing on genomics. J Genet Genomics 38(3):95–109

Zhang L, Xu J, Mi L, Gong H, Jiang S, Yu Q (2012) Multifunctional magnetic–plasmonic nanoparticles for fast concentration and sensitive detection of bacteria using SERS. Biosens Bioelectron 31:130–136

Zhang X, Zhou Q, Shen Z, Li Z, Fei R, Ji EH, Huc S, Hua Y (2015) Quantum dot incorporated Bacillus spore as nanosensor for viral infection. Biosens Bioelectron 74:575–580

Zhao Q, Huang H, Zhang L, Wang L, Zeng Y, Xia X, Liu F, Chen Y (2016) Strategy to fabricate naked-eye readout ultrasensitive Plasmonic Nanosensor based on enzyme mimetic gold nanoclusters. Anal Chem 88:1412–1418

Zhou H, Yang D, Ivleva NP, Mircescu NE, Niessner R, Haisch C (2014) SERS detection of bacteria in water by in situ coating with Ag nanoparticles. Anal Chem 86:1525–1533

Chapter 2
Physical, Chemical and Biochemical Biosensors to Detect Pathogens

Brindha J, Kaushik Chanda, and Balamurali MM

Abstract The development of science and technology has not only improved the comfort of humans but also added a wide variety of hazardous chemicals and life threatening pathogens into the living environment. Surveillance of bacterial pathogens is a daunting task for healthcare industries, food industries and environmental quality control sectors. During the past few decades, pathogens with high virulence have emerged, leading to steady increase in the mortality and morbidity rates, posing burden on the nation's economy. Therefore it becomes necessary to develop devices that can quickly sense pathogens in quantities much lower than pico- and femto-moles. A ideal sensor has short sensing time, low measurable quantities and reliable results.

In this chapter we discuss various types of biosensors for pathogen detection. Optical biosensors have been explored extensively and used as labeled (fluorophores, quantum dots, carbon dots), label free (surface plasmon resonance) and hybrid biosensors for a highly sensitive pathogen detection. Piezoelectric-cantilever biosensors are simple, rapid and as effective as conventional pathogen detection techniques and are notable for detection of food pathogens like *Listeria monocytogenes*. Successful electrochemical biosensors have also been developed with unmodified electrodes and later electrodes were modified with bio-recognition elements such as specific DNA, antibodies or nanoparticles, for detection of pathogens like methicillin resistant *Staphylococcus aureus* and *Salmonella*. Almost all biosensors, including immunosensors, are being improved, by sample enrichment or signal amplification, in order to obtain a simple and rapid pathogen detection tool with lower limits of detection.

B. J • B. MM (✉)
Department of Chemistry, School of Advanced Sciences, VIT University, Chennai campus, Vandalur – Kelambakkam Road, Chennai 600127, Tamil Nadu, India
e-mail: brindha.j@vit.ac.in; balamurali.mm@vit.ac.in

K. Chanda
Department of Chemistry, School of Advanced Sciences, VIT University, Vellore 632014, Tamil Nadu, India
e-mail: chandakaushik1@gmail.com

K M Gothandam et al. (eds.), *Nanotechnology, Food Security and Water Treatment*, Environmental Chemistry for a Sustainable World 11,
https://doi.org/10.1007/978-3-319-70166-0_2

Keywords Biosensors • Immunosensors • Optical sensors • Surveillance tools • Bio-recognition

Contents

2.1 Introduction

Sensors are becoming a vital part in human lives in this fast moving world due to its rapid mode of detection and ease of use. They are used to detect the measurable responses of an analyte in a sample with respect to either physical conditions or chemical interactions. A good sensor is the one that possesses optimum characteristics like reproducibility, sensitivity, selectivity, stability and linearity (Bhalla et al. 2016). Biosensors are analytical devices that work by detecting the presence and/or concentration of a biological analyte, such as biomolecules, microorganism, or any biological structure. It is composed of three major components namely: bio-recognition element that recognizes the analyte and produces a specific signal, transducer that receives and transmits the signal in a readable format which is read by another component, a reader device. With developments in science and technology, there is an increase in exposure to different kinds of hazardous chemicals and life threatening pathogens (Dasgupta et al. 2015, 2017; Shukla et al. 2017; Jain et al. 2016; Ranjan et al. 2014). This urges researchers for the development of a vast range of sensors. Detection of pathogen gains extreme importance in various sectors like health care for clinical diagnosis (Yanase et al. 2014), food industry to ensure food quality (Scognamiglio et al. 2014), and water and environmental quality control (Bereza-Malcolm et al. 2014; Teo and Wong 2014). Most of the available diagnostic methods are time consuming, requires skilled labor and large sample volumes. All these pave way for the researchers to develop sensors that can offer fast response, ease to operate and high sensitivity to detect analytes.

Detection of pathogens in food, water and air has been vital for the researchers due to its crucial effects on the health of people. Though the standard techniques in microbiology using cell culture and plating could confirm the identity of microbial strains (Gracias and McKillip 2004; Monis and Giglio 2006), it usually takes several days to finish the processes. Also, almost all the conventional methods involve complex instrumentation which defies on-site detection process. Conventional methods of pathogen detection are still in use in spite of their longer response times, which is solely due to their high selectivity and sensitivity. Thus, it becomes essential to develop biosensors that could be used for pathogen detection in a rapid and precise manner. Efficient pathogen sensors must satisfy several requisites. Firstly they need to exhibit high sensitivity and lower limits of detection. As the microorganisms like bacteria multiplies at a faster rate, even meager numbers of cells could lead to risk of health in patients (Ward et al. 2014; Thakur et al. 2015). It is also necessary to meet zero tolerance of certain bacterial strains such as *E. coli* O157:H7, *Salmonella*, *Listeria monocytogenes etc.*, in food products to meet the standards of food safety (Batt 2007; Nugen and Baeumner 2008). Second important requirement is faster mode of analysis. This is essential in order to take immediate actions for treating patients and preventing the transmission of pathogens. Next requisite would be high specificity in identifying different bacterial strains. This could be satisfied by means of micro or nano arrays or chips in sensors with a number of target specific probes. Lastly, portability, ease of use and also automation are significant for real time as well as long term monitoring.

A pathogenic biosensor works by transmitting/transducing receptor recognition signals to a specific target pathogen, into a detectable signal. Presently available techniques to detect pathogens include immunosensing or nucleic acid (Deoxyribonucleic acid/aptamers) detection. There are various modes of transduction that includes optical mode involving fluorescence (Li et al. 2015), surface plasmon resonance (Rifat et al. 2015; Liu et al. 2016a), colorimetric (Taneja and Tyagi 2007), mechanical mode involving cantilever (Fritz 2008)/ piezoelectric crystal/quartz crystalline microbalance (Fawcett et al. 1988; Farka et al. 2013) and electrochemical mode consisting of amperometric, voltammetric and impedimetric (Wan et al. 2011; Chen and Shah 2013) and immunosensors (Menti et al. 2016; Sign and Sumana 2016). Immunosensors are based on antigen-antibody interactions, where antibodies specific to antigens are immobilized on sensor surfaces. Antibody development and selection are vital for constructing an immunosensor for pathogen detection. Moreover, as the cells contain lesser concentrations of nucleic acids or other analytes, the biosensors need an enrichment step for amplification of targeted analytes like nucleic acids using polymerase chain reaction (PCR), reverse transcriptase-PCR and various other techniques. Amplification of the transducing signals is also done for confirming the hybridization or affinity binding between the probe and the analyte. The specific target DNA/ribonucleic acid (RNA) can also be identified using different physical bio-sensing methods. Overall, the efficiency of a pathogen sensor is based on the specificity of biochemical reactions, high concentration of analytes under investigation and sensitive/selective detection or transduction methods. The sensor probe ought to be developed with smaller dimensions providing high sensitivity and low detection limit. Here we discuss about the different types of biosensors involved in pathogen detection, brief account on their principles of transduction along with some remarkable research, critically analyzing its construction, specificity, efficiency like limit of detection and ability to detect pathogens.

2.2 Classification of Biosensors

Biosensors can be broadly classified, as physical and chemical biosensors, based on the mode of signal transduction used in detection of biological analytes. The biological components can be enzymes, whole cells, organelles, lipids, peptides, tissues, antibodies or nucleic acids. Biosensor functions by following a biological event (e.g. antibody-antigen interaction) whose signals can be sensed by any physical or physico-chemical transducer as illustrated in Fig. 2.1.

2.2.1 Physical Biosensors

Biosensors that can sense a biological event by following the changes in physical phenomena like mass, resonance frequency, refractive index, fluorescence, etc. of the targeted analyte are classified as physical biosensors. These are further classified into optical and mechanical biosensors.

2.2.1.1 Optical Biosensors

Biosensors that quantify the analyte in a sample by its interaction with photons are categorized as optical biosensors. There are several advanced optical sensing methods available today to overcome the limitations of its preceding version. Few of them are discussed below.

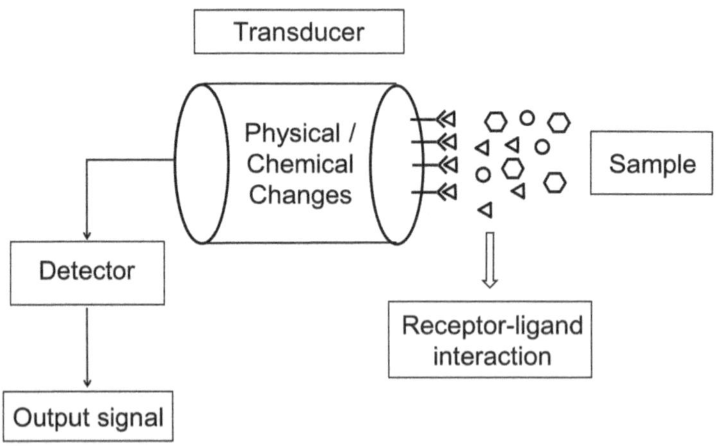

Fig. 2.1 Schematic representation of a Biosensor. Following a biological event, the alterations caused in any of the intrinsic parameters like fluorescence intensity, refractive index, surface tension, viscosity, etc. of the sensor probes can be sensed by a physical or physico-chemical transducer/detector which converts into a readable signal

Labeled Optical Biosensors

Fluorescent Labels

Fluorescence and phosphorescence emissions are about 1000 fold more sensitive with low limits of detection than other available spectrophotometric methods (Guilbault 1990; Yu et al. 2013). A typical fluorescence label method (Li et al. 2015) involves excitation of fluorophore at a particular wavelength followed by emission of light at a different wavelength. When the analytes are present in trace amounts, reporter molecules labeled with fluorescent dyes, are used for sensitive detection. Fluorescence detection has number of advantages (Waggoner 2006) that includes higher level of detection, sensitivity, rapid response times, a selective fluorescence signal, multiple assays using different colored dyes and direct labeling process, which yields specific functional moieties on the analyte.

The reagents used for fluorescent labeling have their key roles for sensitive and reliable mode of detection. Fluorescent labeling agents include organic dyes, nanoparticles like quantum dots, carbon nanotubes and rare-earth elements. Fluorescent biosensor works with optical signals for quantitative or semi-quantitative detection of pathogens. Recently, a number of nanomaterials which can act as efficient fluorescence quenchers have been studied to build biosensing platforms (Jans and Huo 2012; Wang et al. 2015; Wang and Alocilja 2015). Organic dyes were the mostly used labeling molecules in fluorescence biosensors (Panchuk-Voloshina et al. 1999) which included fluorescein, rhodamine dyes, sulfo rhodamine, Alexa dyes (Alexa350, Alexa488, Alexa532 and Alexa594) and cyanine dyes (carbo cyanines Cy3, Cy5 and Cy7). Cyanine dye labels like Cy3, Cy5 and Cy7 were used along with immunological reagents like antibodies and also DNA as bio-recognition elements (Kallioniemi et al. 1992). Sulfonate groups with negative charge, enhances solubility in water, prevents aggregation and minimizes the fluorescence dye-dye interactions bound to the antibodies (Mujumdar et al. 1993). These groups were combined with cyanine dyes in its ring structure, enhancing the intensity of the dyes in aqueous media. Cy3 and Cy5 have been extensively used for DNA and RNA labeling for gene expression studies (Schena et al. 1995). Recently, Cy3 dye bound capillary tube, along with a mesoporous chip was explored in the detection of a carcinoembryonic antigen, a protein available in the blood of some people with certain kinds of cancer (Yu et al. 2013; Yu et al. 2014). Drawbacks faced in using these highly sensitive fluorescent labels for detection includes lower fluorescence intensity, pH-sensitivity, photo-bleaching and non-fluorescent product formation. These limitations restricted their use in a bio-sensing system and paves way for alternatives.

Fluorescence can be directly used as bio-recognition tool/reagent that takes part in the bio assays for the detection of pathogens. Calcein Blue is a fluorophore whose fluorescence is quenched in the presence of iron (III) has been proved to have the potential to directly detect bacteria (Sankaranarayanan et al. 2015). In the presence of an iron chelating molecule (siderophore) released by bacteria, iron is removed from the fluorophore and fluorescence is restored in Calcein Blue. This serves as an easier, sensitive and cheaper method, which could be done in a 96 well plate, to identify and quantify iron chelators, and hence the presence of pathogenic bacteria within 7–8 h of incubation. Formation of siderophores by the Gram positive bacteria such as *Bacillus* species, *Staphylococcus aureus* (Dale et al. 2004; Zawadzka et al. 2009) and Gram negative bacteria like *Escherichia coli*, *Legionella* species, *Proteus* species and *Mycobacterium tuberculosis* (Himpsl et al. 2010; Wells et al. 2013; Adler et al. 2014) have been well studied. Here fluorescence acts as a sensitive indicator of iron sequestration by the virulence marker, siderophores and hence in the detection of the presence of bacterial pathogens in the sample. Fluorescence based antibiogram method was found to be exceptional compared to disc diffusion method or the standard liquid turbidity method. Percentages of individual pathogens susceptible/resistant to different antibiotics are profiled as antibiogram. This technique helps in reporting antibiotic susceptibility, intermediateness and resistance within 6–8 h, allowing rapid selection of suitable antibody against infectious bacteria. Fluorescein diacetate is used as viability indicator along with the antibiotic agent and the bacterial samples that has grown for 7 h (Alagumaruthanayagam and Sankaran 2012). Fluorescence released was measured after half an hour to check for viability of the infectious bacteria in the sample.

Specific antibodies conjugated with fluorescein enriched hollow silica nanospheres (diameter 350 nm) were used along with magnetic probes for selectively detecting *Escherichia coli* O157:H7 (Hu et al. 2016). Here, acidity of the solution is important for the fluorescein release, as the hollow silica nanospheres dissolves only in alkaline solution of sodium hydroxide leading to release of the fluorescent labels from the immune complexes. It involves capture of *Escherichia coli* O157:H7 cells followed by magnetic separation and fluorescein release in a detection solution (pH 10). Quantification of pathogens is based on the fluorescence intensity of the fluorescein released from the fluorescein-enriched hollow silica nanospheres. Here the detection time is only 75 min with limit of detection as low as 3 cfu/ml. The same strategy could be employed in detection for a different pathogen.

Graphene-Fluorophore

Graphene and graphene oxide are used for their functionalization, high volume to surface ratio, desirable physical and electrical properties. Combination of graphene and graphene oxide material with fluorescent labeled probes like aptamer/DNA enabled detection by their fluorescence quenching properties (Zhang et al. 2017). Quenching occurred upon adsorption of the fluorophore conjugated probes on the graphene surface. The analytes present in the sample interacts with the probe leading to its detachment from the graphene surface and generating more fluorescence and hence acting as indicators for detection of the target analytes.

Graphene and graphene oxide can be complemented with different types of sensing mechanisms like optical/fluorescence biosensors and electrochemical sensors (Shao et al. 2010; Liu et al. 2012; Kamali et al. 2015). Based on the sensing mechanism and principle graphene-based biosensors either use its unique properties like high electron mobility, high electron-transfer rates, or high surface to volume ratio for detection of biomolecules (Kuila et al. 2011; Zheng et al. 2014).

Quantum Dots

Nanoparticles in the form of quantum dots and carbon nanotubes have been used for immobilization in bio-recognition applications (Rosi and Mirkin 2005; Ansari et al. 2010; Merkoci 2010). They serve as good alternatives for their biocompatibility, higher distinct surface area, non-toxicity, good electro-catalytic activity, high chemical and thermal stability, and rapid communication by means of electrons. Moreover, nanoparticles associated fluorescence is an emerging research area that enhances sensitivity, ease of use and diversity of the fluorescence-based sensing methods. The development of quantum dots in labeling technology is due to their desirable optical and chemical properties that include broader absorption with narrow photoluminescence spectra, minimum photo-bleaching, high quantum yield and ability to withstand chemical degradation. These are semiconductor nanocrystals that are spherical with size ranging from 1 to 10 nm (Sutherland 2002). An interesting feature of quantum dots is that their fluorescence emission wavelengths can be tuned by varying their particle size and chemical composition. The ability of quantum dots to tune their size has been exploited for labeling and detection of multiple analytes at the same time (Rosenthal 2001). The uninterrupted long term tracking of biological processes is possible with quantum dots due to their low photo degradation rate (Chan and Nie 1998; Wu et al. 2003). Quantum dots are considered to be more advantageous than traditional organic dyes as fluorescent probes (Vinayaka and Thakur 2010), as they are 20 folds brighter and 100 folds more resistant to photo-bleaching (Chan and Nie 1998). But it also has three major limitations, synthesis from toxic elements, no inherent solubility and aggregation in water leading to lower quantum yields, which restricts its use for biological applications (Shen et al. 2012).

Carbon Dots

Carbon-based nanomaterials, fluorescent carbon dots with similar optical properties like quantum dots have emerged which possesses photoluminescence properties. By means of purification of single-walled carbon nanotubes through preparative electrophoresis carbon dots of size below 10 nm, were first formed in 2004 (Xu et al. 2004). Carbon nanotubes offer high surface to volume ratio, strong adsorption ability, desirable electronic and mechanical properties (Davis et al. 1998). It is commonly used as, single-wall carbon nanotubes and multi-wall carbon nanotubes. Due to its high surface to volume ratio and good adsorption, carbon nanotubes are employed for adsorption of bacteria and enrichment of various types of pathogens (Srivastava et al. 2004; Deng et al. 2008; Upadhyayula et al. 2009). It also serves as a material with high antimicrobial nature (Kang et al. 2007).

Combination of Quantum Dots with Bio-recognition Elements

One major development in fluorescence based detection technique includes immunoassays that uses bio-recognition elements along with fluorescent labels (Yang and Li 2006). Quantum dots are bound to biomolecules like antibodies by strategies like covalent binding, electrostatic binding, non-covalent biotin-avidin binding and nickel-based histidine tagging (Xing et al. 2007; Lee and Kang 2009; Zhang et al. 2010). Semiconductor quantum dots with different emission wavelengths (525 nm and 705 nm) were bound to anti-*Escherichia coli* O157 and anti-*Salmonella* antibodies by streptavidin and biotin coupling and used as fluorescent labels in immunoassay for detection of *E. coli* O157:H7 and *Salmonella* (Yang and Li 2006).

Fluorescence-Based Fiber Optic and Planar Waveguide Biosensors

Fluorescence based optical fiber and planar waveguide consists of an optical transmitter for the bio-recognition and for transporting the signal responses to a photodetector which converts the photons into an electrical signal. The optical biosensors offer advantages like low cost, compact size, ease of use and ability for real time monitoring or detection of specific species in test samples as well as quantification of the binding events (Taitt et al. 2016). These sensors find their use in detection of pathogen or contamination in food samples, environment and also in clinical diagnosis (Ligler and Taitt 2008; Benito-Pena et al. 2016). Fluorescence based transduction is most common in optical biosensors (White and Errington 2005; Borisov and Wolfbeis 2008). Parameters like fluorescence intensities measured at a particular excitation/emission wavelengths and decay time as a function of intensities could be used for bio-sensing (Demchenko 2008). It offers various ways to improve the performance, selectivity and sensitivity of the optical biosensors. Optical waveguides are made up of dielectric structures which transmit energy between ultra violet-visible and infra-red regions in the electromagnetic spectrum. Based on their configuration they are categorized as cylindrical and planar wave guides (Fig. 2.2). Optical fibers are composed of cylindrical central dielectric core covered by a lower refractive index material, whereas a planar waveguide consists of block of dielectric core placed in between two layers with reduced refractive indexes (Banica 2012).

A fiber-optic biosensor, shown schematically in Fig. 2.3, used to detect *Staphylococcus aureus* was established in 1996 (Hsin Chang et al. 1996). The fluorescein derivative fluorescein isothiocyanate was bound to the anti-Protein A immunoglobulin G, produced by *Staphylococcus aureus* that emits fluorescent signals of the antigen-antibody reaction with lower detection time (24 h). Similarly, the above method was used to detect Staphylococcal enterotoxin B with cyanine dye Cy5 as fluorescent label (Rowe-Taitt et al. 2000). A planar waveguide attracted the Staphylococcal enterotoxin B present in samples, a small diode laser excited the fluorophore Cy5, the pattern of fluorescent biochemical assay was recorded and extraction procedure lasted for less than 20 min. Another fiber optic biosensor

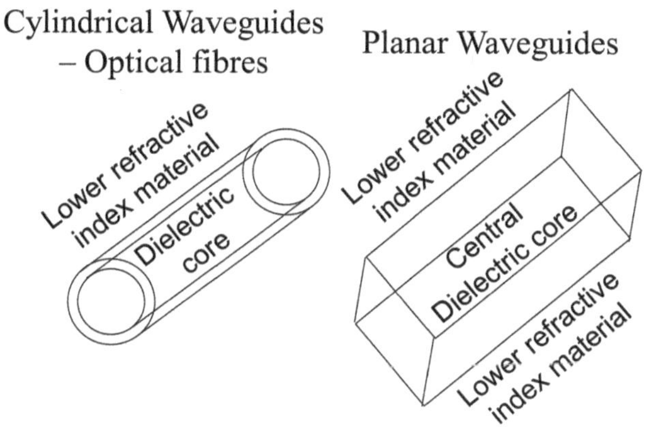

Fig. 2.2 Schematic of Optical waveguides. Cylindrical optical waveguides with central cylindrical dielectric core covered by low refractive index material (*Left*). Planar waveguides with block of central dielectric core placed between two layers of reduced refractive index materials (*Right*)

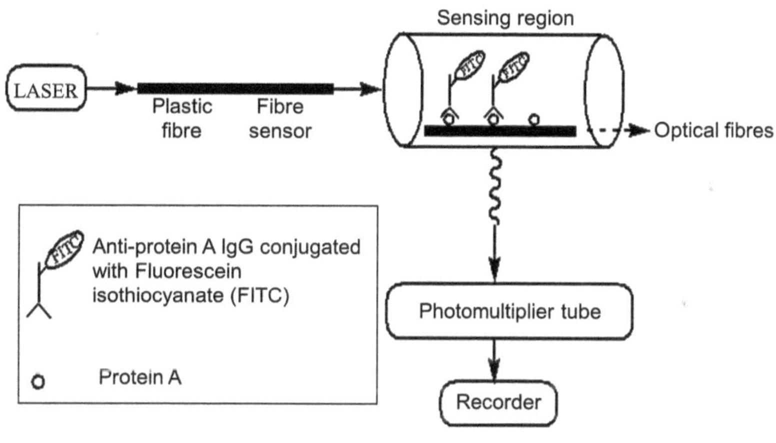

Fig. 2.3 Schematic representation of a fluorescence-based fiber optic biosensor. In this illustration, a laser source coupled with fiber optics is used in the detection of protein A from *Staphylococcus aureus*. This antigen-antibody reaction induces fluorescence signals from fluorescein isothiocyanate (FITC) conjugated anti-protein A immunoglobulin G (IgG), when complexed with protein A which then can be detected by a conventional optical set up

(Ko and Grant 2006) based on fluorescence resonance energy transfer with donor fluorophore as Alexa Fluor 546 and Alexa Fluor 594 as acceptor fluorophore due to its high spectral overlap and energy transfer (Panchuk-Voloshina et al. 1999) for the detection of *Salmonella typhimurium* in the samples of ground pork was established. A glass fibre based lateral flow DNA biosensor was reported (Chua et al. 2011) for the detection of food borne *Vibrio cholerae* by conjugation of the capture reagents to carrier beads and fluorescein labeled detector reagent bound to gold nanoparticles.

Label Free Optical Biosensors

Surface Plasmon Resonance Biosensors

While sensing biological analytes it becomes necessary to analyze them in their naturally occuring form. Early optical sensors required a fluorescent probe to be tagged to the analyte. Such tagging may interefere with the native interpretations. Therefore it becomes necessary to identify techniques that can go label-free, background-free and detect heterogeneity among interactions. In this context, surface plasmon resonance biosensors follow a label-free technique. They have been developed for sensitive and specific real time analyte detection. These surface plasmon resonance instruments work by optically monitoring the changes that occur on its surface, (usually metals like gold, copper or silver), with respect to the analyte flowing over the surface. A surface plasmon occurs at the metal-dielectric interface, as a charge density wave also called as surface plasmon wave (Liedberg et al. 1995). Surface plasmon resonance biosensors employ an optical phenomenon where the surface plasmon wave undergoes changes corresponding to the changes in the refractive index of the dielectric, and due to the binding of the analyte to the surface (Homola 2003). The most commonly used surface plasmon resonance sensing method involves the prism-coupling system (Fig. 2.4) where an incident light at the metal/glass interface at the bottom is coupled to the surface plasmon waves on the metal or liquid interface at the top. The light rays with an incident angle bends due to high refractive index of the prism which then interacts with the surface plasmon dispersion curve at the interface (Pi et al. 2016). Here the change in signal will be solely due to the variation in the surface refractive index at the top interface.

Some examples and improved methodologies employed for pathogen detection along with their limitations are listed below. Surface plasmon resonance has proven

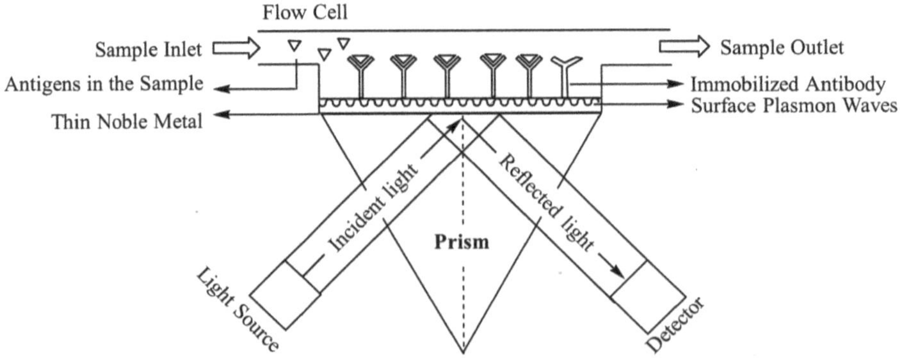

Fig. 2.4 Prism coupled-surface plasmon resonance biosensor with immobilized antibodies to detect specific antigens in the sample. The antigen-antibody reaction, causes a change in the refractive index at the interface (thin noble metal), which is being monitored as variations in output optical signals by the detector

to have the potential to identify a particular target bacterium in a sample containing closely-related species. Differentiation of spores of *Bacillus globigii*, a simulant of the bio-warfare agent *Bacillus anthracis*, from a sample mixture of various other *Bacillus* species, was made possible with a portable surface plasmon resonance biosensor. Detection was performed by direct capture using anti-*Bacillus*. However this technique was impractical to use due to its higher limit of detection (10^7 spores/ml) compared to other detection methods (Byrne et al. 2009) and the lethal dose of *Bacillus anthracis* (100 spores) (Housewright and Glassman 1966). This also signifies the need for a biosensor to have lower limits of detection for pathogens in order to detect them sensitively.

Hybrid Optical Biosensors

Integration of surface plasmon resonance biosensors with other detection elements/techniques like, fluorophore, immunolabels, magnetic nano particles, PCR, etc., was found to enhance sensitivity and reduce limit of detection.

An improved version of surface plasmon resonance sensor was developed by integrating with immunolabels, for Salmonella detection in food samples with lower limit of detection of 10^3 cfu.ml^{-1}(Farka et al. 2016). The surface plasmon resonance chip was immobilized with capture antibody, to which the samples containing Salmonella will be added. Salmonella if present in the sample will bind to the immobilized capture antibody followed by binding to secondary/detection antibody conjugated to Horse Radish Peroxidase. Here surface plasmon resonance detection is enhanced by Horse Radish Peroxidase catalysed reaction where 4-chloro-1-naphthol gets converted into insoluble benzo-4-chlorocyclohexadienone as shown in Fig. 2.5. This has been found to increase the detection signal and 40 times enhancement of sensitivity in detecting Salmonella compared to label-free detection. Enzymatic reaction oriented bio-recognition along with antibodies have proved to enhance the ability of surface plasmon resonance biosensor in pathogen detection compared to direct detection with antibodies or amplification with secondary antibodies. Use of antibody functionalized with Fe_3O_4 (iron oxide) magnetic nanoparticles (immuno magnetic nanoparticles) in surface plasmon resonance sensors based on sandwich immunoassay for Salmonella detection (Fig. 2.6) was found to be more sensitive with limit of detection as low as 14 cfu/ml (Liu et al. 2016a). These immune magnetic nanoparticles work as vehicles to selectively detect, isolate and deliver Salmonella from a sample mixture onto the sensor surface. Moreover, it also serves as labels that amplify sensor signal by the enhancement of refractive index variations for particular analyte pathogen. Sensitivity of the surface plasmon resonance sensors employing these immuno magnetic nanoparticles were proved to have four times more sensitivity compared to direct detection surface plasmon resonance sensors. Surface plasmon resonance biosensors with labeled antibodies, (be it enzyme/magnetic nanoparticles), that were based on sandwich immunoassay were found to have high sensitivity compared to other direct detection surface plasmon

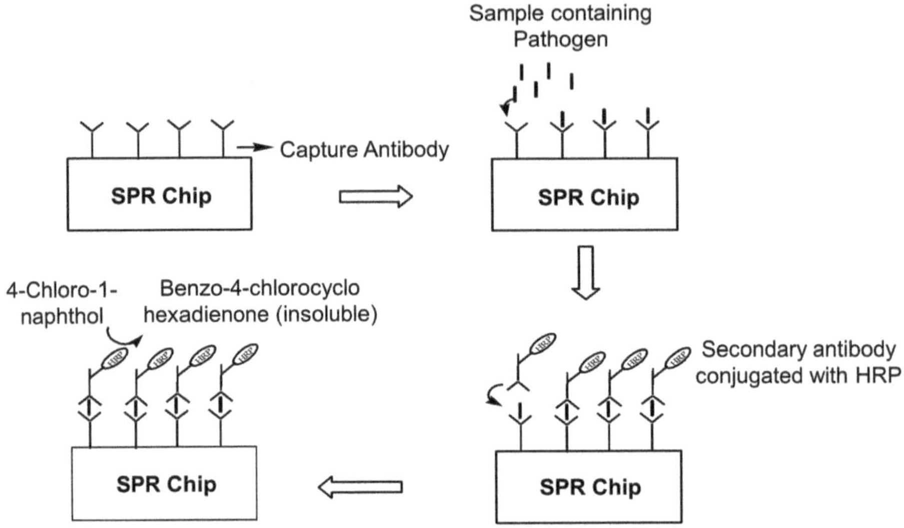

Fig. 2.5 Schematic representation of surface plasmon resonance chip that can detect pathogenic antigens in the sample, by surface plasmon coupled sandwich immunoassay, enhanced by Horse radish peroxidase labeled antibodies catalyzing the formation of insoluble benzo-4-chlorocyclohexadienone; SPR- Surface plasmon resonance, HRP- Horse radish peroxidase

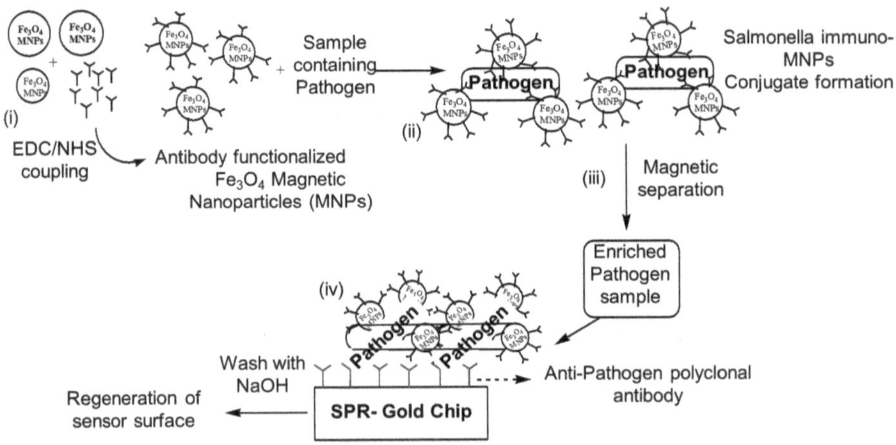

Fig. 2.6 (i) Formation of antibody functionalized iron oxide magnetic nanoparticles by EDC/NHS coupling; (ii) Binding of specific antibodies conjugated iron oxide magnetic nanoparticles with pathogenic antigens; (iii) Enrichment of pathogens complexed with antibody-iron oxide nanoparticles from sample mixture by magnetic separation; (iv) Pathogen detection by surface plasmon resonance based gold chip coated with anti-pathogen polyclonal antibody. SPR - Surface plasmon resonance, EDC/NHS - 1-Ethyl-3-(3-dimethylaminopropyl)-carbodiimide/N-Hydroxysuccinimide, Fe_3O_4 MNPs - iron oxide magnetic nanoparticles

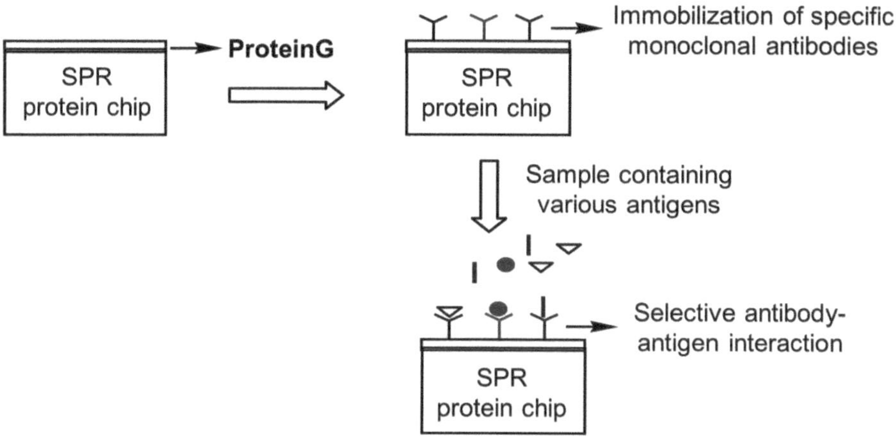

Fig. 2.7 Illustration of a biosensor chip that can selectively detect different pathogens by specific antigen-antibody interaction together with surface plasmon resonance. Here protein G coated on surface of the chip helps to preserve the orientation of immobilized monoclonal antibodies.

resonance sensors. With appropriate antibodies immobilized on gold surface of surface plasmon resonance sensors, pathogens in food samples could be detected in less than 5 h (Sharma and Mutharasan 2013). All these prove the ability of surface plasmon resonance sensors to be a sensitive, selective, quicker and label-free detection tool. Moreover, these sensor surfaces can be regenerated and reused by washing with NaOH (5 mM). A surface plasmon resonance based protein chip with Protein G immobilized on its surface was developed to selectively detect different pathogens (Fig. 2.7) like *Escherichia coli* O157:H7, *Salmonella typhimurium*, *Legionella pneumophila*, and *Yersinia enterocolitica* (Oh et al. 2005) that exist in the contaminated environment. The immobilised Protein G preserves the orientation of the monoclonal capture antibodies on sensor surface. It enhances the sensitivity by associating specifically with the fragment crystallizable portion of antibody immunoglobulin G (Boyle and Reis 1987) and improving the efficiency of antigen-antibody binding.

As production of specific antibodies is expensive, time consuming, complicated and are prone to lose its activity due to external conditions (Arya et al. 2011) DNA based surface plasmon resonance biosensors were developed. Here the carboxylated dextran fixed onto the gold surface of sensor chip was coated with streptavidin covalently linked to biotinylated single stranded oligonucleotide probe as shown in Fig. 2.8 (Zhang et al. 2012). The probe was designed in such a way that it is specific to a highly conserved gene, a potential target for pathogenicity in pathogens like Salmonella which undergoes hybridization that lasts only for 15 min.. Detection of Salmonella cells upto 10^2 cfu/ml could be performed in 4.5 h using this surface plasmon resonance DNA biosensor. Regeneration of the sensor surface makes it possible to reuse the sensor for atleast 300 assay cycles, thereby reducing the cost of pathogen detection. By standardizing the sample preparation procedures to extract

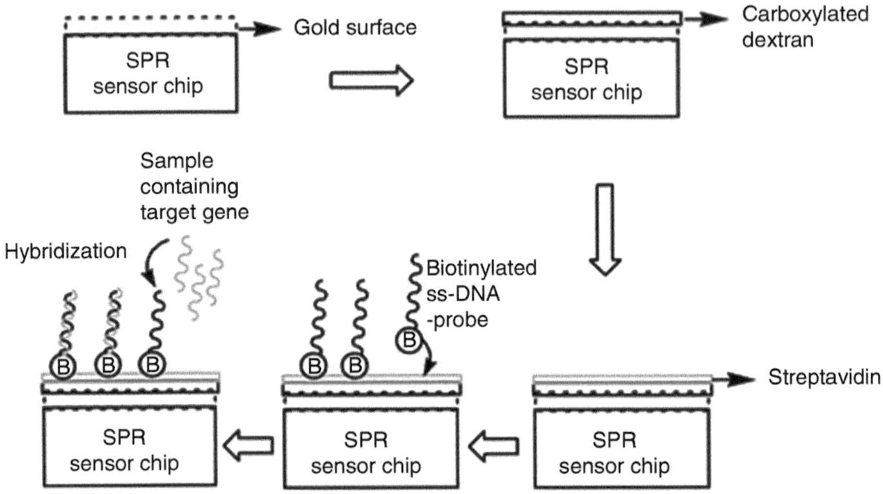

Fig. 2.8 Schematic representation of a surface plasmon resonance - DNA biosensor chip, coated with gold, carboxylated dextran and streptavidin on top that binds biotin conjugated ss-DNA probe, followed by hybridization with complementary pathogenic target gene from the sample. SPR - surface plasmon resonance, ss-DNA - single stranded-deoxyribonucleic acid

DNA followed by its amplification, this sensor could be used as an efficient tool to detect and monitor pathogens in real samples.

Surface plasmon resonance biosensors based on grating coupled long range surface plasmons were employed along with magnetic nanoparticles to detect bacterial pathogens like *Escherichia coli* O157:H7 at concentrations as low as 50 cfu/ml (Wang et al. 2012a). Here the limit of detection is four times better than that obtained by the regular surface plasmon resonance with direct detection. Long range surface plasmons provide reduction in losses compared to the normal surface plasmons and exhibits narrow resonance, thereby rendering accurate measurements of changes in refractive index. Long range surface plasmons based biosensors could probe higher distances from the surface (Dostalek et al. 2007) and possess several fold enhanced refractive index resolution compared to regular grating-coupled surface plasmons suitable for analyzing large analytes (Huang et al. 2011; Vala et al. 2009).

Further PCR microchip hybridized with optical fiber surface plasmon resonance sensor with bimetallic coating (Ag/Al) (Nguyen et al. 2017) serves as a label-free, reusable, point of care diagnostic instrument by genetic analysis. In this, DNA of the pathogen to be detected is amplified within 30 min on the PCR microchip followed by detection with the surface plasmon resonance optical fibre sensor without the aid of any kind labels. Here the pathogen detection is free from sample volume restriction and labeling, until the DNA sample is in close contact to the surface of the sensor. It has paved way for further miniaturization and point of care diagnosis by genetic analysis towards effective pathogen detection without much human intervention.

2.2.1.2 Mechanical Biosensors

Sensors that sense the analytes by monitoring the change in mass associated with it during the interaction process. They are more advantageous (Arlett et al. 2011) than any other biosensors as they do not require any sample preparation. The commonly available mechanical sensors follow either quartz crystal microbalance or cantilever technology for the detection of analytes. Moreover, these sensors work as label-free unlike other optical techniques.

Piezoelectric Quartz Crystal Biosensors

In piezoelectric quartz crystal biosensors, the detection of an analyte is usually based on adsorbate identification. Here corresponding to the selective binding, mass change occurs and is detected by change in electrical/acoustic parameters of a piezoelectric quartz crystal. Piezoelectric effect is well observed in crystals on applying pressure to it, thereby distorting the crystal lattice and inducing dipole moment in the crystal molecules (Alder and McCallum 1983). Quartz crystal is the most commonly used (Deakin and Buttry 1989), due to its desirable electrical, mechanical and chemical properties. The piezoelectric quartz crystal is placed in between two metal (gold/silver) electrodes and when a potential difference is applied, the crystal lattice gets distorted and undergoes oscillation with a characteristic resonance frequency. This oscillating piezoelectric quartz crystal has a characteristic frequency dependent on its own physical properties and other phases adjacent to it. This particular proportional relationship between the resonance frequency and the mass of the crystal is exploited from this technique for pathogen detection.

Piezoelectric biosensors are being used for rapid detection (Farka et al. 2013) of pathogens either by *active mode* or *passive mode*. In *active mode*, the piezoelectric crystal oscillates and the corresponding resonance frequency is measured by means of a frequency counter (Arnau 2008). The passive mode is an expensive one where the resonators are monitored for their impedance (Zhang et al. 2002). *Passive mode piezoelectric biosensor* employs one equipment that can recognize the changes in mass and viscosity (Itoh and Ichihashi 2008). These sensors detect the change in resonance frequency that is caused by the binding of an analyte or increase in the mass on its surface. A more stable and sensitive *series piezoelectric quartz crystal* sensors (Shen et al. 1993; Jang et al. 2009) were designed by interconnecting several piezoelectric quartz crystal with conducting electrodes successively, which responded with resonance frequencies to every change in the electrical parameters (He et al. 2003; Lakshmanan et al. 2014). With the aid of series piezoelectric quartz crystal having multiple channels several samples could be analyzed at the same time (Tong et al. 2014) and hence can be used for high-throughput analyses.

To further increase the sensitivity, *series piezoelectric quartz crystal with interdigital electrodes* where a series of cathodes and anodes are connected with micro gap in between were designed. This enables detection of even minute variations on its surface (He et al. 2007; Ren et al. 2010). Graphene coated interdigital gold electrodes are connected in series with piezoelectric quartz crystal to develop a series piezoelectric quartz crystal sensor upon which the probes like pathogen specific aptamers or antibodies, are impregnated as assemblies for the corresponding pathogen detection. This was demonstrated initially by binding *Staphylococcus aureus* aptamers to grapheme by π-π stacking of DNA bases (Lian et al. 2015). Upon introducing analyte/target DNA (*Staphylococcus aureus*), the aptamers bind strongly with the target DNA. The force of interaction of aptamer with graphene becomes relatively weaker and the aptamer-target DNA complex gets itself freed from the electrode surface, thereby causing changes in electrical parameters of electrode surface with corresponding shift in oscillation frequencies. These sensors have limit of detection of pathogens upto 41 cfu/ml. Similarly interdigital electrodes impregnated with pleurocidin, an antimicrobial peptide with broad spectrum of antimicrobial activity, help in rapid detection of microbes in clinical blood samples in a time span of 15 min has also been established (Shi et al. 2017). This sensor has found its importance in clinical diagnosis (Jordana-Lluch et al. 2013) and food safety applications (Farahi et al. 2012). In this the transducing element is an integrated single walled carbon nanotubes that crosslinks pleurocidin with interdigital electrodes connected to a multichannel series piezoelectric quartz crystal. This sensor works by specific binding of pleurocidin with microbe which causes detachment of pleurocidin from the single walled carbon nanotubes associated with the electrodes. The response in the form of frequency changes in multichannel series piezoelectric quartz crystal are detected by the transducer which is proportional to the microbes present in the sample. It fulfills the need to detect the microbes which is the first step in the microbial blood stream infection tests (Gonsalves and Sakr 2010), followed by identification and drug susceptibility of microbes.

As a further step towards increasing the sensitivity, a sensor holding a similar transducer as above is used along with gold nanoparticles (He et al. 2016), aptamers specific for the complex of 10 kDa culture filtrate protein and 6 kDa early secreted antigen target (Renshaw et al. 2002; Philip et al. 2005), an early phase antigen, secreted by the pathogenic *Mycobacterium tuberculosis*, is used as probe for detection of *Mycobacterium tuberculosis*. These aptamers along with complementary DNA bound gold nanoparticles over the interdigital electrodes showed more sensitivity (Zheng et al. 2007; Li et al. 2009) when the target protein in the test sample interacts with the aptamer. The DNA bound gold nanoparticles detached from the electrode surface followed by electrical changes on the electrode due to the substitution of conductive DNA-gold nanoparticles by non-conductive target proteins. This had led to a more sensitive detection of frequency changes by the multichannel series piezoelectric quartz crystal with a detection time of 96.3 h. This sensor serves as a rapid and early detection tool for detection of pathogenic *M. tuberculosis* which is not possible with other sensors based on antigen-antibody interactions (Ren et al. 2008; Kirsch et al. 2013).

Piezoelectric – Cantilever Biosensors

Cantilevers are rigid structural elements held at one end to a rigid support. Micro-cantilever sensors have bio-receptors bound to it and have characteristic resonance frequencies of oscillation upon interaction with bio-analytes (Ahmed et al. 2014). The cantilever undergoes mechanical stress due to increase in mass on the sensor surface and induces variation in the resonance frequencies. This property has been used in the detection of various bacteria like *Escherichia coli* O157:H7 (Zhang and Hai-Feng 2004; Campbell and Mutharasan 2005a), *Salmonella typhimurium* (Zhu et al. 2007), *Vibrio cholerae* (Sungkanak et al. 2010), and the biowarfare agent *Francisella tularensis* (Ji et al. 2004). Antibody functionalized cantilevers of the size of a millimeter excited by piezoelectric effect has the ability to detect various bacteria like *Escherichia coli* cells (Campbell and Mutharasan 2007), *Listeria monocytogenes* cells in milk (Sharma and Mutharasan 2013). By using an impedance analyzer the changes in the resonance frequencies when the analyte binds specifically to the antibody at the cantilever tip could be measured. In a 1 ml sample cell, *Escherichia coli* O157:H7 upto 700 cells/ml has been detected with a cantilever functionalized with monoclonal antibody specific to *Escherichia coli* O157:H7 (Campbell and Mutharasan 2005b). Amplitude ratio and phase angle variations with respect to mass changes at the cantilever were measured using an impedance analyzer. Further, flow cells wherein analyte solution flows at a particular rate have been used along with the functionalized piezoelectric-cantilever to detect pathogens more sensitively (Campbell and Mutharasan 2006).

Piezoelectric cantilever sensors employing specific gene based detection of pathogenic bacteria have also been developed. A food pathogen, *Listeria monocytogenes* could be detected with one such sensor employing a probe specific for the target virulence hemolysin gene, hlyA, within 90 min (Sankaranarayanan et al. 2015). The hybridization response of the specific virulence gene from the sample genome extract of the pathogens, with the probe on the cantilever was detected by different means. It includes, a fluorescent indicator to detect the hybridized double stranded DNA, a secondary single stranded DNA to detect the unhybridized portion of the target DNA and also gold nanoparticles tagged secondary single stranded DNA to amplify the target hybridization. It is necessary to extract the genomic DNA of the test sample prior to detection with the cantilever. It proved to be a rapid technique compared to other conventional detection techniques with almost same detection limit.

2.2.2 *Chemical Biosensors*

Chemical biosensors monitor the changes in the chemical phenomenon of the analytes upon interaction with the sensor probes. Further discussion on this will be carried out under two categories as electrochemical and biochemical sensors.

2.2.2.1 Electrochemical Biosensors

Here the variations in electrical properties like current, potential or impedance at the electrode surface upon binding to the bio-analytes in the test sample will be examined. It can be further classified as label dependent and label free electrochemical biosensors (Xu et al. 2017) based on the technique used for detection. Labels like enzymes, metal particles, etc. are specifically tagged to the target analyte in case of label dependent sensors. Label free technique involves the binding of bacterial cells on the electrode surface that elicits changes in electrical parameters which could then be measured (Sang et al. 2016). With respect to the methods used for measurement of electrical parameters, electrochemical biosensors can be sub divided as amperometric, potentiometric, voltametric, conductometric and impedimetric.

Amperometric and Potentiometric Detection

Pathogen detection by amperometric or potentiometric methods (Monzo et al. 2015) depends on the variations in current or potentials in which the electrode is kept at a constant potential or current, relative to the reference electrode (Bard and Faulkner 2001). This type of sensor has biological receptors like aptamers/antibodies that can specifically adsorb microbes coupled to an enzyme transducer. Upon binding of microbes, the enzymatic reactions are initiated, which either produces or takes up a species, followed by its detection, by a selective electrode (Barlett 2008). It proves to be a sensitive method as it is dependent on logarithmic concentration. Recently, the use of a semiconductor device, field effect transistors, for microbial detection enhanced the signal of the sensors ensuring increased sensitivity (Grieshaber et al. 2008; Lin et al. 2008). It is also a rapid and cost effective method (Barlett 2008; Wei et al. 2009), but these amperometric sensors tend to have poor selectivity. This is because all the constituents in the solution with a standard potential less than operating potential will add to the faradaic current, generated by the reduction or oxidation of those constituents at the electrode surface (Monzo et al. 2015).

Impedimetric Detection

Here the impedance changes caused by the voltage signal when the analyte binds, relative to the frequency of an electrode with dielectric properties, are measured (Bard and Faulkner 2001; Barlett 2008). Two approaches are employed for the identification of pathogens using impedimetric techniques. One involves the use of unmodified electrodes, or electrodes bound to biological receptors, for the measurement of the variations in impedance due to unspecific or specific attachment of bacteria to the electrode, respectively (Wang et al. 2012b). Another approach is the direct detection of the metabolites or toxins that are released by the pathogens during their growth and is considered to be an important tool in detecting infected samples (Felice et al. 1999; Gomez et al. 2002). The association of the microbes or the metabolites with the electrode can be measured by differences in the capacitance magnitude at the junction of electrode or variations in impedance. Upon adsorption of either metabolites or microbes, reduction in total electrode surface area occurs which further shifts the impedance to a higher level. Sensitivity and selectivity are found to be high in impedance spectroscopy towards detecting biological elements (Felice et al. 1999). Carbon electrode modified with reduced graphene oxide has been reported in the detection of methicillin resistant *Staphylococcus aureus* by means of impedance spectroscopy (Wang et al. 2011). Single stranded probe DNA bound to the functionalized electrode undergoes hybridization with the complementary target DNA present in the test sample, followed by measuring the change in impedance values. This proves to be a more sensitive and selective method in the detection of pathogen with the help of DNA as bio-receptor. A rapid (45 min) and label free approach was demonstrated for the detection of Salmonella by using a combination of polypyrrole based polymer, poly [pyrrole-co-3-carboxyl-pyrrole] copolymer and a specific aptamer (Sheikhzadeh et al. 2016). Impedimetric measurements upon interaction of the aptamer with the target, cause changes in electrical parameters of the polypyrrole based copolymer. It is considered to be a cost effective and rapid detection technique that could be used for analyzing the contaminants in food samples and also in the environment. As an alternative to the above, gene/immune based approach with engineered synthetic peptides specific for pathogens have been used while monitoring the changes in impedance (Liu et al. 2016b).

Voltammetric Detection

Voltammetric method for pathogen detection depends on the variations in the potential at the junction of the electrode surface and the analyte solution with respect to time, along with the measure of current (Bard and Faulkner 2001). Cyclic sweep voltammetry is a commonly used technique for acquiring information like oxidation or reduction potentials, reaction mechanisms and kinetics (Bard and Faulkner 2001; Compton and Banks 2011). Here the voltage scan is performed from a minimum to a maximum level of potential at a specific scan rate. The response in the form of current is recorded with respect to the voltage instead of time. Voltammetric sensing is done either with the help of biological receptors or by direct detection of the metabolite species undergoing oxidation or reduction on a modified or unmodified electrode. Several advanced voltammteric detection methods employing pulse strategies are available. These include differential pulse voltammetry, normal pulse voltammetry and square wave voltammetry. For better time resolution and high frequency operation, differential pulse voltammetry and square wave voltammetry are more preferred for electroanalysis (Chen and Shah 2013). Genomic DNA of the targeted pathogens like Salmonella or *Escherichia coli* cultures without any PCR amplification could be detected by means of sandwich like assay consisting of magnetic nanoparticles, the genomic target DNA and gold nanoparticles imprinted on carbon electrodes (Blow 2015).

An antibody based sensor with an immunoelectrode containing Graphene oxide-silver nanoparticles immobilized with antibodies of the pathogen *Salmonella typhimurium* utilizes cyclic voltammetry for the detection the same pathogen (Sign and Sumana 2016). This could be extended to detect various other pathogens with corresponding specific antibodies.

2.2.2.2 Biochemical Sensors

Immunosensors

Immunosensors are solid-state devices that combine immunochemical reactions and transducers. Here the sensitivity and selectivity depends on specific ligand-receptor interactions. A typical immunosensor is composed of two elements: a bio-recognition element and a transducer. Here the bio-recognition element is formed by impregnation of specific antigens/antibodies, and their interaction is converted into a measurable signal by the transducer.

The immunoassay strategies involve direct or indirect methods of detection. In the direct method, the immunochemical reaction is directly quantified with respect to the physical changes that occur due to the antigen-antibody interaction. Indirect method involves the use of a label in association with the antibody or antigen.

Immunosensors are combined with magnetic particles or gold nanoparticles for enrichment or signal amplification in the detection of pathogens (Wang and Alocilja 2015). An *in-situ* immuno-gold nanoparticle integrated network-based enzyme-linked immunosorbent assay biosensor together with initial sample enrichment using immuno-magnetic separation have been developed for the detection of pathogens with high sensitivity (Cho and Irudayaraj 2013). It involves the formation of immuno-gold nanoparticle network onto the antigenic site at the bacterial outer membrane surface, followed by analytical validation using microtiter immunoassay. Here magnetic and gold nanoparticles are coupled with antibodies specific to the target bacterium. Initially separation and enrichment of bacteria will be done by immuno-magnetic separation followed by secondary antibodies functionalized with gold nanoparticles (30 nm in size) and are able to bind to complementary targets on the cell surface to form a network structure that can grow with time amplifying the signal in the network structure (Lee and Irudayaraj 2009; Lee et al. 2011). It proved to be highly sensitive to detect pathogens at extremely lower concentration like 3 cells/ml of *Escherichia coli* O157:H7 and *Salmonella typhimurium* in buffer, and 3 cfu/ml of *Escherichia coli* O157: H7 and 15 cfu/ml of *Salmonella typhimurium* in real sample conditions within 2 h of inoculation. The ability to detect and monitor target bacteria with enhanced analytical sensitivity compared to other current techniques makes it suitable as a tool for routine monitoring and improved food safety. Based on the measurement of the amount of specific virulence factors formed by the pathogenic organism, the pathogen can be detected. Immunochemical approaches have been developed to detect *Pseudomonas aeruginosa* infections by quantification of a specific virulence factor pyocyanin secreted exclusively by these organisms (Pastells et al. 2016). Antibodies specific to 1-hydroxyphenazine, the major metabolite of pyocyanin virulence factor enables us to quantify both 1-hydroxyphenazine and pyocyanin in clinical test samples in 20 min. This assay completes in 2 h and offers simultaneous detection of several samples. It proves to be a remarkable development in diagnosis of infections in test samples.

Immunoassays/ Immunochemical Methods to Detect Pathogenic Infections

Antibodies suitable for immunoassays are chosen based on factors like, the assay format where these monoclonal/polyclonal antibodies will be employed according to antigen specificity. These assays are integrated with sensors to form immunosensors. After establishing the ability of antibodies to recognize particular antigens, it becomes easier to classify, detect and ultimately prevent pathogenic infections. Several immunoassays have emerged through the years, to detect pathogens in the clinical stool samples or industrial food products. Immunoassays have been developed for food industry to detect and quantify various food components like protein, enzymes, vitamins and contaminants like microorganism, toxins, pesiticides, hormones and others (Fukal and Kas 1989). Immunoassay format can be designed for identification by means of understanding the immunochemical reaction between antigen-antibody complex that is formed during an immunological reaction between the analyte (antigen) and the reagent antibody (Kas et al. 1986). Immunoassays serve as a tool for analyzing the presence or concentration of an antigen. Initially the agglutination of the antigen-antibody complex was visualized followed by visible clumping of cells and antibodies (Burnet 1934; Pauling 1940), which would be discussed in the latter part of this chapter. This was fine-tuned with the establishment of radial immune diffusion assays commonly referred to as the double disk diffusion assays (Ouchterlony 1949). This immune diffusion is based on the principle that molecules with similar structure will migrate with similar rates through the agarose gel, without any restriction of movement of the molecules (Lam and Mutharia 1994). Double disk diffusion assay is used in the determination of minimum concentration of antigen required for precipitation of the antibody-antigen complex. Here the center part of the agarose gel consists of a well or disk with an antibody surrounding which antigens are placed at equal distances around the center forming a streak or band like precipitate or the precipitin line would be formed, ensuring the binding of antigen with the specific antibody. This type of double disk diffusion assay was used to identify various components of *Escherichia coli* like enzymes (Lee and Englesberg 1962), surface proteins (Guinee et al. 1976; Isaacson 1977), enterotoxins (Honda et al. 1981; Tsuji et al. 1985) and Shiga toxins (Oku et al. 1989).

Immunoassays work by the hydrophobic interaction of an antibody and antigen, where either component could be bound to hydrophobic surfaces to quantify the relationship between antigen and antibody. Immunoassays were developed further with multiple screening of samples in a microtiter plate (96-well plates). Upon binding of the antibodies to solid phase surfaces that are mostly hydrophobic, quantification of these antibodies could be done by attachment of labels to antibody which can be detected with a transducer. The important approaches include radio immuno assay, enzyme-linked immunosorbant assay and latex agglutination.

Radio Immunoassay

Radio immunoassay pioneered in incorporating labels (Lequin 2005) like the radioisotopes ^{125}I and ^{3}H were commonly used (Soergel et al. 1982; De Boever et al. 1983). Here the radio labelled antibodies are allowed to interact with the analytes and the screening was carried out with the help of a liquid scintillation counter, which quantifies the particles emitted due to radioactive decay (Chase 1980). It was then followed by the emergence and usage of non-radioactive labels like enzyme-conjugated chromophore systems that indirectly forms a detectable label in an immune assay reaction (Lequin 2005).

Enzyme-Linked Immunosorbent Assay

Several enzyme associated formats of immunoassays emerged with the successful implementation of enzyme labels for detection of antigens (Lequin 2005). Enzyme linked immunosorbent assay development could be seen as a way of improving the detection methods in clinical and public health setting, where cell culture or radio immune assay was commonly used (Downes et al. 1989). Enzyme linked immunosorbent assay has evolved into different forms for antigen detection and quantification as illustrated in the Fig. 2.9.

A sandwich assay (Skinner et al. 2013) was developed based on enzyme-linked immunosorbent assay for detection of *Escherichia coli* producing Shiga toxin Stx2 subtype known as Stx2f. Although the Stx2f toxin was not involved in any critical human disease (Melton-Celsa 2014), the assay proved to be useful in detecting the presence of phages and plasmids in *Escherichia coli* isolates (Skinner et al. 2013). Prior to the development of immunoassay for the Stx2f toxin, monoclonal antibodies were generated by murine immunization with a Stx2f A subunit and fusion of the mice spleenocytes with the myeloma cells. It was followed by screening of

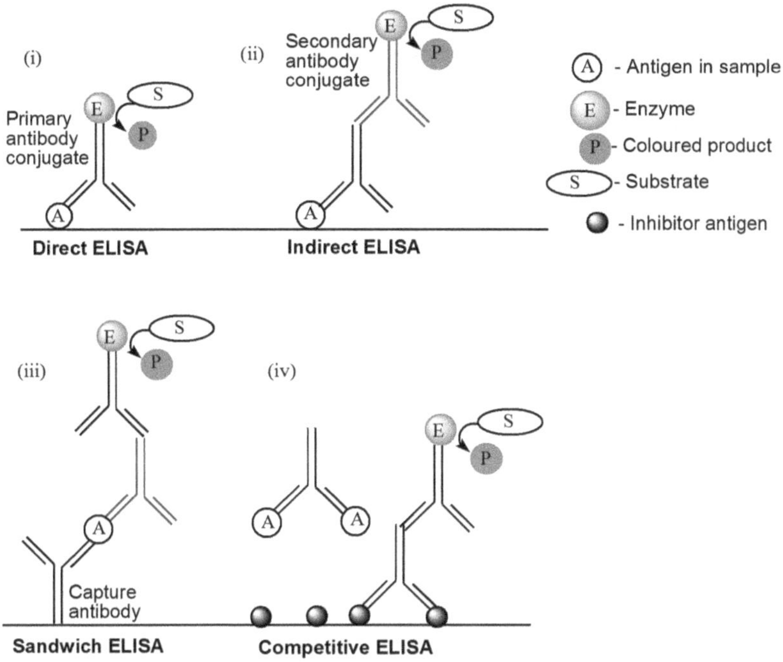

Fig. 2.9 Illustration of Enzyme labeled immunosorbent assay (ELISA), (i) Direct ELISA - antigen adsorbed on plastic/microtiter plate binds to the enzyme-antibody complex added (ii) Indirect ELISA - the enzyme-antibody complex uses an antibody against the isotype of antibody that is used to detect the antigen (iii) Sandwich ELISA - antigens in sample bind to capture antibody followed by binding with monoclonal antibody and secondary antibody conjugated with enzyme, (iv) Competitive ELISA - more the antigen present in the sample, the less antibody will be present to bind to the antigen coated in the microtiter well, followed by the enzyme-antibody conjugate that binds the isotype of antibody that is used to detect the antigen; Addition of enzyme's substrate produces colored product

antibodies for immune reactions with the antigen Stx2f toxin, where four unique antibodies (Stx2f-1, Stx2f-2, Stx2f-3 and Stx2f-4) were found to be specific for Stx2f toxin.

Latex Agglutination

This method uses latex beads immobilized with specific antibodies and is one of the widely used technique to sense the pathogens by their specificity towards antibodies (Hegde et al. 2012). It is the common technique employed in clinical laboratories for O antigen identification (Atkinson et al. 2012). Latex-bound antibodies generates complexes with any antigen present in the sample and forms visually detectable precipitates (Boer and Heuvelink 2000).

In addition, an optical immunoassay has been developed for visually identifying microorganisms. The *BioStar optical immunoassay SHIGATOX kit* (Inverness

Medical Professional Diagnostics, Inc.) serves as a visual identification tool to detect the presence of shiga toxins Stx1 and Stx2, but the drawback is that it does not differentiate the two toxins (Teel et al. 2007). Here a mixture of anti-Stx antibodies (either Stx1 or Stx2 specific) is bound to a silicon wafer. The principle is based on the reflection of light from the silicon wafer, which appears gold in color when bare and purple as the thickness increases due to the binding of shiga toxins from the pathogen (Teel et al. 2007).

Antibiogram

Antibiogram was employed as a surveillance tool for the detection of early microbial growth. It was initially developed as a 96 well plate format which was later transferred to a portable, low cost point of care resazurin based polymethylmethacrylate microfluidic chip (Elavarasan et al. 2013) for live cells. Resazurin is a blue colored water soluble dye that undergoes two stages of reduction in the presence of viable cells (Fig. 2.10), first stage resulting in an irreversible pink colored resorufin, formed by loss of one oxygen atom and a reversible second stage resulting in colorless hydroresorufin (Sarker et al. 2007). It serves as an indicator of cell viability, growth and toxicity (Palomino et al. 2002). Here, in this immunoassay, blue color indicates the blank/antibiotic susceptibility of the microbial sample where there is no cell growth; pink/colorless indicates antibiotic resistance of the sample and violet coloration shows intermediate to poor growth of the sample due to partial/complete reduction of resazurin by cell growth. This method is used for testing contamination of milk with bacteria and detection of multi drug resistant microbes. This microfluidic chip can be used as a one-time disposable device and thereby no cross contamination.

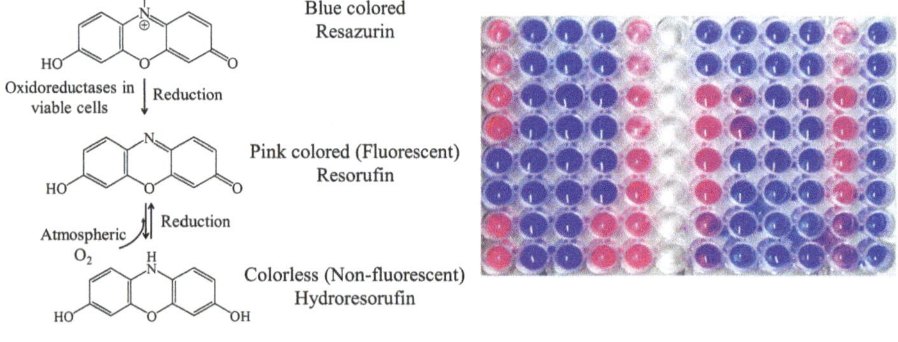

Fig. 2.10 Immunoassay based on Resazurin reduction reaction (*Left*), blue colored resazurin loses one oxygen atom in the presence of oxidoreductases in viable cells to form pink colored resorufin followed by further reduction to a colourless hydroresorufin. Resazurin dye reduction antibiogram in 96 well plate format (*Right*), blue color- blank/antibiotic susceptibility, pink/colorless- antibiotic resistance, violet- intermediate to poor growth of microbes in the sample

2.3 Conclusion

Researchers working on pathogenic biosensors focus mainly on lowering the time taken for detection, the limit of detection, the need for skilled labor and sample volume required for pathogen detection. Pathogen surveillance tools including microchips and other diagnostic kits, based on surface plasmon resonance or immunochemical reactions are remarkable. New trends involve the integrated use of micro and nano fabrication techniques along with sample enrichment using techniques like magnetic separation, followed by signal amplification and detection using various transducers in the area of biosensors for pathogen detection (Zuo et al. 2013; Kim et al. 2014; Hsieh et al. 2015; Altobelli et al. 2016). Progress has been witnessed in label free technologies like surface plasmon resonance/piezoelectric quartz crystal based optical, mechanical and biochemical biosensors, in spite of the improvements in label-based detection techniques which includes fluorophores, quantum/carbon dots based techniques. As a next step, these significant pathogen detection tools ought to be made available in the market for timely detection of pathogens in no minute at a cheaper cost, thereby preventing epidemic conditions.

Acknowledgement The authors gratefully acknowledge VIT University, Vellore for the support through Seed Grant for Research.

References

Adler C, Corbalan NS, Peralta DR, Pomares MF, de Cristóbal RE, Vincent PA (2014) The alternative role of enterobactin as an oxidative stress protector allows Escherichia coli colony development. PLoS One 9:e84734. https://doi.org/10.1371/journal.pone.0084734

Ahmed A, Rushworth JV, Hirst NA, Millner PA (2014) Biosensors for whole-cell bacterial detection. Clin Microbiol Rev 27:631–646. https://doi.org/10.1128/CMR.00120-13

Alagumaruthanayagam A, Sankaran K (2012) High-throughput fluorescence-based early antibiogram determination using clinical Escherichia coli isolates as case study. Microb Drug Resist 18:586–596. https://doi.org/10.1089/mdr.2011.0238

Alder JF, McCallum JJ (1983) Piezoelectric crystals for mass and chemical measurements. A review. Analyst 108:1169–1189. https://doi.org/10.1039/AN9830801169

Altobelli ER, Mohan KE, Mach ML, Sin Y, Anikst V, Buscarini M, Wong PK, Gau V, Banaei N, Liao JC (2016) Integrated biosensor assay for rapid uropathogen identification and phenotypic antimicrobial susceptibility testing. Eur Urol Focus. https://doi.org/10.1016/j.euf.2015.12.010

Ansari AA, Alhoshan M, Alsalhi MS, Aldwayyan AS (2010) Prospects of nanotechnology in clinical immunodiagnostics. Sensors 10:6535–6581. https://doi.org/10.3390/s100706535

Arlett J, Myers E, Roukes M (2011) Comparative advantages of mechanical biosensors. Nat Nanotechnol 6:203–215. https://doi.org/10.1038/nnano.2011.44

Arnau A (2008) A review of interface electronic systems for AT-cut quartz crystal microbalance applications in liquids. Sensors 8:370–411. https://doi.org/10.3390/s8010370

Arya SK, Singh A, Naidoo R, Wu P, McDermott MT, Evoy S (2011) Chemically immobilized T4-bacteriophage for specific Escherichia coli detection using surface plasmon resonance. Analyst 136:486–492. https://doi.org/10.1039/c0an00697a

Atkinson R, Besser J, Bopp C, Carlson C, Crandall C, George K, Gerner-Smidt P, Gladbach S, Gould LH and Hartley C (2012) Guidance for public health laboratories on the isolation and characterization of Shiga toxin-producing Escherichia coli (STEC) from clinical specimens. Silver Spring (MD): APHL:1–102

Banica FG (2012) Chemical sensors and biosensors: fundamentals and applications. Wiley. https://doi.org/10.1002/9781118354162

Bard AJ, Faulkner LR (2001) Fundamentals and applications. Electrochem Methods 2. 978-0-471-04372-0

Barlett P (2008) Bioelectrochemistry: fundamentals. Wiley, Experimental techniques and applications. 978-0-470-84364-2

Batt CA (2007) Food pathogen detection. Science 316:1579–1580. https://doi.org/10.1126/science.1140729

Benito-Peña E, Valdés MG, Glahn-Martínez B, Moreno-Bondi MC (2016) Fluorescence based fiber optic and planar waveguide biosensors. A review. Anal Chim Acta 943:17–40. https://doi.org/10.1016/j.aca.2016.08.049

Bereza-Malcolm LT, Mann GI, Franks AE (2014) Environmental sensing of heavy metals through whole cell microbial biosensors: a synthetic biology approach. ACS Synth Biol 4:535–546. https://doi.org/10.1021/sb500286r

Bhalla N, Jolly P, Formisano N, Estrela P (2016) Introduction to biosensors. Essays Biochem 60:1–8. https://doi.org/10.1042/EBC20150001

Blow M (2015) Streamlining a DNA-based biosensor detection system for the detection of non-PCR amplified genomic pathogenic DNA targets in food. 2015 annual meeting-international association for food protection, Portland, Oregon

Boer E, Heuvelink A (2000) Methods for the detection and isolation of Shiga toxin-producing Escherichia coli. J Appl Microbiol 88. doi:https://doi.org/10.1111/j.1365-2672.2000.tb05341.x

Borisov SM, Wolfbeis OS (2008) Optical biosensors. Chem Rev 108:423–461. https://doi.org/10.1021/cr068105t

Boyle M, Reis K (1987) Bacterial Fc receptors. Nature Biotechnol 5:697–703. https://doi.org/10.1038/nbt0787-697

Burnet F (1934) Antigenic differences between related bacterial strains: a criticism of the mosaic hypothesis. Br J Exp Pathol 15:354

Byrne B, Stack E, Gilmartin N, O'Kennedy R (2009) Antibody-based sensors: principles, problems and potential for detection of pathogens and associated toxins. Sensors 9:4407–4445. https://doi.org/10.3390/s90604407

Campbell GA, Mutharasan R (2005a) Detection of pathogen Escherichia coli O157: H7 using self-excited PZT-glass microcantilevers. Biosens Bioelectron 21:462–473. https://doi.org/10.1016/j.bios.2004.11.009

Campbell GA, Mutharasan R (2005b) Escherichia coli O157: H7 detection limit of millimeter-sized PZT cantilever sensors is 700 cells/mL. Anal Sci 21:355–357. https://doi.org/10.2116/analsci.21.355

Campbell GA, Mutharasan R (2006) Detection of Bacillus anthracis spores and a model protein using PEMC sensors in a flow cell at 1mL/min. BiosensBioelectron 22:78–85. https://doi.org/10.1016/j.bios.2005.12.002

Campbell GA, Mutharasan R (2007) A method of measuring Escherichia coli O157: H7 at 1 cell/mL in 1 liter sample using antibody functionalized piezoelectric-excited millimeter-sized cantilever sensor. Environ Sci Technol 41:1668–1674. https://doi.org/10.1021/es061947p

Chan WC, Nie S (1998) Quantum dot bioconjugates for ultrasensitive nonisotopic detection. Science 281:2016–2018. https://doi.org/10.1126/science.281.5385.2016

Chase GD (1980) Applications of liquid scintillation counting to radioimmunoassay. In: Peng C-T, Horrocks DL, Alpen EL (eds) Liquid scintillation counting, recent applications and development: sample preparation and applications, pp 489–502. ISBN:9780323146128

Chen A, Shah B (2013) Electrochemical sensing and biosensing based on square wave voltammetry. Anal Methods 5:2158–2173. https://doi.org/10.1039/C3AY40155C

Cho IH, Irudayaraj J (2013) In-situ immuno-gold nanoparticle network ELISA biosensors for pathogen detection. Int J Food Microbiol 164:70–75. https://doi.org/10.1016/j.ijfoodmicro.2013.02.025

Chua A, Yean CY, Ravichandran M, Lim B, Lalitha P (2011) A rapid DNA biosensor for the molecular diagnosis of infectious disease. Biosens Bioelectron 26:3825–3831. https://doi.org/10.1016/j.bios.2011.02.040

Compton RG and Banks CE (2011) Understanding voltammetry, World.Scientific. ISBN:978-1-84816-585-4.

Dale SE, Doherty-Kirby A, Lajoie G, Heinrichs DE (2004) Role of siderophore biosynthesis in virulence of Staphylococcus aureus: identification and characterization of genes involved in production of a siderophore. Infect Immun 72:29–37. https://doi.org/10.1128/IAI.72.1.29-37.2004

Dasgupta N, Ranjan S, Mundra S, Ramalingam C, Kumar A (2015) Fabrication of food grade vitamin E nanoemulsion by low energy approach, characterization and its application. Int J Food Prop 19(3):700–708

Dasgupta N, Ranjan S, Ramalingam C (2017) Applications of nanotechnology in agriculture and water quality management. Environ Chem Lett 15(4):591–605

Davis JJ, Green ML, Hill HAO, Leung YC, Sadler PJ, Sloan J, Xavier AV, Tsang SC (1998) The immobilisation of proteins in carbon nanotubes. Inorg Chim Acta 272:261–266. https://doi.org/10.1016/S0020-1693(97)05926-4

De Boever J, Kohen F, Vandekerckhove D (1983) Solid-phase chemiluminescence immunoassay for plasma estradiol-17 beta during gonadotropin therapy compared with two radioimmunoassays. Clin Chem 29:2068–2072

Deakin MR, Buttry DA (1989) Electrochemical applications of the quartz crystal microbalance. Anal Chem 61:1147A–1154A. https://doi.org/10.1021/ac00195a001

Demchenko AP (2008) Introduction to fluorescence sensing. Springer Science & Business Media. ISBN:978-1-4020-9002-8.

Deng S, Upadhyayula VK, Smith GB, Mitchell MC (2008) Adsorption equilibrium and kinetics of microorganisms on single-wall carbon nanotubes. IEEE Sensors J 8:954–962. https://doi.org/10.1109/JSEN.2008.923929

Dostálek J, Kasry A, Knoll W (2007) Long range surface plasmons for observation of biomolecular binding events at metallic surfaces. Plasmonics 2:97–106. https://doi.org/10.1007/s11468-007-9037-8

Downes FP, Green J, Greene K, Strockbine N, Wells J, Wachsmuth I (1989) Development and evaluation of enzyme-linked immunosorbent assays for detection of shiga-like toxin I and shiga-like toxin II. J Clin Microbiol 27:1292–1297

Elavarasan T, Chhina SK, Parameswaran M, Sankaran K (2013) Resazurin reduction based colorimetric antibiogram in microfluidic plastic chip. Sensor Actuat B: Chem 176:174–180. https://doi.org/10.1016/j.snb.2012.10.011

Farahi R, Passian A, Tetard L, Thundat T (2012) Critical issues in sensor science to aid food and water safety. ACS Nano 6:4548–4556. https://doi.org/10.1021/nn204999j

Farka Z, Kovář D, Přibyl J, Skládal P (2013) Piezoelectric and surface plasmon resonance biosensors for Bacillus atrophaeus spores. Int J Electrochem Sci 8:100–112

Farka ZK, Juřík TS, Pastucha MJ, andSkládal P (2016) Enzymatic precipitation enhanced surface Plasmon resonance Immunosensor for the detection of salmonella in powdered milk. Anal Chem 88:11830–11836. https://doi.org/10.1021/acs.analchem.6b03511

Fawcett NC, Evans JA, Chen LC, Drozda KA, Flowers AN (1988) A quartz crystal detector for DNA. Anal Lett 21:1099–1110

Felice CJ, Madrid RE, Olivera JM, Rotger VI, Valentinuzzi ME (1999) Impedance microbiology: quantification of bacterial content in milk by means of capacitance growth curves. J Microbiol Methods 35:37–42. https://doi.org/10.1016/S0167-7012(98)00098-0

Fritz J (2008) Cantilever biosensors. Analyst 133:855–863. https://doi.org/10.1039/B718174D

Fukal L, Kas J (1989) The advantages of immunoassay in food analysis. TrAC Trends Anal Chem 8:112–116. https://doi.org/10.1016/0165-9936(89)85010-1

Gómez R, Bashir R, Bhunia AK (2002) Microscale electronic detection of bacterial metabolism. Sensor Actuat B: Chem 86:198–208. https://doi.org/10.1016/S0925-4005(02)00175-2

Gonsalves MD, Sakr Y (2010) Early identification of sepsis. Curr Infect Dis Rep 12:329–335. https://doi.org/10.1007/s11908-010-0122-3

Gracias KS, McKillip JL (2004) A review of conventional detection and enumeration methods for pathogenic bacteria in food. Can J Microbiol 50:883–890. https://doi.org/10.1139/w04-080

Grieshaber D, MacKenzie R, Voeroes J, Reimhult E (2008) Electrochemical biosensors-sensor principles and architectures. Sensors 8:1400–1458. https://doi.org/10.3390/s80314000

Guilbault GG (1990) Practical fluorescence. Press, CRC. 9780824783501

Guinée P, Jansen W, Agterberg C (1976) Detection of the K99 antigen by means of agglutination and immunoelectrophoresis in Escherichia coli isolates from calves and its correlation with entertoxigenicity. Infect Immun 13:1369–1377

He F, Zhao J, Zhang L, Su X (2003) A rapid method for determining Mycobacterium tuberculosis based on a bulk acoustic wave impedance biosensor. Talanta 59:935–941. https://doi.org/10.1016/S0039-9140(02)00643-4

He F, Ren J, Liu Z (2007) The study and application of a new IDE–PQC sensor. Sensor Actuat B Chem 123:1057–1063. https://doi.org/10.1016/j.snb.2006.11.017

He F, Xiong Y, Liu J, Tong F, Yan D (2016) Construction of Au-IDE/CFP10-ESAT6 aptamer/DNA-AuNPs MSPQC for rapid detection of Mycobacterium tuberculosis. Biosens Bioelectron 77:799–804. https://doi.org/10.1016/j.bios.2015.10.054

Hegde NV, Cote R, Jayarao BM, Muldoon M, Lindpaintner K, Kapur V, DebRoy C (2012) Detection of the top six non-O157 Shiga toxin–producing Escherichia coli O Groups by ELISA. Foodborne Pathog Dis 9:1044–1048. https://doi.org/10.1089/fpd.2012.1231

Himpsl SD, Pearson MM, Arewang CJ, Nusca TD, Sherman DH, Mobley HL (2010) Proteobactin and a yersiniabactin-related siderophore mediate iron acquisition in Proteus mirabilis. Mol Microbiol 78:138–157. https://doi.org/10.1111/j.1365-2958.2010.07317.x

Homola J (2003) Present and future of surface plasmon resonance biosensors. Anal Bioanal Chem 377:528–539. https://doi.org/10.1007/s00216-003-2101-0

Honda T, Tsuji T, Takeda Y, Miwatani T (1981) Immunological nonidentity of heat-labile enterotoxins from human and porcine enterotoxigenic Escherichia coli. Infect Immun 34:337–340

Housewright RD, Glassman HN (1966) Concluding remarks of the second international conference on aerobiology (airborne infection). Bacteriol Rev 30:696

Hsieh K, Ferguson BS, Eisenstein M, Plaxco KW, Soh HT (2015) Integrated electrochemical microsystems for genetic detection of pathogens at the point of care. Acc Chem Res 48:911–920. https://doi.org/10.1021/ar500456w

Hsin Chang Y, Chang TC, Kao EF, Chou C (1996) Detection of protein A produced by Staphylococcus aureus with a fiber-optic-based biosensor. Biosci Biotechnol Biochem 60:1571–1574. https://doi.org/10.1271/bbb.60.1571

Hu RR, Yin ZZ, Zeng YB, Zhang J, Liu HQ, Shao Y, Ren SB, Li L (2016) A novel biosensor for Escherichia coli O157: H7 based on fluorescein-releasable biolabels. Biosens Bioelectron 78:31–36. https://doi.org/10.1016/j.bios.2015.11.018

Huang CJ, Dostalek J, Sessitsch A, Knoll S (2011) Long-range surface plasmon-enhanced fluorescence spectroscopy biosensor for ultrasensitive detection of E. Coli O157: H7. Anal Chem 83:674–677. https://doi.org/10.1021/ac102773r

Isaacson R (1977) K99 surface antigen of Escherichia coli: purification and partial characterization. Infect Immun 15:272–279

Itoh A, Ichihashi M (2008) A frequency of the quartz crystal microbalance (QCM) that is not affected by the viscosity of a liquid. Meas Sci Technol 19:075205. https://doi.org/10.1088/0957-0233/19/7/075205

Jain A, Ranjan S, Dasgupta N, Ramalingam C (2016) Nanomaterials in food and agriculture: an overview on their safety concerns and regulatory issues. Crit Rev Food Sci Nutr 15:1–21

Jang HD, Chang KS, Lee YG, Lee SJ and Hsu CL (2009) Quantitative determination of Lactobacillus spp. in Milk using a series piezoelectric quartz crystal sensor, vol 229. European Food Research and Technology, pp 349–355. doi:https://doi.org/10.1007/s00217-009-1047-7

Jans H, Huo Q (2012) Gold nanoparticle-enabled biological and chemical detection and analysis. Chem Soc Rev 41:2849–2866. https://doi.org/10.1039/C1CS15280G.

Ji HF, Yang X, Zhang J, Thundat T (2004) Molecular recognition of biowarfare agents using micromechanical sensors. Expert Rev Mol Diagn 4:859–866. https://doi.org/10.1586/14737159.4.6.859

Jordana-Lluch E, Carolan HE, Giménez M, Sampath R, Ecker DJ, Quesada MD, Mòdol JM, Arméstar F, Blyn LB, Cummins LL (2013) Rapid diagnosis of bloodstream infections with PCR followed by mass spectrometry. PLoS One 8:e62108. https://doi.org/10.1371/journal.pone.0062108

Kallioniemi OP, Kallioniemi A, Kurisu W, Thor A, Chen LC, Smith HS, Waldman FM, Pinkel D, Gray JW (1992) ERBB2 amplification in breast cancer analyzed by fluorescence in situ hybridization. Proc Natl Acad Sci 89:5321–5325. https://doi.org/10.1073/pnas.89.12.5321

Kamali KZ, Pandikumar A, Sivaraman G, Lim HN, Wren SP, Sun T, Huang NM (2015) Silver@graphene oxide nanocomposite-based optical sensor platform for biomolecules. RSC Adv 5:17809–17816. https://doi.org/10.1039/C4RA11356J

Kang S, Pinault M, Pfefferle LD, Elimelech M (2007) Single-walled carbon nanotubes exhibit strong antimicrobial activity. Langmuir 23:8670–8673. https://doi.org/10.1021/la701067r

Kas J, Fukal L, Rauch P (1986) Immunochemical assay of enzymes. TrAC Trends Anal Chem 5:205–209. https://doi.org/10.1016/0165-9936(86)80014-0

Kim TH, Park J, Kim CJ, Cho YK (2014) Fully integrated lab-on-a-disc for nucleic acid analysis of food-borne pathogens. Anal Chem 86:3841–3848. https://doi.org/10.1021/ac403971h

Kirsch J, Siltanen C, Zhou Q, Revzin A, Simonian A (2013) Biosensor technology: recent advances in threat agent detection and medicine. Chem Soc Rev 42:8733–8768. https://doi.org/10.1039/c3cs60141b

Ko S, Grant SA (2006) A novel FRET-based optical fiber biosensor for rapid detection of Salmonella typhimurium. Biosens Bioelectron 21:1283–1290. https://doi.org/10.1016/j.bios.2005.05.017

Kuila T, Bose S, Khanra P, Mishra AK, Kim NH, Lee JH (2011) Recent advances in graphene-based biosensors. Biosens Bioelectron 26:4637–4648. https://doi.org/10.1016/j.bios.2011.05.039

Lakshmanan RS, Efremov V, Cullen SM, Killard AJ (2014) Measurement of the evolution of rigid and viscoelastic mass contributions from fibrin network formation during plasma coagulation using quartz crystal microbalance. Sensor Actuat B: Chemical 192:23–28. https://doi.org/10.1016/j.snb.2013.10.081

Lam JS, Mutharia LM (1994) Antigen-antibody reactions. Methods for general and molecular bacteriology. American Society for Microbiology, Washington, DC, pp 104–132. https://doi.org/10.1128/9781555817497.ch08

Lee N, Englesberg E (1962) Dual effects of structural genes in Escherichia coli. Proc Natl Acad Sci 48:335–348

Lee K, Irudayaraj J (2009) Periodic and dynamic 3-D gold nanoparticle-DNA network structures for surface-enhanced Raman spectroscopy-based quantification. J Phys Chem C 113:5980–5983. https://doi.org/10.1021/jp809949v

Lee KR, Kang IJ (2009) Effects of dopamine concentration on energy transfer between dendrimer–QD and dye-labeled antibody. Ultramicroscopy 109:894–898. https://doi.org/10.1016/j.ultramic.2009.03.012

Lee K, Drachev VP, Irudayaraj J (2011) DNA-gold nanoparticle reversible networks grown on cell surface marker sites: application in diagnostics. ACS Nano 5:2109–2117. https://doi.org/10.1021/nn1030862

Lequin RM (2005) Enzyme immunoassay (EIA)/enzyme-linked immunosorbent assay (ELISA). Clin Chem 51:2415–2418. https://doi.org/10.1373/clinchem.2005.051532

Li G, Li X, Wan J, Zhang S (2009) Dendrimers-based DNA biosensors for highly sensitive electrochemical detection of DNA hybridization using reporter probe DNA modified with Au nanoparticles. Biosens Bioelectron 24:3281–3287. https://doi.org/10.1016/j.bios.2009.04.022

Li B, Yu Q, Duan Y (2015) Fluorescent labels in biosensors for pathogen detection. Crit Rev Biotechnol 35:82–93. https://doi.org/10.3109/07388551.2013.804487

Lian Y, He F, Wang H, Tong F (2015) A new aptamer/graphene interdigitated gold electrode piezoelectric sensor for rapid and specific detection of Staphylococcus aureus. Biosens Bioelectron 65:314–319. https://doi.org/10.1016/j.bios.2014.10.017

Liedberg B, Nylander C, Lundström I (1995) Biosensing with surface plasmon resonance—how it all started. Biosens Bioelectron 10:i–ix. doi:https://doi.org/10.1016/0956-5663(95)96965-2

Ligler FS, Taitt CR (2008) Fluorescence lifetime biosensing: entering the mainstream. Optical Biosensors: Today and Tomorrow 287. 978-0-444-53125-4

Lin YH, Chen SH, Chuang YC, YC L, Shen TY, Chang CA, Lin CS (2008) Disposable amperometric immunosensing strips fabricated by Au nanoparticles-modified screen-printed carbon electrodes for the detection of foodborne pathogen Escherichia coli O157: H7. Biosens Bioelectron 23:1832–1837. https://doi.org/10.1016/j.bios.2008.02.030

Liu Y, Dong X, Chen P (2012) Biological and chemical sensors based on graphene materials. Chem Soc Rev 41:2283–2307. https://doi.org/10.1039/C1CS15270J

Liu X, Hu Y, Zheng S, Liu Y, He Z, Luo F (2016a) Surface plasmon resonance immunosensor for fast, highly sensitive, and in situ detection of the magnetic nanoparticles-enriched Salmonella enteritidis. Sensor Actuat B Chem 230:191–198. https://doi.org/10.1016/j.snb.2016.02.043

Liu X, Marrakchi M, Xu D, Dong H, Andreescu S (2016b) Biosensors based on modularly designed synthetic peptides for recognition, detection and live/dead differentiation of pathogenic bacteria. Biosens Bioelectron 80:9–16. https://doi.org/10.1016/j.bios.2016.01.041

Melton-Celsa AR (2014) Shiga toxin (Stx) classification, structure, and function. Microbiol Spectr 2. https://doi.org/10.1128/microbiolspec.EHEC-0024-2013

Menti C, Henriques JA, Missell FP, Roesch-Ely M (2016) Antibody-based magneto-elastic biosensors: potential devices for detection of pathogens and associated toxins. Appl Microbiol Biotechnol 100:6149–6163. https://doi.org/10.1007/s00253-016-7624-3

Merkoçi A (2010) Nanoparticles-based strategies for DNA, protein and cell sensors. Biosens Bioelectron 26:1164–1177. https://doi.org/10.1016/j.bios.2010.07.028

Monis PTand Giglio S (2006) Nucleic acid amplification-based techniques for pathogen detection and identification. Infect Genet Evol 6:2–12. https://doi.org/10.1016/j.meegid.2005.08.004

Monzo J, Insua I, Fernandez-Trillo F, Rodriguez P (2015) Fundamentals, achievements and challenges in the electrochemical sensing of pathogens. Analyst 140:7116–7128. https://doi.org/10.1039/c5an01330e

Mujumdar RB, Ernst LA, Mujumdar SR, Lewis CJ, Waggoner AS (1993) Cyanine dye labeling reagents: sulfoindocyanine succinimidyl esters. Bioconjug Chem 4:105–111. https://doi.org/10.1021/bc00020a001

Nguyen TT, Trinh KTL, Yoon WJ, Lee NY, Ju H (2017) Integration of a microfluidic polymerase chain reaction device and surface plasmon resonance fiber sensor into an inline all-in-one platform for pathogenic bacteria detection. Sensor Actuat B Chem 242:1–8. https://doi.org/10.1016/j.snb.2016.10.137

Nugen SR, Baeumner A (2008) Trends and opportunities in food pathogen detection. Anal Bioanal Chem 391:451. https://doi.org/10.1007/s00216-008-1886-2

Oh BK, Lee W, Chun BS, Bae YM, Lee WH, Choi JW (2005) The fabrication of protein chip based on surface plasmon resonance for detection of pathogens. Biosens Bioelectron 20:1847–1850. https://doi.org/10.1016/j.bios.2004.05.010

Oku Y, Yutsudo T, Hirayama T, O'Brien AD, Takeda Y (1989) Purification and some properties of a Vero toxin from a human strain of Escherichia coli that is immunologically related to Shiga-like toxin II (VT2). Microb Pathog 6:113–122. https://doi.org/10.1016/0882-4010(89)90014-4

Ouchterlony O (1949) Antigen-antibody reactions in gels. Acta Pathologica Microbiologica Scandinavica 26:507–515. https://doi.org/10.1111/j.1699-0463.1949.tb00751.x

Palomino JC, Martin A, Camacho M, Guerra H, Swings J, Portaels F (2002) Resazurin microtiter assay plate: simple and inexpensive method for detection of drug resistance in Mycobacterium tuberculosis. Antimicrob Agents Chemother 46:2720–2722. https://doi.org/10.1016/j.ejcdt.2013.05.008

Panchuk-Voloshina N, Haugland RP, Bishop-Stewart J, Bhalgat MK, Millard PJ, Mao F, Leung WY, Haugland RP (1999) Alexa dyes, a series of new fluorescent dyes that yield exceptionally bright, photostable conjugates. J Histochem Cytochem 47:1179–1188. https://doi.org/10.1177/002215549904700910

Pastells C, Pascual N, Sanchez-Baeza F, Marco MP (2016) Immunochemical determination of pyocyanin and 1-hydroxyphenazine as potential biomarkers of Pseudomonas aeruginosa infections. Anal Chem 88:1631–1638. https://doi.org/10.1021/acs.analchem.5b03490.

Pauling L (1940) A theory of the structure and process of formation of antibodies. J Am Chem Soc 62:2643–2657. https://doi.org/10.1021/ja01867a018.

Philip A, Odabasi Z, Matiuzzi G, Paetznick VL, Tan SW, Warmington J, Rex JH, Ostrosky-Zeichner L (2005) Syscan3, a kit for detection of anti-Candida antibodies for diagnosis of invasive candidiasis. J Clin Microbiol 43:4834–4835. https://doi.org/10.1128/JCM.43.9.4834-4835.2005

Pi S, Zeng X, Zhang N, Ji D, Chen B, Song H, Cheney A, Xu Y, Jiang S, Sun D (2016) Dielectric-grating-coupled surface plasmon resonance from the back side of the metal film for ultrasensitive sensing. IEEE Photon J 8:1–6. https://doi.org/10.1109/JPHOT.2015.2509870

Ranjan S, Dasgupta N, Chakraborty AR, Melvin Samuel S, Ramalingam C, Shanker R, Kumar A (2014) Nanoscience and nanotechnologies in food industries: opportunities and research trends. J Nanopart Res 16(6):1–23

Ren J, He F, Yi S, Cui X (2008) A new MSPQC for rapid growth and detection of Mycobacterium tuberculosis. Biosens Bioelectron 24:403–409. https://doi.org/10.1016/j.bios.2008.04.018

Ren J, He F, Zhang L (2010) The construction and application of a new PPY-MSPQC for l-asparaginase activity assay. Sensor Actuat B Chem 145:272–277. https://doi.org/10.1016/j.snb.2009.12.006

Renshaw PS, Panagiotidou P, Whelan A, Gordon SV, Hewinson RG, Williamson RA, Carr MD (2002) Conclusive evidence that the major T-cell antigens of the mycobacterium tuberculosis complex ESAT-6 and CFP-10 form a tight, 1: 1 complex and characterization of the structural properties of ESAT-6, CFP-10, and the ESAT-6· CFP-10 complex implications for pathogenesis and virulence. J Biol Chem 277:21598–21603. https://doi.org/10.1074/jbc.M201625200

Rifat AA, Mahdiraji GA, Sua YM, Shee YG, Ahmed R, Chow DM, Adikan FRM (2015) Surface Plasmon resonance photonic crystal fiber biosensor: a practical sensing approach. IEEE Photon Technol Lett 27:1628–1631. https://doi.org/10.1109/LPT.2015.2432812

Rosenthal SJ (2001) Bar-coding biomolecules with fluorescent nanocrystals. Nature Biotechnol 19:621–623. https://doi.org/10.1038/90213

Rosi NL, Mirkin CA (2005) Nanostructures in biodiagnostics. Chem Rev 105:1547–1562. https://doi.org/10.1021/cr030067f

Rowe-Taitt CA, Golden JP, Feldstein MJ, Cras JJ, Hoffman KE, Ligler FS (2000) Array biosensor for detection of biohazards. Biosens Bioelectron 14:785–794. https://doi.org/10.1016/S0956-5663(99)00052-4

Sang S, Wang Y, Feng Q, Wei Y, Ji J, Zhang W (2016) Progress of new label-free techniques for biosensors: a review. Crit Rev Biotechnol 36:465–481. https://doi.org/10.3109/07388551.2014.991270

Sankaranarayanan R, Alagumaruthanayagam A, Sankaran K (2015) A new fluorimetric method for the detection and quantification of siderophores using Calcein Blue, with potential as a bacterial detection tool. Appl Microbiol Biotechnol 99:2339–2349. https://doi.org/10.1007/s00253-015-6411-x

Sarker SD, Nahar L, Kumarasamy Y (2007) Microtitre plate-based antibacterial assay incorporating resazurin as an indicator of cell growth, and its application in the in vitro antibacterial screening of phytochemicals. Methods 42:321–324. https://doi.org/10.1016/j.ymeth.2007.01.006

Schena M, Shalon D, Davis RW, Brown PO (1995) Quantitative monitoring of gene expression patterns with a complementary DNA microarray. Science 270:467. https://doi.org/10.1126/science.270.5235.467

Scognamiglio V, Arduini F, Palleschi G, Rea G (2014) Biosensing technology for sustainable food safety. TrAC Trends Anal Chem 62:1–10. https://doi.org/10.1016/j.trac.2014.07.007

Shao Y, Wang J, Wu H, Liu J, Aksay IA, Lin Y (2010) Graphene based electrochemical sensors and biosensors: a review. Electroanalysis 22:1027–1036. https://doi.org/10.1002/elan.200900571

Sharma H, Mutharasan R (2013) Rapid and sensitive immunodetection of Listeria monocytogenes in milk using a novel piezoelectric cantilever sensor. Biosens Bioelectron 45:158–162. https://doi.org/10.1016/j.bios.2013.01.068

Sheikhzadeh E, Chamsaz M, Turner A, Jager E, Beni V (2016) Label-free impedimetric biosensor for Salmonella typhimurium detection based on poly [pyrrole-co-3-carboxyl-pyrrole] copolymer supported aptamer. Biosens Bioelectron 80:194–200. https://doi.org/10.1016/j.bios.2016.01.057

Shen D, Zhu W, Nie L, Yao S (1993) Behaviour of a series piezoelectric sensor in electrolyte solution: part I. Theory. Anal Chim Acta 276:87–97. https://doi.org/10.1016/0003-2670(93)85042-I

Shen J, Zhu Y, Yang X, Li C (2012) Graphene quantum dots: emergent nanolights for bioimaging, sensors, catalysis and photovoltaic devices. Chem Commun 48:3686–3699. https://doi.org/10.1039/C2CC00110A

Shi X, Zhang X, Yao Q, He F (2017) A novel method for the rapid detection of microbes in blood using pleurocidin antimicrobial peptide functionalized piezoelectric sensor. J Microbiol Methods 133:69–75. https://doi.org/10.1016/j.mimet.2016.12.005

Shukla A, Dasgupta N, Ranjan S, Singh S, Chidambram R (2017) Nanotechnology towards prevention of anaemia and osteoporosis: from concept to market. Biotechnol Biotechnol Equip 31(5):863–879

Sign C, Sumana G (2016) Antibody conjugated graphene nanocomposites for pathogen detection. J Phys Conf Ser IOP Publishing. https://doi.org/10.1088/1742-6596/704/1/012014

Skinner C, Patfield S, Stanker L, He X (2013) Development of monoclonal antibodies and immunoassays for sensitive and specific detection of Shiga toxin Stx2f. PLoS One 8:e76563. https://doi.org/10.1371/journal.pone.0076563

Soergel ME, Schaffer FL, Blank HF (1982) Solid-phase radioimmunoassay for detection of plague antigen in animal tissue. J Clin Microbiol 16:953–956

Srivastava A, Srivastava O, Talapatra S, Vajtai R, Ajayan P (2004) Carbon nanotube filters. Nat Mater 3:610–614. https://doi.org/10.1038/nmat1192

Sungkanak U, Sappat A, Wisitsoraat A, Promptmas C, Tuantranont A (2010) Ultrasensitive detection of Vibrio cholerae O1 using microcantilever-based biosensor with dynamic force microscopy. Biosens Bioelectron 26:784–789. https://doi.org/10.1016/j.bios.2010.06.024

Sutherland AJ (2002) Quantum dots as luminescent probes in biological systems. Curr Opin Solid State Mater Sci 6:365–370. https://doi.org/10.1016/S1359-0286(02)00081-5

Taitt CR, Anderson GP, Ligler FS (2016) Evanescent wave fluorescence biosensors: advances of the last decade. Biosens Bioelectron 76:103–112. https://doi.org/10.1016/j.bios.2015.07.040

Taneja NK, Tyagi JS (2007) Resazurin reduction assays for screening of anti-tubercular compounds against dormant and actively growing Mycobacterium tuberculosis, Mycobacterium bovis BCG and Mycobacterium smegmatis. J Antimicrob Chemother 60:288–293. https://doi.org/10.1093/jac/dkm207

Teel LD, Daly JA, Jerris RC, Maul D, Svanas G, O'Brien AD, Park CH (2007) Rapid detection of Shiga toxin-producing Escherichia coli by optical immunoassay. J Clin Microbiol 45:3377–3380. https://doi.org/10.1128/JCM.00837-07

Teo SC, Wong LS (2014) Whole cell-based biosensors for environmental heavy metals detection. Annual Res Rev Biol 4:2663–2674. https://doi.org/10.1016/S0003-2670(98)00725-9

Thakur B, Amarnath CA, Mangoli S, Sawant SN (2015) Polyaniline nanoparticle based colorimetric sensor for monitoring bacterial growth. Sensor Actuat B Chem 207:262–268. https://doi.org/10.1016/j.snb.2014.10.045

Tong F, Lian Y, Zhou H, Shi X, He F (2014) Multichannel series piezoelectric quartz crystal cell sensor for real time and quantitative monitoring of the living cell and assessment of cytotoxicity. Anal Chem 86:10415–10421. https://doi.org/10.1021/ac502926k

Tsuji T, Honda T, Miwatani T, Wakabayashi S, Matsubara H (1985) Analysis of receptor-binding site in Escherichia coli enterotoxin. J Biol Chem 260:8552–8558

Upadhyayula VK, Deng S, Smith GB, Mitchell MC (2009) Adsorption of Bacillus subtilis on single-walled carbon nanotube aggregates, activated carbon and NanoCeram™. Water Res 43:148–156. https://doi.org/10.1016/j.watres.2008.09.023

Vala M, Etheridge S, Roach J, Homola J (2009) Long-range surface plasmons for sensitive detection of bacterial analytes. Sensor Actuat B Chem 139:59–63. https://doi.org/10.1016/j.snb.2008.08.029

Vinayaka A, Thakur M (2010) Focus on quantum dots as potential fluorescent probes for monitoring food toxicants and foodborne pathogens. Anal Bioanal Chem 397:1445–1455. https://doi.org/10.1007/s00216-010-3683-y

Waggoner A (2006) Fluorescent labels for proteomics and genomics. Curr Opin Chem Biol 10:62–66. https://doi.org/10.1016/j.cbpa.2006.01.005

Wan Y, Lin Z, Zhang D, Wang Y, Hou B (2011) Impedimetric immunosensor doped with reduced graphene sheets fabricated by controllable electrodeposition for the non-labelled detection of bacteria. Biosens Bioelectron 26:1959–1964. https://doi.org/10.1016/j.bios.2010.08.008

Wang Y, Alocilja EC (2015) Gold nanoparticle-labeled biosensor for rapid and sensitive detection of bacterial pathogens. J Biol Eng 9:16. https://doi.org/10.1186/s13036-015-0014-z

Wang Z, Zhang J, Chen P, Zhou X, Yang Y, Wu S, Niu L, Han Y, Wang L, Boey F (2011) Label-free, electrochemical detection of methicillin-resistant staphylococcus aureus DNA with reduced graphene oxide-modified electrodes. Biosens Bioelectron 26:3881–3886. https://doi.org/10.1016/j.bios.2011.03.002

Wang Y, Knoll W, Dostalek J (2012a) Bacterial pathogen surface plasmon resonance biosensor advanced by long range surface plasmons and magnetic nanoparticle assays. Anal Chem 84:8345–8350. https://doi.org/10.1021/ac301904x

Wang Y, Ye Z, Ying Y (2012b) New trends in impedimetric biosensors for the detection of foodborne pathogenic bacteria. Sensors 12:3449–3471. https://doi.org/10.3390/s120303449

Wang R, Xiang Y, Zhou X, Liu LH, Shi H (2015) A reusable aptamer-based evanescent wave all-fiber biosensor for highly sensitive detection of Ochratoxin A. Biosens Bioelectron 66:11–18. https://doi.org/10.1016/j.bios.2014.10.079

Ward AC, Connolly P, Tucker NP (2014) Pseudomonas aeruginosa can be detected in a polymicrobial competition model using impedance spectroscopy with a novel biosensor. PLoS One 9:e91732. https://doi.org/10.1371/journal.pone.0091732

Wei D, Bailey MJ, Andrew P, Ryhanen T (2009) Electrochemical biosensors at the nanoscale. Lab Chip 9:2123–2131. https://doi.org/10.1039/B903118A

Wells RM, Jones CM, Xi Z, Speer A, Danilchanka O, Doornbos KS, Sun P, Wu F, Tian C, Niederweis M (2013) Discovery of a siderophore export system essential for virulence of Mycobacterium tuberculosis. PLoS Pathog 9:e1003120. https://doi.org/10.1371/journal.ppat.1003120.

White NS, Errington RJ (2005) Fluorescence techniques for drug delivery research: theory and practice. Adv Drug Deliv Rev 57:17–42. https://doi.org/10.1016/j.addr.2004.08.003

Wu X, Liu H, Liu J, Haley KN, Treadway JA, Larson JP, Ge N, Peale R, Bruchez MP (2003) Immunofluorescent labeling of cancer marker Her2 and other cellular targets with semiconductor quantum dots. Nature Biotechnol 21:41–46. https://doi.org/10.1038/nbt764

Xing Y, Chaudry Q, Shen C, Kong KY, Zhau HE, Chung LW, Petros JA, O'regan RM, Yezhelyev MV, Simons JW (2007) Bioconjugated quantum dots for multiplexed and quantitative immunohistochemistry. Nat Protoc 2:1152–1165. https://doi.org/10.1038/nprot.2007.107

Xu X, Ray R, Gu Y, Ploehn HJ, Gearheart L, Raker K, Scrivens WA (2004) Electrophoretic analysis and purification of fluorescent single-walled carbon nanotube fragments. J Am Chem Soc 126:12736–12737. https://doi.org/10.1021/ja040082h

Xu M, Wang R, Li Y (2017) Electrochemical biosensors for rapid detection of Escherichia coli O157:H7. Talanta 162:511–522. https://doi.org/10.1016/j.talanta.2016.10.050

Yanase Y, Hiragun T, Ishii K, Kawaguchi T, Yanase T, Kawai M, Sakamoto K, Hide M (2014) Surface plasmon resonance for cell-based clinical diagnosis. Sensors 14:4948–4959. https://doi.org/10.3390/s140304948

Yang L, Li Y (2006) Simultaneous detection of Escherichia coli O157: H7 and Salmonella typhimurium using quantum dots as fluorescence labels. Analyst 131:394–401. https://doi.org/10.1039/B510888H

Yu X, Xia HS, Sun ZD, Lin Y, Wang K, Yu J, Tang H, Pang DW, Zhang ZL (2013) On-chip dual detection of cancer biomarkers directly in serum based on self-assembled magnetic bead patterns and quantum dots. Biosens Bioelectron 41:129–136. https://doi.org/10.1016/j.bios.2012.08.007

Yu Q, Wang X, Duan Y (2014) Capillary-based three-dimensional immunosensor assembly for high-performance detection of carcinoembryonic antigen using laser-induced fluorescence spectrometry. Anal Chem 86:1518–1524. https://doi.org/10.1021/ac402973n

Zawadzka AM, Kim Y, Maltseva N, Nichiporuk R, Fan Y, Joachimiak A, Raymond KN (2009) Characterization of a Bacillus subtilis transporter for petrobactin, an anthrax stealth siderophore. Proc Natl Acad Sci 106:21854–21859. https://doi.org/10.1073/pnas.0904793106

Zhang J, Hai-Feng J (2004) An anti E. coli O157: H7 antibody-immobilized microcantilever for the detection of Escherichia coli (E. coli). Anal Sci 20:585–587. https://doi.org/10.2116/analsci.20.585

Zhang J, Su X, O'shea S (2002) Antibody/antigen affinity behavior in liquid environment with electrical impedance analysis of quartz crystal microbalances. Biophys Chem 99:31–41. https://doi.org/10.1016/S0301-4622(02)00109-6

Zhang Y, Zeng Q, Sun Y, Liu X, Tu L, Kong X, Buma WJ, Zhang H (2010) Multi-targeting single fiber-optic biosensor based on evanescent wave and quantum dots. Biosens Bioelectron 26:149–154. https://doi.org/10.1016/j.bios.2010.05.032.

Zhang D, Yan Y, Li Q, Yu T, Cheng W, Wang L, Ju H, Ding S (2012) Label-free and high-sensitive detection of salmonella using a surface plasmon resonance DNA-based biosensor. J Biotechnol 160:123–128. https://doi.org/10.1016/j.jbiotec.2012.03.024

Zhang H, Zhang H, Aldalbahi A, Zuo X, Fan C, Mi X (2017) Fluorescent biosensors enabled by graphene and graphene oxide. Biosens Bioelectron 89:96–106. https://doi.org/10.1016/j.bios.2016.07.030

Zheng J, Lin L, Cheng G, Wang A, Tan X, He P, Fang Y (2007) Study on an electrochemical biosensor for thrombin recognition based on aptamers and nano particles. Sci China Ser B Chem 50:351–357. https://doi.org/10.1007/s11426-007-0062-4

Zheng Q, Wu H, Wang N, Yan R, Ma Y, Guang W, Wang J, Ding K (2014) Graphene-based biosensors for biomolecules detection. Curr Nanosci 10:627–637. https://doi.org/10.2174/1573413710666140422231701

Zhu Q, Shih WY, Shih WH (2007) In situ, in-liquid, all-electrical detection of Salmonella typhimurium using lead titanate zirconate/gold-coated glass cantilevers at any dipping depth. Biosens Bioelectron 22:3132–3138. https://doi.org/10.1016/j.bios.2007.02.005

Zuo P, Li X, Dominguez DC, Ye BC (2013) A PDMS/paper/glass hybrid microfluidic biochip integrated with aptamer-functionalized graphene oxide nano-biosensors for one-step multiplexed pathogen detection. Lab Chip 13:3921–3928. https://doi.org/10.1039/c3lc50654a

Chapter 3
Nanotechnology in the Food Industry

Arun G. Ingale and Anuj N. Chaudhari

Abstract Nanotechnology delivers emerging applications in functional food by engineering biological and synthetic molecules toward functions that are exceptionally changed from those they have originally. Nanotechnology has enhanced the superiority of foods by making them flavoured, nutritive and more healthier. Nanotechnology generates also novel food products, better packaging, coating and shelf storage techniques. Applications in food also improve shelf life, food quality, safety and fortification. Biosensors in food packaging are designed to detect contaminated or spoiled food. Nanotechnology improve food processes that use enzymes to confer nutrition and health benefits. This report reviews applications of nanotechnology in agriculture, and food science and technology. Furthermore, risk assessment, safety concerns and social implications are discussed.

Keywords Food nanotechnology • Agriculture nanotechnology • Food processing • Food packaging • Nanotechnology in food supplements

Contents

A.G. Ingale (✉) • A.N. Chaudhari
Department of Biotechnology, School of Life Sciences, North Maharashtra University, Jalgaon, Maharashtra, India
e-mail: agibiotech@gmail.com; anujchaudhari01@gmail.com

K M Gothandam et al. (eds.), *Nanotechnology, Food Security and Water Treatment*, Environmental Chemistry for a Sustainable World 11,
https://doi.org/10.1007/978-3-319-70166-0_3

3.1 Introduction

Nanotechnologists are being hopeful that nanotechnology will able to transform the entire scenario of food industry by bringing significant changes in the various processes in the food industry: production, processing, packaging, transportation and consumption of food and food products. Increase and exploitation of nanotechnology in these processes ensures the safety of food and food products and creates a healthy food culture which dominantly enhancing the nutritional quality of foods which is the need of an hour (deAzeredo 2009; Ravichandran 2010; Dasgupta et al. 2017; Shukla et al. 2017; Jain et al. 2016). Moreover smart food packaging systems can be developed using nanotechnology that in turn boosts the shelf-life of food products by developing active antifungal and antimicrobic surfaces, improving properties like heat-resistance and mechanical, modifying the permeation behaviour of foils simultaneously detecting and signalling biochemical and microbiological changes (Neethirajan and Jayas 2011). Implication of nanotechnology in current food processing is creating an incredible impact on the advanced development of interactive and functional foods which deliver nutrients and respond to the body's requirements in an efficient manner (Dunn 2004). Nanocapsules which are added into food products with the aim to deliver nutrients and nanoparticles when added to food and food products increase the absorption of nutrients. Organic and polymeric nanoparticles are being used to deliver vitamins or other nutrients in food and beverages without affecting the taste and appearance of the product. The nutrients are encapsulated by the nanoparticles and carried all the way through the stomach and reaches into the bloodstream. This method applied for delivery of many vitamins in higher percentage of availability of the nutrients to be used by the body because, when nutrients are not encapsulated by the nanoparticles, few of the nutrients would be lost in stomach (Ezhilarasi et al. 2013). Researchers are also engaged in development of nanocapsules containing nutrients that would be released when nanosensors detect a deficiency in of particular nutrient in body. Fundamentally the phenomenon behind this is to construct a super vitamin storage system in body that gives only what body need and when body need. Dominantly nanomaterials are being developed to improve the overall quality viz. taste, color, and texture of foods. Taken an example "interactive" foods are continuously being developed that would allow consumer to choose which flavor and color a piece of food has (Dunn 2004). Nanosensors are enormously developed and improved that can prior detect bacteria and other contaminants on the surfaces of food at a packaging plants. Resultantly this will allow for frequent testings at a much lower cost than usually is acquire by sending samples to a lab for analysis and wait till the results will arrive. This particular packaging testing point is conducted more properly and carefully, has great potential to considerably reduce the chance of contaminated food reaching grocery store shelves.

In August 2006, the US Food and Drug Administration (FDA) formed a Nanotechnology Task Force with goals that include; (i) evaluate the effectiveness of the agency's regulatory approaches and authorities to meet any unique challenge that may be presented by the use of nanotechnology materials in FDA-regulated products. (ii) Explore opportunities to foster innovation using nanotechnology materials to develop safe and effective drugs, biologics, and devices, and to develop safe foods, feeds, and cosmetics. (The US Department of Agriculture 2015) Focusing this goals nanotechnology is having a significant impact on numerous facet of the food industry, from how food is grown to how it is travelled, processed and packaged. Industries are developing nanomaterials that will make a difference not only in the taste of food, but also in food safety, and the health benefits food going to delivers when consumed. Increasing developments in micro- and nanotechnologies are growing at a rapid rate and strongly offers the functional potential to not only improve the products in terms of textural and sensory qualities, stability and health benefits, but also develop new products or manufacturing processes for the food industry. Today food industry is the one of the largest manufacturing sector in the world, with counting an annual turnover approximating US $4 trillion. Worldwide, a large amount of foods are consumed after only minimum processing e.g., fresh fruits, vegetables, nuts, some cereals and with additionally high post-harvest losses considerably with fruits and vegetables (Fig. 3.1) (U.S. FDA 2014).

Nanotechnology offering wide ranges of opportunities for the development and improvement of innovative products and applications in food manufacturing system. Recently developed functional foods, neutraceuticals, bioactive compounds, enriched farmafoods, etc. are highlighted examples of it.

Majorly of the sector where nanotechnologies are settling to make a difference is in meat food processing in near future like, intelligent packaging of meat and meat products, meat derived bioactive peptides, pro- and pre-biotics inclusion in processed meat products, fat based nanoemulsions, nanosensors and nanotracers for meat biosecurity tracing and nanostructured meat food products with distinct functions (Table 3.1). Complex set of engineering techniques and occurrence of scientific challenges in the food and bioprocessing industries for manufacturing high quality and safe food through efficient and sustainable means can proven to be solved through nanotechnology. Identification of contaminating bacteria and monitoring of food quality using highly precised biosensors; intelligent, active, and smart food packaging systems; and nano-encapsulation of bioactive food compounds are strong establishing applications of nanotechnology for the next generation food industry. Nanotechnologies are not new and researchers have been making various polymers based on nanoscale subunits for many years. The result of new, previously unknown, properties attributed to engineered nanoparticles (NP) many inventive consumer products containing these Nanoparticles have been launched to the market recently. Application of nanotechnologies in

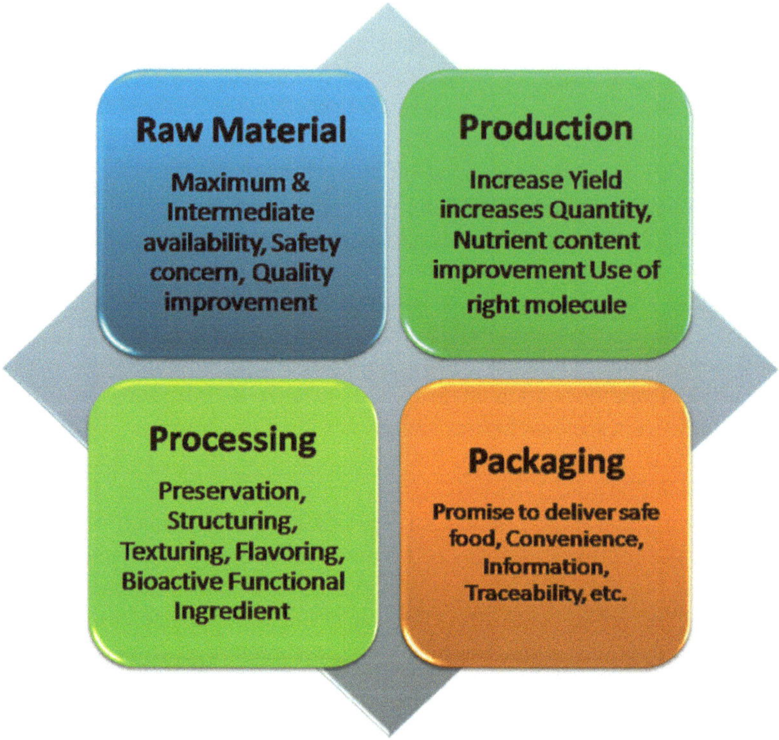

Fig. 3.1 Aims of next generation food industry of the future

electronics, medicine, textiles, defence, food, agriculture, cosmetics, and other areas are already a reality and fruits of it applications are beginning to impact the next generation of food production and processing industries (Chen et al. 2006). Within food and agricultural sector nanotechnologies covering many aspects, such as food security, packaging materials, disease treatment, delivery systems, bio-availability, new tools for molecular and cellular biology and new materials for pathogen detection (Maynard et al. 2006; Jasińska et al. 2010). The latent profit for consumers and producers of these innovative products are widely emphasized.

3.2 Role of Nanotechnology in Agriculture

There is huge potential in nanoscience and technology in the stipulation of state-of-the-art key for various challenges faced by and opportunities missed by agriculture development society today and in the future. Concerning to climate change, increase in urbanization, elevated use of natural resources and environmental issues like runoff and continuous accumulation of pesticides and fertilizers are the burning

Table 3.1 Applications of various nanoparticles in active functions

Type of nanoparticles	Deliverable application	Active functions
Metal nanoparticles (Silver, ZnO)	Food additive/supplement	Enhanced gastrointestinal uptake of metal
	Packaging materials/storage, Food preparation devices	Increase barrier properties Clean surface
	Refrigerators, storage containers	Anti-bacterial coating
	Water purification/soil cleaning	Removal/catalysation/oxidation of contaminants
Sprays	Refrigerators, storage containers	Anti-bacterial coating
Complex nanostructures	Nanosensors in packaging	Detection of food deterioration
Hand-held devices	Storage conditions evaluation	Monitoring of contaminants
Incorporated active nanoparticles	Migration out of packaging materials	Oxygen scavenging, prevention of growth of pathogens
Filters with nano-pores contaminants	Water purification	Removal pathogens
Nano-sized nutrients/foods	Food additives/supplements	Claimed enhanced uptake
Delivery systems (nano-encapsulates)	Food additives/supplements	Protecting and (targeted) delivery of content

issues for today's agriculture sector and the researcher community, who are waiting for the recommendations of many nanotechnological strategies for the advancement of scientific and technological knowledge currently being examined. The future, demand for food will increase tremendously while natural resources such as land, water and soil fertility are actually limited. Utilization of nanotechnology in materials science and biomass conversion technologies applied in agriculture are the starting point of providing food, feed, fiber, fire and fuels in the agriculture sector. With reference to the cost of production inputs like chemical fertilizers and pesticides is increasing at an alarming rate due to the limited reserves of fuel such as natural gas and petroleum. To overcome these constraints, precision farming is proving a better option to reduce the initial production costs and to maximize agricultural product output. Possible through implicational advancement in nanotechnology, a number of techniques exist for the improvement of precision farming practices which will allow precise control at nanometer scale (Ingale and Chaudhari 2013).

Nanoencapsulation researchers are working on applications of pesticides encapsulated in nanoparticles; these only release active pesticide in a target insect's stomach, which minimizes the possibility of contamination of plants themselves. A further development being looked at is a network of nanosensors, nanotracers and dispensers throughout a food crop. The sensors and tracers have ability to recognize when a particular plant needs nutrients or water, before crop grower could see any sign that the plant is deficient of the nutrients and water. Dispensers release fertilizer, nutrients, or water as needed demand, optimizing the growth of each plant in the field one by one.

3.2.1 Precision Farming

The precision farming is process that maximizes the crop yield and minimizing the excessive usage of pesticides, fertilizers and herbicides through efficient monitoring. Precision farming applies advanced remote sensing devices, computers and global satellite positioning systems to precisely analyze various environmental conditions in order to determine the growth behavior of plants under the conditions and identify problems related to crops and their growing environments. Mainly this helps in determine timely development of plant, soil conditions, usage and take up of water and chemicals, fertilizers and seeding and consequently controls environmental pollution to a minimum extent by helping in reducing agricultural waste. The implementation of nanotechnology in the form of sensors and monitoring devices will anticipate creating a positive impact on the future use of precision farming methodologies. Nanosensors enabling systems help in growing the use of autonomous sensors that are linked to GPS systems to provide proficient monitoring services which focusing on crop growth and soil conditions. The use of smart sensors in precision farming will results in incredibly increased agricultural crop productivity by providing accurate information that will enable the farmers to make accurate decisions related to plant growth and soil suitability. Applying precision farming tools like, centralized data storage and collection system to determine soil conditions and plant development, seeding, fertilizer, chemical and water use can be accurately fine-tuned to lowers the production costs and certainly increase production which all the way benefiting the hard worker farmer (Rickman et al. 1999).

Precision farming is also helping to reduce the massive generation of agricultural waste and thus keeping environmental pollution to a minimum extent. But still not fully put into practice yet, small sensors and monitoring systems enabled by nanotechnology certainly will have a huge impact on future precision farming methodologies. One of the major functionality of nanotechnology-enabled devices will be the increased use of autonomous sensors linked into a GPS system for real-time monitoring of the implemented field and its environmental conditions. These nanosensors are distributed throughout the field where they can monitor soil conditions and crop growth.

The use of nanotechnology in development of precision sensors will create equipment of increased sensitivity, allowing an earlier response to environmental changes. For example: Nanosensors which are utilizing carbon nanotubes (Fujii et al. 2005) or nano-cantilevers (Vashist 2007) are fairly small enough to trap and measure individual proteins or even small molecules of farmer interest. Nanoparticles or nanosurfaces can be specifically engineered to trigger a specific electrical or chemical signal in the existence of a active contaminant such as bacteria. While other nanosensors work by triggering a running enzymatic reaction or by using nanoengineered branching molecules called dendrimers as detecting probes to bind to target chemicals and proteins (Ruiz-Altisenta et al. 2010). Sooner or later, future precision farming, with the help of advanced smart sensors, will allow enhanced productivity in agriculture by providing accurate information, thus helping farmers to make better decisions and improving the profit folds.

Particular Nanobiosensors are successfully used for sensing of a wide variety of chemical fertilizers, herbicide, pesticide, insecticide, pathogens, moisture, soil pH, and their controlled use can support sustainable development of agriculture for enhancing crop quality and productivity (Rai et al. 2012). This technology provides farmers a better fertilization management, reduction of inputs, and better management of time and the environment. This system could help in the efficient use of available agricultural natural resources like water, nutrients, and chemicals through precision farming. Operational nanosensors which are dispersed in the field can also successfully detect the presence of plant viruses and other crop pathogens, and the level of soil nutrients in the defined area (Jones 2006; Brock et al. 2011). Levels of existing environmental pollution can be evaluated quickly by nano-smart dust (the use of tiny wireless sensors and transponders) and gas sensors (Mousavi and Rezaei 2011). Application of nanobarcodes and nanoprocessing could also be used to monitor the quality of agricultural produce (Li et al. 2005). By keeping an eye on nanotechnology-based plant regulation of hormones such as auxin helps scientists to understand how the plant roots adapt to their environment variations, especially to marginal soils where the variations are too flexible (McLamore et al. 2010). The biosensors based on specific interactions creats atomic force spectroscopy more effective in detecting enzyme-inhibiting herbicides. A nanobiosensor based on an atomic force microscopy tip functionalized with the acetolactate synthase enzyme was successfully detected for the herbicide metsulfuron-methyl, an acetolactate synthase inhibitor, through the acquisition of force curves (daSilva et al. 2013). Bionanosensors also allow the more quantification and rapid detection of bacteria and viruses, thereby increasing the safety of the food for the customer (Otles and Yalcin 2010).

It is difficult to predict the long-term and broad applications of nanotechnology in agriculture and agricultural process and product development. Within agriculture, precision farming is settling a promising objective. Applications like smart sensors for early warning of changing climatic and crop conditions, and the use of nanocapsules with pesticides that are able to respond to the raised different conditions (Fig. 3.2).

3.2.2 *Agrochemicals*

Extensive amount of work is also being carried out in the development of various nanosized agrochemicals, such as fertilizers, pesticides and veterinary medicines. The aim to use of nanosized active ingredients has been suggested to offer

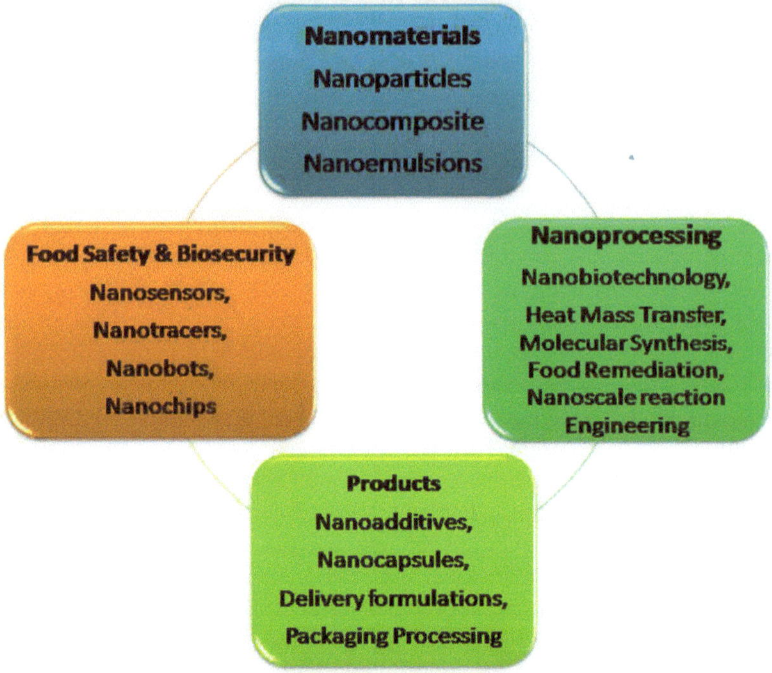

Fig. 3.2 Possible application areas of nanotechnology in food science and technology

improved and targeted delivery of agrochemicals in the only required area of field, better efficacy of pesticides in terms of minimal use and better control over necessary dosing of veterinary products. Nano-encapsulated and solid lipid nanoparticles have been widely discovered for the delivery of agrochemicals (Frederiksen et al. 2003); applicably it includes slow- or controlled-release of fertilizers and pesticides within the stipulated time and required area. Combination of fertilizer with pesticide formulation encapsulated in nanoclay for the slow release of growth stimulants and biocontrol agents, has been tested (Chung et al. 2010).

The progress of nano-emulsions (water/polyoxyethylene) non-ionic surfactant (methyl decanoate) containing the pesticide beta-cypermethrin has been characteristically described by Wang et al. (2007a, b) and similarly, the manufacturing of organic–inorganic nanohybrid material for controlled release of the herbicide 2,4-dichlorophenoxyacetate has been described by Bin Hussein et al. (2005). Porous hollow silica nanoparticles, developed for the controlled delivery of the water-soluble pesticide validamycin with a high loading capacity (36 wt%), have been shown to have a multistaged release pattern (Liu et al. 2006).

Use of zinc–aluminium layered double hydroxide to host the herbicide active ingredient by self-assembly. A few fertilizers claimed to contain nanosized micronutrients which are mainly oxides and carbonates of zinc, calcium, magnesium, molybdenum, etc. are available till date. For remineralisation of soil a micronized (volcanic) rock dust is available from a variety of sources. Product, which comprises sulphates of iron, cobalt, aluminium, magnesium, manganese, nickel and silver, is available for treatment of seed and bulbs prior to planting. Research and development into slow- or controlled-release fertilizers is continuously being carried out in China and India in large extent. The use of nanoforms of agrochemicals offers a number of potential benefits in terms to significantly reduced use of toxic chemicals, but may also raise numerous concerns over exposure of agricultural workers, and elevated contamination of agri-food products. Apart from the intended use and application of nanotechnologies in agriculture and food sectors, there may be seen instances where engineered nanomaterials can get into food and drinks through environmental contamination. Researchers identified possible route of exposure through environmental contamination from the manufacture, use and disposal of consumer products containing advanced engineered nanomaterials. In the product list the major share is contributed by cosmetics and personal care products (TiO_2, ZnO, fullerene (C60), Fe_2O_3, Ag, Cu, Au), catalysts, lubricants and fuel additives (CeO_2, Pt, MoS3), paints and coatings (TiO_2, SiO_2, Ag, quantum dots), water treatment and environmental remediation (Fe, Fe–Pd, polyurethane), agrochemicals (porous SiO_2 carriers and other nanosized agrochemicals), food packaging (Ag, nanoclay, TiO_2, ZnO, TiN), nanomedicine and carriers (silver, Fe, magnetic nanoparticles) (Dasgupta et al. 2016; Boxall et al. 2007).

3.3 Role of Nanotechnology in Food Processing

Consumers increasing demand of fresh, authentic, textured, convenient and flavourful food products in competitive market is keeping nanotechnology to leadership in the food and food processing industry. The next generation food industry belongs to innovative products manufactured by novel processes, with the ambition to enhancing the performance of the product, prolonging the shelf life, keeping the freshness, improving the safety and quality of food product. The advance processing of food products has been asserting to give new tastes; improved textures, consistency and stability of used emulsions, compared with conventionally produced and processed food products (Nandita et al. 2016; Walia et al. 2017). An increasing demand of the heath conscious society, benefit of this technology results in form of a low-fat containing food product that is as creamy as the full-fat alternative, and hence offers a healthy option to the consumer. The food product which is oil in water emulsion that contains nanodroplets of water inside the oil droplets is in the pipeline. This offers different taste and texture attributes similar to the full-fat equivalent, but with a substantial reduction in fat intake by the consumer (Kaiser 2004) (Table 3.2).

Another area of application of nanotechnology in food processing involves the development of nanostructures also termed nanotextures in foodstuffs. Mostly

Table 3.2 Application of nanotechnology for the food and bioprocessing industries

Technology	Description	Benefits
Nanostructures of food ingredients	Nanosized ingredients, additives	Improved texture, flavor, taste; Reduction in the amount of salt and sugar; enhanced bioavailability
Nanoparticle-based intelligent inks; reactive nanolayers	Nanolithography depositions	Traceability, authentication, prevention of adulteration
Nanoencapsulaton of supplements based on micelles and liposomes	Delivery systems for supplements	Taste masking; protection from degradation during processing
Membrane	Filtration effective separation of target material from food	Higher quality food products and fluids
Nanoparticle form of additives and supplements	Nano-engineered particulate additives	Antimicrobial; health benefits; enhanced bioavailability of nutrients
Nutrient delivery	Enzymatic structure, modification, emulsion and foams	Targeted delivery of nutrients, increased bioavailability of nutrients
Improved and active nanocomposites, intelligent and smart packaging	Food packaging	Improve flexibility, durability, temperature/moisture stability, barrier properties
Surface disinfectant	Engineering nanoparticles	Non-contaminated foods, protection from pathogens

mechanisms commonly used for preparing nanostructured food products include nano-emulsions, surfactant micelles, emulsion bilayers, double or multiple emulsions and reverse micelles (Weiss et al. 2006). Examples of nanotextured foodstuffs include spreads, mayonnaise, cream, yoghurts, ice creams, etc. a further area of application involves the use of nanosized or nano-encapsulated food additives. The broaden application is expected to exploit a much huge segment of the health food sector, and include colours, preservatives, flavourings and supplements. Efforts are taking to make better dispersion of water-insoluble additives in food products without the use of excessive fat or surfactants, and enhanced tastes and flavours due to the enlarged surface area of nanosized additives, compared with conventional additives forms. Quantities of consumer products containing nanosized additives are already available in some food markets, including foods, health foods, supplements and nutraceuticals. This includes minerals, antimicrobials, vitamins, antioxidants, etc. nearly all of these products are claiming to have improved absorption and bioavailability in the body compared with their conventional equivalents.

3.3.1 Monitoring Food Quality

Researchers are importantly keeping eye on quality assurance in food production and processing industry because consumers demands of more safe and wholesome food products in addition to the governments imposing to strictly follow stringent regulations to ensure food products safety and feed hygiene. For this sensor or advanced detection systems for rapid detection of spoilage of product and its components, for quality control, and for ignorance in detection at source and during production chain is possible through nanotechnology. In monitoring of food quality the analytical methods for contamination detection must have the flexibility to detect different analytes as well as the specificity to distinguish between different bacteria, and the sensitivity to detect bacteria directly and on-line in real samples without pre treatment to meet consumer expectations. Biosensor technology is holding a promise with manufacturing of inexpensive and simple devices to satisfy these requirements (Palchetti and Mascini 2008; Ozimek et al. 2010). Mainly biosensors can be an inherent alternative to the traditional methods for the detection of toxins and pathogens in food and food product (Bogue 2005, 2008; Connolly 2008).

3.3.1.1 Nanosensors for Bacteria Identification

Most commonly found food born *Campylobacter jejuni* are bacteria on infection which cause savoir abdominal cramps and diarrhea in humans (Ingale and Goto 2013). The campylobacter infections can possibly be traced from poultry meat products which have been contaminated with intestinal contents of the livestock during processing. Stutzenberger et al. (2007) group worked to tackle this food safety problem, they have developed a novel strategy that utilizes bioactive nanoparticles in the chicken feed which is specifically designed to bind to the biomolecular structures on the surfaces of campylobacters. This antibiotic enriched feed with functioning nanocarbohydrate particles successively binds with the bacterium's surface to remove it through the animals excretes.

Another biosensor developed by Fu et al. (2008) utilized fluorescent dye particles which attaches to anti-salmonella antibodies on a silicon or gold nanorod array. On testing the nanosized dye particles on the sensor become visible when the salmonella bacteria present in the food. In contrast to the time-consuming conventional laboratory testings that are based on bacterial cultures maintenance and labour consuming process, this biosensor can detect the salmonella in food instantly.

An analytical technology called reflective interferometry have been developed by Horner et al. (2006), which provides specific, rapid, and label-free optical detection of biomolecules in complex mixtures. This new technology has provided food quality assurance by detecting *Escherichia coli* (*E. coli*) bacteria in a food sample by measuring and detecting light scattering by cell mitochondria. This sensor works on the principle of this sensor is that a protein of a known and characterized bacterium is set on a silicon chip and can bind with any other *E. coli* bacteria present in the food sample. A nanosized light scattering is emitted by this binding and detected by analysis of digital images.

Industry named Agromicron Ltd, located in Hong Kong has developed a low cost Nano Bioluminescent Spray (Plexus Institute 2006), which can react with the pathogen strain on food and produce a visual glow for easy detection. The spray contains nanoparticles and would work based on its adherence reaction with the bacteria. The higher the number of adherence between bacteria and molecules, the more intense the glow produced by the particles. A broad range of food related pathogens are identified by this spray, such as *Salmonella and E. coli*. Cheng et al. (2009) demonstrated detection of *E. coli* in food using biofunctional magnetic nanoparticles (about 20 nm in diameter) in combination with adenosine triphosphate bioluminescence. Zhao et al. (2004) successfully developed an ultrasensitive immunoassay for in situ pathogen quantification in spiked ground beef samples using antibody-conjugated silica fluorescent nanoparticles (about 60 nm in diameter).

3.3.2 Nanoencapsulation

Recently role of food materials has progressed from being only a source of energy and nutrients to actively contributing to the health conscious consumers. The nutrients such as enzymes can be sensitive to proteases and other denaturing compounds to protect this nutrients there is need to immobilise it on different tailored carriers this may also improved nutrients stability to pH and temperature changes. Hence the protection as well as controlled release of bioactive compounds at the right time and the right place can be implemented by encapsulation. Nanoencapsulation remains to be the one of the most promising technologies having the feasibility to entrap bioactive compounds offers targeted site-specific delivery and efficient absorption through cells. Encapsulation is mostly carried out by physical and chemical techniques such as emulsification, coacervation, inclusion, complexation nanoprecipitation, emulsification–solvent evaporation, and supercritical fluid for food ingredients, drying techniques such as spray drying and freeze drying for stabilization of nanoparticles describes in Table 3.3.

Microencapsulation of foods components or products is well established technique, microencapsulated fish oil has been added to bread for a health benefit which masks the unpleasant taste of fish oil (Chaudhry et al. 2008a, b) and this bread is marketed successfully. The nanoencapsulation of food components and additives is a coherent advancement of the encapsulation technology to provide protective barriers, flavour and taste masking, increased bioavailability, increased potency, controlled release and better dispersion in aqueous systems for water-insoluble food ingredients and additives (Chaudhry et al. 2008a, b; Mozafari et al. 2006).

The chief protein found in corn, zein, has received attention in food nanotechnologies because it has the potential to form a mesh like tubular network resistant to

Table 3.3 Encapsulation techniques used for various bioactive compounds utilized in food and bioprocessing industries

Techniques used for encapsulation	Purposes for encapsulations	Functional compounds used	Active emulsifiers and wall materials used	References
Emulsification	Enhancement of anti-inflammation Activity	Curcumin (L)	Tween-20	Wang et al. (2008a, b)
	Protection of the droplets from recoalescence	D-Limonene (L)	Maltodextrin; emulsifiers: modified starch (Hi-Cap 100)	Jafari et al. (2007b)
	Increase the oxidative stability	Salmon oil (L)	Other materials: marine lecithin, α-tocopherol, quercetin, chloroform, methanol, diethylic ether, hexane	Belhaj et al. (2010)
	Improvement in physical stability and commercial application	β-Carotene (L)	Tween-20, Tween-40, Tween-60, and Tween-80	Yuan et al. (2008b)
	Optimization of the conditions to produce nanoemulsion	Sunflower oil (L)	Tween-80, Span-80, and sodium dodecyl sulfate	Leong et al. (2009)
	Improving the physical stability	β-Carotene (L)	Tween-20	Yuan et al. (2008a)
	Improving the stability for use in food or pharmaceutical industry	MCT (L)	OSA starch, chitosan, and lambda-carrageenan	Preetz et al. (2008)
	Optimizing operating conditions to prevent the droplet from coalescence and cavitational bubble cloud formation	Flax seed oil (L)	Emulsifiers: Tween-40	Kentish et al. (2008)
Coacervation	Masking its pungent odor and improving the stability	Capsaicin (L)	Gelatin, acacia, and tannins; emulsifiers: Tween-60; other material: glutaraldehyde	Jincheng et al. (2010)
	Improving the efficiency and delaying the release property	Capsaicin (L)	Gelatin, acacia, and hydrolysable tannins; emulsifiers: hydroxylethyl, cellulose; other material: glutaraldehyde	Xing et al. (2004)
	Masking the pungent odor, giving biocompatibility and biodegradation	Capsaicin (L)	Gelatin, maltodextrin and tannins; emulsifiers: Tween-60; other material: glutaraldehyde	Wang et al. (2008a, b)

(continued)

Table 3.3 (continued)

Techniques used for encapsulation	Purposes for encapsulations	Functional compounds used	Active emulsifiers and wall materials used	References
Nanoprecipitation	Improving the bioavailability, bioactivity, encapsulation efficiency and enhancing the cellular uptake	Curcumin (L)	Poly (lactide-co-glycolide); emulsifiers: poly-ethylene glycol-5000	Anand et al. (2010)
	Improving physical, chemical stability and bioavailability	β-carotene (L)	Poly(D,L-lactic acid) and poly(D,L-lactic-coglycolic acid); emulsifiers: gelatin or Tween-20	Ribeiro et al. (2008)
	Improving the solubility	Curcumin (L)	Monomethoxy poly (ethylene glycol)-poly (3-caprolactone) micelles	Gou et al. (2011)
	Improving the solubility and bioavailability	Astaxanthin (L)	Poly (ethylene oxide)-4-methoxycinnamoylphthaloylchitosan,poly (vinylalcohol-co-vinyl-4-methoxycinnamate), poly(vinylalcohol), and ethyl cellulose	Tachaprutinun et al. (2009)
Emulsification–solvent evaporation	For controlled release	Curcumin (L)	Chitosan cross-linked with tripolyphosphate; emulsifiers: Span-80 and Tween-80; other materials: acetic acid and ethanol	Sowasod et al. (2008)
	Enhance absorption and prolong the rapid clearance of curcumin	Curcumin (L)	Hydroxyl propyl methyl cellulose and polyvinyl pyrrolidone; emulsifiers: D-α-Tocopheryl polyethylene glycol 1000 succinate, Tween-80, Tween-20, cremophor-RH 40, pluronic-F68, pluronic-F127	Dandekar et al. (2010)
	Improving the controlled release and encapsulation efficiency	Quercetin	Poly-D,L-lactide and polyvinyl alcohol	Kumari et al. (2010)
	Improving the reproducibility, stability and target drug loading yield	Coenzyme Q10 (L)	Poly (-methyl methacrylate) and polyvinyl alcohol	Kwon et al. (2002)
	Optimize the operating conditions and reduce phytosterol loss	Phytosterol (L)	Tween-20; other materials: hexane, isopropyl alcohol, ethanol, and acetone	Leong et al. (2011)
	Minimizing the recoalescence improve the physical stability and solubility	α-Tocopherol (L)	Tween-20,	Cheong et al. (2008)

	Optimizing the processing condition and improving bioavailability	Astaxanthin	Sodium caseinate	Anarjan et al. (2011)
	Improving the physical stability	β-carotene (L)	Tween-20	Silva et al. (2011)
	It gives smooth, spherical PLGA nanospheres formation, high yield, drug entrapment efficiency with a narrow size range and sustained delivery	Curcumin (L)	Poly(D,L-lactide-co-glycolide) and polyvinyl alcohol; other materials: chloroform and ethanol	Mukerjee and Vishwanatha (2009)
Spray drying	Increasing the stability, protecting from oxidation and incorporation into beverages	Catechin (H)	Carbohydrate matrix and maltodextrin; other materials: acetone	Ferreira et al. (2007)
	Improving dispersibility, coloring strength and bioavailability	β-carotene (L)	Modified n-octenyl succinatestarch; other materials: ethyl acetate (droplet size); 12 um (particle size)	De Paz et al. (2012)
	Increasing the retention, stability during process	D-Limonene (L)	Maltodextrin; emulsifiers: Hi-Cap, whey protein concentrate, and Tween-20 (emulsion droplet size); 21–53 μm (dried particle size)	Jafari et al. (2007a)
	Minimizing the un-encapsulated oil at the surface and maximizing encapsulation efficiency	Fish oil (L)	Maltodextrin; emulsifiers: modified starch (Hi-Cap)/whey protein concentrate (droplet size); 25–41 μm (particle size)	Jafari et al. (2008)
Freeze drying	Preventing oxidation and masking the odor	Fish oil (L)	β- cyclodextrin, polycaprolactone; emulsifiers: pluronic F68; other materials: ethyl acetate	Choi et al. (2010)
	Increasing the oxidative stability and encapsulation efficiency	Fish oil (L)	Poly-ε-caprolactone	Bejrapha et al. (2010)
	Improving the stability and Rehydrating to study the dispersion characteristics and gel network formation prevents the denaturation	Capsicum oleoresin (L)	Poly-ε-caprolactone and gelatin; emulsifiers: pluronic F68	Nakagawa et al. (2011)
	Effect of excipients on the stability and particle size of nanocapsules	Capsicum oleoresin (L)	Poly-ε-caprolactone; emulsifiers: pluronic F68; other materials: trehalose, D-sucrose, D-mannitol, dextrose, D-sorbitol gelatin and κ-carrageenan	Bejrapha et al. (2011)

(continued)

Table 3.3 (continued)

Techniques used for encapsulation	Purposes for encapsulations	Functional compounds used	Active emulsifiers and wall materials used	References
	Extending the shelf-life, minimizing environmental stress and can apply to the food products	Capsicum oleoresin (L)	Poly-ε-caprolactone; emulsifiers: pluronic F68	Surassamo et al. (2010)
	Improving stability and protecting from environmental factor	Tocopherol (L)	Chitosan, zein; emulsifiers: Tween-20 α-	Luo et al. (2011)
	Increasing the stability, retention percentage and extending the shelf-life	Vitamin E (L)	Polyethylene glycol; emulsifier: Tween-80	Zhao et al. (2011)
	Improving the stability	Curcuminoids (L)	Dioctyl sodium sulfosuccinate, poloxamer 188, glyceryl monostearate	Tiyaboonchai et al. (2007)
	Protecting catechin from degradation	Catechin (H)	Chitosan and sodium tripolyphosphate	Dube et al. (2010)
Inclusion complexation	Formation of transparent solution, improve the colloidal stability, protection against degradation and useful for enrichment of acid drinks	DHA (L)	Beta-lactoglobulin and low methoxyl pectin	Zimet and Livney (2009)
	Improving the thermal stability	Linoleic acid (L)	α- and β-cyclodextrin	Hadaruga et al. (2006)
Supercritical antisolvent precipitation	Bioactivity, promote to food industry and to avoid thermal/light degradation	Lutein (L)	Hydroxylpropyl methyl cellulose phthalate	Heyang et al. (2009)

microorganisms. The zein nanomaterial is widely as a vehicle for flavouring ingredients of the food product and the nanoencapsulation of dietary supplements has been explored (Sozer and Kokini 2009). Just like carbon nanotubes, the nanotubes of a-lactalbumin have a cavity diameter ranging of 8–10 nm which may enable the binding of food components such as vitamins and enzymes (Srinivas et al. 2010), cavities could also be used to encapsulate neutraceuticals or to cover undesirable flavour or aroma compounds (Graveland-Bikker and de Kruif 2006). Nanotubes can be obtained from milk protein by appropriately conditioning the partial hydrolysis of milk with a specific protease, a-lactalbumin will self assemble into nanotubes (Graveland-Bikker and de Kruif 2006). As the origin of these nanomaterials used for encapsulation is milk protein or in the case of zein, corn protein, they are considered to be food grade material and so their introduction to the market should be relatively easy for a nano ingredient. The food grade association of these proteins may facilitate widespread applications in nanoencapsulating nutrients, supplements and pharmaceuticals.

Thus the size and the structure of food is influenced the functionality of foods by providing the taste, texture, and stability properties that desired by the consumer. Here nanotechnology can prove to play a vital role in controlling the size and structure of food to a greater extent to make the desirable texture of food. It serves healthier foods (lower fat, lower salt) with desirable sensory properties; ingredients with improved neutraceutic properties; and the potential for removal of certain additives without loss of spreadability and stability (Garti and Benichou 2004).

3.3.3 Ultrafiltration in Food Processing

Filtration process has been widely applied in numerous foods processing industries for the last two decades due to its operational advantages over conventional ingredient separation processes such as gentle product treatment, high selectivity, and lower energy consumption (Mohammad et al. 2012). Ultrafiltration becomes an essential part in food technology as a tool for separation and increase the concentration. Same time membranefouling compromises the benefits of ultrafiltration as fouling significantly reduces the performance and hence increases the cost of ultrafiltration resulted in overall increase in product cost. Recently various advanced intensive studies carried out to improve ultrafiltration, focusing on membrane fouling control and cleaning of fouled membranes.

Membrane filters are extensively used in dairy processing industries. The dairy industry has been one of the pioneers in the development of equipment and techniques of ultrafiltration based on the practice gained from its application in the dairy processing field (Daufin et al. 2001; Fox et al. 2004; Moresi and Lo Presti 2003; Pouliot 2008; Rosenberg 1995; Saxena et al. 2009). Ultrafiltration has found a major appliance in the making of cheese, during cheese production, whey was discharged to the sewer due to its high salt and lactose content, causing the direct use as a food supplement difficult, but now whey can be processed to obtain additional food values through a newer process using ultrafiltration membrane by increasing the fraction of milk proteins used as cheese or some other useful products and reduce the waste disposal problem represented by whey (Saxena et al. 2009). Membrane filtration technology is documented as a standard tool in the food processing and beverage industry (Cheryan 1998). It is being in use for processing a variety of fruit and vegetable juices (lemon, orange, grapefruit, tangerine, tomato, pomegranate, sweetlemon, cucumber, carrot, and mushroom) (Echavarria et al. 2011). For clarification of juices, ultrafiltration can be used to separate juices into fibrous concentrated pulp and a clarified fraction free of spoilage microorganisms. The clarified fraction can then undergo non-thermal membrane concentration and eventually whole juice reconstitution by combination with pasteurized pulp, in order to obtain a product with improved organoleptic qualities (Cassano et al. 2008). In addition, a better quality clarified fruit juice could able to stand in new market areas, such as clear juice blends, liqueur and related food juice products such as carbonated soft drinks (de Barros et al. 2003). Ultrafiltration is also applied in the concentration process in fruit juice production and processing industry. To recover bioactive components in fruit juice ultrafiltration were employed; bioactive compounds of the depectinized kiwifruit juice were recovered in the clarified fraction of the ultrafiltration process (Galaverna et al. 2008). Ultrafiltration used for fractionation and recovery of waste in fish processing industry. To improve the bioactivity of a saithe protein hydrolysates the ultrafiltration process were employed, by fractionating or concentrating some specific molecular weight peptide classes it increases the concentration of the protein in the filtrate (Chabeaud et al. 2009). Using cross-flow membrane ultrafiltration and nanofiltration the protein recovery from fish meal effluents were made technically and economical feasible (Afonso et al. 2004).

3.4 Role of Nanotechnology in Food Packaging

Increasing the shelf life of food by avoiding spoilage, bacteria, or the loss of food nutrient can be achieved by smart packaging. Nanotechnology offering advanced hopes in food packaging by promising its longer shelf life, safer packaging, better traceability of food products, and does providing healthier food. Intelligent, smart, and active packaging systems produced by nanotechnology would be able to repair the tears and leakages, and respond to environmental conditions. In addition polymer nanocomposite technology also holds the key to future advances in flexible, intelligent, and active packaging (Cushena et al. 2012).

Smart food packaging can detect when its contents are spoiling, and alert the consumer prior to start or in the early stages of the spoiling, as active packaging will release a preservative such as antimicrobials, flavors, colors, or nutritional supplements into the food when it begins to spoil (Ranjan et al. 2014). Nanotechnology provide solutions for food packaging by modifying the permeation behavior of foils, increasing barrier properties (mechanical, chemical, and microbial), providing antimicrobial properties, and by improving heat-resistance properties (Brody et al. 2008; Chaudhry et al. 2008a, b).

3.4.1 Nanoparticles for Food Packaging

The application of bionanocomposites for food packaging protects the food and increases its shelf life, and also be considered a more eco friendly solution because it reduces the requirement to use plastics as toxic, nondegradable packaging materials. Conventional packaging materials are made from nondegradable materials, which increase environmental pollution in addition it consuming restricted fossil fuels for their production. This current alternative of biodegradable films exhibit poor barrier and mechanical properties and these properties need to be improved considerably before they could replace traditional plastics (Suyatma et al. 2004; Tharanathan 2003) and thus help to manage the global waste problem (Sorrentino et al. 2007. The use of inorganic particles, such as clay, into the biopolymeric matrix enhances the biodegradability of a packaging material and can also be controlled with surfactants that are used for the modification of layered silicate. The inorganic particles use also makes it possible to introduce multiple functionalities, which might help to improve the delivery of fragile micronutrients within edible capsules (Bharadwaj et al. 2002; Alexandre and Dubois 2000).

3.4.2 Improved Food Storage

Storage of food is the major concern in the food industry as there are numerous reasons which affect the food storage. The main cause for food deterioration inside food packaging is oxygen; as a result of it oxidation of fats and oils and growth of microorganisms develops in the package. It also accelerates the processes inside food packaging leading to discoloration, changes in texture, rancidity and off-odor, and flavor trouble. Nanotechnologies effectively produce oxygen scavengers for sliced and processed meat, beer, beverages, cooked pastas, and ready-to-eat snacks; moisture absorber sheets for fresh meat, poultry, and fish; and ethylene-scavenging bags for packaging of fruit and vegetables. A functional packaging film for selective control of oxygen transmission through the package and aroma affecting enzymes has been developed by using the nanotechnology approach (Rivett and Speer 2009). The modification of the surface of nanosized materials by dispersing agents can act as substrates for the oxidoreductase enzymes based on reactions catalyzed by food grade enzymes oxygen absorbing s packaging system are also commercially available in the market. Packaging film supplemented with silicate nanoparticles produced, reduces the entrance of oxygen and other gasses, and the exit of moisture and can prevent the food from spoilage. The clay nanoparticles embedded in the plastic bottles strengthen the packaging, reducing gas permeability, and minimizes the loss of carbon dioxide from the beer and the ingress of oxygen to the bottle, keeping the beverage fresher and increases the shelf life (Avella et al. 2005).

3.4.3 Antimicrobial Packaging

Antimicrobial packaging systems holds the impression of being significant for the food industry and the consumers sight because these systems can help extend the product shelf life and maintain food safety by reducing or merely inhibiting the growth rate of microorganisms. Antimicrobial nanoparticle covering in the matrix of the packaging material can reduce the growth of bacteria on or near the food product, inhibiting the microbial growth on nonsterilized foods and maintain the sterility and quality of pasteurized foods by preventing the post manufacture contamination (Table 3.4). Sophisticated techniques of antimicrobial packaging systems contain adding an antimicrobial nanoparticle sachet into the package, dispersing bioactive antimicrobial agents in the packaging; coating of bioactive agents on the surface of the packaging material or utilizing antimicrobial macromolecules with film forming properties or edible matrices (Coma 2008). Foods such as cheese, sliced meat, and bakery that are prone to spoiling on the surface can be protected by contact packaging imbued with antimicrobial nanoparticles, a typical antimicrobial coating nanopackaging film was developed (Buonocore et al. 2005). Paper having active antifungal properties developed for packaging by Rodriguez et al. (2008) which incorporating cinnamon oil with solid wax paraffin using nanotechnology as an active coating was shown to be used as an effective packaging material for numerous bakery products. Working with oregano oil and apple puree, Rojas-Grau et al. (2006) have created edible food films that are able to kill *E. coli* bacteria.

Nanoparticles posing antimicrobial property have been synthesized and tested for applications in antimicrobial packaging and food storage boxes which include silver oxide nanoparticles (Sondi and Salopek-Sondi 2004), zinc oxide, and magnesium oxide nanoparticles (Jones et al. 2008) and nisin particles produced from the fermenation of a bacteria (Gadang et al. 2008). Many antimicrobials are hypothetically proposed to be used in the formulation of edible films and coatings in order to inhibit the spoilage flora and to decrease the risk of pathogens. There is a trend to select the antimicrobials from natural sources and to use generally recognized as safe compounds so as to meet consumer demands for healthy foods, free of harmful chemical additives (Devlieghere et al. 2004).

The most commonly used antimicrobials are organic acids, the polysaccharide chitosan, some polypeptides as nisin, the lactoperoxidase system, and some plant extracts and its essential oils among others. Organic acids such as lactic, acetic, malic, and citric acids, among others, are present in the ingredients of many foods and are broadly used for preservation. The efficacy of antimicrobial activity is

Table 3.4 Food borne pathogens causing illness, their major sources and time taken for action

Pathogen	Sources	Symptoms	Incubation and duration
Campylobacter jejuni	Raw milk, untreated water, raw and undercooked meat, poultry, or shellfish	Diarrhea (sometimes bloody), stomach cramps, fever, muscle pain, headache, and nausea.	Generally 2 to 5 days after eating contaminated food
Clostridium botulinum	Home-canned and prepared foods, vacuum-packed and tightly wrapped food, meat products, seafood, and herbal cooking oils	Dry mouth, double vision followed by nausea, vomiting, and diarrhea. Later, constipation, weakness, muscle paralysis, and breathing problems may develop. Botulism can be fatal.	12 to 72 h after eating contaminated food (in infants 3 to 30 days)
Clostridium perfringens	Meat and meat products	Abdominal pain, diarrhea, and sometimes nausea and vomiting.	8 to 16 h after eating contaminated food
Pathogenic *Escherichia coli (E. coli)*	Meat (undercooked or raw hamburger), uncooked produce, raw milk, unpasteurized juice, and contaminated water	Severe stomach cramps, bloody diarrhea, and nausea. It can also manifest as non-bloody diarrhea or be symptomless. Must-know: *E.coli*0157:H7 can cause permanent kidney damage which can lead to death in young children.	Usually 3 to 4 days after ingestion, but may occur from 1 to 10 days after eating contaminated food.
Listeria monocytogenes	Refrigerated, ready-to-eat foods (meat, poultry, seafood, and dairy – unpasteurized milk and milk products or foods made with unpasteurized milk)	Fever, headache, fatigue, muscle aches, nausea, vomiting, diarrhea, meningitis, and miscarriages.	9 to 48 h after ingestion, but may occur up to 6 weeks after eating contaminated food.
Salmonella Enteritidis	Raw and undercooked eggs, raw meat, poultry, seafood, raw milk, dairy products, and produce	Diarrhea, fever, vomiting, headache, nausea, and stomach cramps must-know: Symptoms can be more severe in people in at-risk groups, such as pregnant women.	12 to 72 h after eating contaminated food
Salmonella Typhimurium	Raw meat, poultry, seafood, raw milk, dairy products, and produce	Diarrhea, fever, vomiting, headache, nausea, and stomach cramps must-know: Symptoms can be more severe in people in the at-risk groups, such as pregnant women.	12 to 72 h after eating contaminated food

(continued)

Table 3.4 (continued)

Pathogen	Sources	Symptoms	Incubation and duration
Shigella	Salads, milk and dairy products, raw oysters, ground beef, poultry, and unclean water	Diarrhea, fever, stomach cramps, vomiting, and bloody stools	1 to 2 days after eating contaminated food
Staphylococcus aureus	Dairy products, salads, cream-filled pastries and other desserts, high-protein foods (cooked ham, raw meat and poultry), and humans (skin, infected cuts, pimples, noses, and throats)	Nausea, stomach cramps, vomiting, and diarrhea	Usually rapid – within 1 to 6 h after eating contaminated food
Vibrio cholerae	Raw and undercooked seafood or other contaminated food and water.	Often absent or mild. Some people develop severe diarrhea, vomiting, and leg cramps. Loss of body fluids can lead to dehydration and shock. Without treatment, death can occur within hours.	6 h to 5 days after eating contaminated food
Vibrio parahaemolyticus	Raw or undercooked fish and shellfish	Diarrhea, stomach cramps, nausea, vomiting, headache, fever, and chills	4 to 96 h after eating contaminated food
Vibrio vulnificus	Raw fish and shellfish, especially raw oysters	Diarrhea, stomach pain, nausea, vomiting, fever, and sudden chills. Some victims develop sores on their legs that resemble blisters.	1 to 7 days after eating contaminated food or exposure to organism
Yersinia enterocolitica	Raw meat and seafood, dairy products, produce, and untreated water	Fever, diarrhea, vomiting, and stomach pain Must-know: Symptoms may be severe for children.	1 to 2 days after eating contaminated food

Table 3.5 Antimicrobial compounds used in nanocomposites, films and coatings for food packaging showing antimicrobial activity

Antimicrobial compounds used in the nanocomposite	Active compounds	Microorganism target	Assay performed	Inhibitory observations	Reference
Acidulants					
Citric, lactic, malic, tartaric acids in combination with nisin	Soy protein	*L. monocytogenes, E. coli O157:H7, S. gaminara*	Film disk agar diffusion assay; total plate count of survivors	L. monocytogenes was inhibited by all acids Salmonella	Eswaranandam et al. (2004)
Sodium lactate	Sodium caseinate	*L. monocytogenes*	Plate count of mo population from inoculated agar systems in contact with antimicrobial films	A slight inhibition was observed by addition of 40% w/w	Kristo et al. (2008)
Malic, citric, lactic acids	Whey protein	*L. monocytogenes*	Film disk agar diffusion assay	Antimicrobial activity in increasing order: lactic < citric < malic	Pintado et al. (2009)
Lipophilic acids					
Sorbic acid or paraminobenzoic acid	Whey protein isolate	*L. monocytogenes, E. coli O157:H7, S. typhimurium*	Film disk agar diffusion assay	L. monocytogenes and E. coli were inhibited for all levels of both antimicrobials	Cagri et al. (2001)
Potassium sorbate	Xantham gum and tapioca starch	*Z. bailii*	Plate count of mo population from inoculated film disk in contact with agar plates	A microbiostactic effect was observed; xanthan gum exert a negative effect on inhibition	Flores et al. (2010)
	Alginate	*Total aerobic bacteria*	Plate count of microbial population in coated potato samples	A delay in microbial growth was observed	Mitrakas et al. (2008)
	Sweet potato starch	*E. coli, S. aureus*	Film disk agar diffusion assay	E. coli growth was inhibited by a 15% w/w KS	Shen et al. (2010)
	Tapioca starch	*Z. bailii, Lactobacillus spp.*	Plate count of microbial population from inoculated film disk in contact with agar plates	No inhibition of Lactobacillus spp. was observed Z. bailii population decreased 2 log cycles after 48 h	Vásconez et al. (2009)

(continued)

Table 3.5 (continued)

Antimicrobial compounds used in the nanocomposite	Active compounds	Microorganism target	Assay performed	Inhibitory observations	Reference
Chitosan					
Chitosan	Chitosan	*P. aeruginosa*, *S. aureus* *L. monocytogenes*	Plate count of mo population from film disk in contact with surface inoculated agar plates	S. aureus and L. monocytogenes surface growth was inhibited by chitosan film; P. aeruginosa was not inhibited	Coma et al. (2003)
	Chitosan with thyme, clove and cinnamon essential oils	*S. aureus*, *L. monocytogenes*, *P. aeruginosa*, *S. enteriditis*	Film disk agar diffusion assay	Chitosan alone did not inhibit the bacteria, thyme essential oil showed the highest antimicrobial efficacy	Hosseini et al. (2009)
	Sweet potato starch	*E. coli S. aureus*	Film disk agar diffusion assay	E. coli and S. aureus growth was inhibited by chitosan	Shen et al. (2010)
	Chitosan	*A. niger* *A. alternata* *R. oryzae*	Measurement of radial growth form film disk in contact with surface inoculated agar plate	Molds growth was decreased by films and coatings, effectiveness varied with the type of mold	Ziani et al. (2009)

based on pH reduction, disruption of substrate transport, and reduction of proton motive force. The most common acidulant agents are acetic, lactic, and malic acids. They are obtained by fermentation and are effective against the main pathogen bacteria encountered in foods (Samelis and Sofos 2003).

In whey-protein-based packaging films containing of glycerol as plasticizer, the use of formic, acetic, and fumaric acids or citric acid produced films of extreme brittleness. In the case of use of acetic acid, whey proteins precipitated since pH was close to isoelectric pH of proteins; as a consequence, gels formed were thick and could not form films (Pintado et al. 2009). Different molecular weighted chitosan can be extracted from shell wastes with different degree of deacetylation, due to difference in molecular weight different functional properties and biological activities were exerted by these chitosan (No et al. 2007). The antimicrobial activity is related to its positively charged amino group which interacts with negatively charged microbial cell membrane promoting an increase in their permeability in the cell and causing disruptions that lead to cell death (Ziani et al. 2009). Chitosan inhibit the growth of many spoilage causing, yeast, molds and pathogenic bacteria (No et al. 2007; Roller 2003). The antimicrobial activity of chitosan is depends on the type of chitosan, degree of acetylation, its molecular weight, the target microorganism, pH of the medium, and presence of other additives or food components (Aider 2010). Researchers have reported that efficacy of chitosan activity depends on the application technique used; in a coating solution it is more available to act as a preservative than when the preservative is forming the film (Vásconez et al. 2009; Zivanovic et al. 2005). Addition of other antimicrobials to chitosan films and coatings generally enhances the antimicrobial activity and also modified physical and mechanical properties of films and coatings. Taking into consideration the mentioned trend, addition of another antimicrobial agent such as potassium sorbate, nisin, and essential oils, to enhance chitosan antimicrobial action explained in Table 3.5 (Hosseini et al. 2009; Pranoto et al. 2005; Vásconez et al. 2009). Combination of compounds from aromatic plant, clove, and cinnamon essential oils to chitosan films, in general, inhibited the growth of *L. monocytogenes, S. aureus, Salmonella enteriditis, and Pseudomonas aeruginosa* (Hosseini et al. 2009). Essential oil exhibited the greatest inhibitory action on contaminating bacteria and also modifies physical and mechanical properties of films and coatings. It has seen in essential-oil-free films, inhibition of bacterial growth was not observed, suggesting that chitosan is unable to diffuse through the agar layer and pointed out the necessary addition of other antimicrobial in the film to exert chitosan its influence (Pranoto et al. 2005). Zinc oxide nanoparticles have been incorporated in different materials including glass, low density polyethylene, polypropylene, polyurethane, paper and chitosan using different incorporation methods (Espitia et al. 2012). Antimicrobial activity if zinc oxide nanocomposites material has been tasted by agar diffusion test, direct contact with culture broth contained microorganisms followed by colony counting (Applerot et al. 2009; Jin et al. 2009; Vicentini et al. 2010). Antimicrobial activity of ZnO nanocomposites performed against Gram-negative bacteria such as *E. coli* as well as Gram-positive bacteria such as *B. subtilis, S. aureus* and *L. plantarum* (Applerot et al. 2009; Emamifar et al.

2010; Jin and Gurtler 2011). Paper coated with ZnO nanoparticles has shown antimicrobial activity against *E. coli* (Ghule et al. 2006).

3.4.4 Green Packaging

Use of natural biopolymer, bio-nanocomposites-based functional packaging materials have generated great potential for enhancing food quality, safety, and stability as smart packaging and processing technology. Researchers are taking efforts to manufacture biodegradable and fully compostable bioplastics packaging (CSIRO 2006), made from organic corn starch using aspects of nanotechnology. Use of bio degradable biopolymer in food packaging material also provide enhanced organoleptic characteristics such as appearance, odor, and flavor (Zhao et al. 2008). The exceptional advantages of the use of natural biopolymer packaging are that these can easily handle particulate foods, can act as carriers for functionally active components, and provide nutritional supplements (Rhim and Ng 2007). A natural polymer and a main component of lobster shells called chitin is used for the making of biodegradable green food packaging using electrospinning technique. The electrospinning technique involves dissolving chitin in a solvent and drawing it through a tiny hole with applied electricity to produce nanoslim fiber spins (Kriegel et al. 2009).

3.4.5 Edible Films and Coatings

Development of edible films and coatings has been possible due to the film forming capacity of natural biopolymers. Hydrocolloids have good ability to form a uniform and cohesive matrix with controlling mechanical properties (Bourtoom 2008, 2009). This ability of hydrocolloids is related to the chemical structure of these compounds, which allows the association through hydrogen bonding of their polymeric chains. The most common biopolymers used for edible antimicrobial film production are polysaccharides, proteins single or mixtures from different sources, and combination of carbohydrates and proteins. While lipids such as waxes and fatty acids are main constituent of edible films and coatings, they do not possess a stand-alone film making nature. For this reason, lipids are often supported on a polysaccharide matrix to provide a film with mechanical strength (Bourtoom 2009). Incorporation of lipids in hydrocolloid-based film formulations facilitates to improve their water barrier characteristics or change their visual appearance (Karbowiak et al. 2010; Maftoonazad et al. 2007) (Table 3.5).

Polysaccharides Polysaccharides make transparent and homogeneous edible films with moderate mechanical properties. The application of these films is limited by their water solubility and poor permeability. To solve this issue, the blending of this with different biopolymers (Xu et al. 2005), addition of hydrophobic materials like oils or waxes (Anker et al. 2001; Ayranci and Tunc 2003; García et al. 2000), or chemical modification of polymer structure have been proposed (Marques et al. 2006).

Cellulose and Derivatives Cellulose is the major structural material of plant cell walls and it is composed of linear chains of (1 → 4)-β-D-glucopyranosyl units. Chemical substitution of some hydroxyl groups along the chain gives origin to ionic (carboxymethylcellulose, CMC) and nonionic cellulose ethers (methylcellulose, MC; hydroxypropylcellulose, HPC; hydroxypropyl methylcellulose, HPMC). Cellulose derivatives films are tough, flexible, totally transparent, and highly sensible to water presence but resistant to fats and oils (Lin and Zhao 2007; Vargas et al. 2008). Crosslinking treatments can be used to decrease the water solubility of cellulose ethers (Coma et al. 2003).

Chitosan Chitosan is a natural carbohydrate polymer derived by deacetylation of chitin [poly-β-(1 → 4)-N-acetyl-D-glucosamine]. It is a high molecular weight cationic polysaccharide that exhibits antibacterial, antifungal activity and film-forming properties (Fernandez-Saiz et al. 2009; Ziani et al. 2009: Arvanitoyannis 2008; Sebti et al. 2005). Numerous information has been reported about chitosan potential to act as a food preservative, function that was evaluated either on the basis of in vitro trials or through direct application of chitosan on real complex matrix foods (Durango et al. 2006; Han et al. 2004; Park et al. 2004; Ribeiro et al. 2007; Vásconez et al. 2009). Because of the good film-forming capacity of chitosan, it is broadly used to protect, improve quality and extend the shelf life of fresh and processed foods. Only chitosan coating was successfully applied on silver carp (Fan et al. 2009) and ready-to-eat roast beef coating (Beverlya et al. 2008); chitosan coatings incorporated with cinnamon oil retained the good quality characteristics as well as extended the shelf life during the refrigerated storage of rainbow trout (Ojagh et al. 2010); modified atmosphere packaging in combination with chitosan edible coating maintained quality and enhanced phenolic content in carrot sticks (Simões et al. 2009) and coatings based on selectively high molecular weight chitosan alone (Han et al. 2005) or combined with oleic acid extended strawberry shelf life (Vargas et al. 2006).

3.4.6 Nanocomposites for Food Packaging

Efficient nanocomposite materials for food packaging are developed with nanotechnology can provide better solutions to food industry challenges concerning to product safety and performance as well as economic and environmental advantages (Silvestre et al. 2011; Ingale 2014). In preservation of food, implications of nanotechnology can extends and improve functional packaging, which like been inhibition and protection from contaminants, preservation, marketing and communication, leading to a active food packaging system. In synthesis of nanocomposite, the composites are made of a polymeric matrix act as continuous phase and a discontinuous phase known as filler, fibres, platelets and particles have been widely used as fillers, so as to improve the mechanical properties and heat resistance of polymers hence it also enhances the overall properties of nanocomposites. (Ajayan et al. 2003; Chaudhry et al. 2008a, b; Arora and Padua 2010). Zinc oxide (ZnO) is an inorganic compound enormously used in everyday applications and also facilitates in formation of nanocomposites since currently listed as a generally recognized as safe (GRAS) material by the Food and Drug Administration and hence is used as element of food additive. ZnO nanoparticles have shown antimicrobial properties and been incorporated in polymeric matrices in order to fabricate nanocomposite which provide antimicrobial activity to the packaging material and improve packaging properties (Espitia et al. 2012). Polyurethane films incorporated nanoparticles have shown antimicrobial activity against *E. coli* and *B. subtilis*, amongst *E. coli* being more sensitive to the developed nanocomposite material, eventually this may be the result of a strong affinity of the nanoparticles with *E. coli* cells and consider that the antibacterial activity of ZnO is due to the generation of H_2O_2 in nanoparticle surface (Li et al. 2009).

3.5 Role of Nanotechnology in Food Supplements

Food supplements are generally considered to include vitamins, minerals, fiber, fatty acids, or amino acids, among or within other food substances. Nutrients are essential to the sustainability of a body, the bioactive compounds are not essential since the body can functions without them but bioactive compounds can have an influence on health and can believe to act as alternative of direct drug use external to the body. Bioactive compounds are known to be considering are flavonoids, caffeine, carotenoids, carnitine, choline, coenzyme Q, creatine, dithiolthiones, phytosterols, phytoestrogenss, glucosinolates, polyphenols, anthocyanins (Golmohamadi et al. 2013).

3.5.1 Bioactive Compounds

Bioactive compounds defined as extra nutritional constituents that normally occur in small quantities in foods, include beta-carotene from carrots, lycopene from tomato, beta-glucan from oats, omega-3 acid from salmon oil, conjugated linoleic acid from cheese, *Lactobacillus* from yogurt, and isoflavones from soybeans, etc. There is evidence to recommend consuming food sources rich in bioactive compounds. From a practical perspective, this translates to recommending a diet rich in a variety of fruits, vegetables, whole grains, legumes, oils, and nuts (Kris-Etherton et al. 2002). Nanotechnology has made known better potential in improving the efficiency and delivery of nutraceuticals and bioactive compounds in functional foods ultimately to improve human health. Enhanced solubility; improve bioavailability and protection of the stability of micronutrients and bioactive compounds during processing, storage and distribution results by efforts of nanoencapsulation (Chen et al. 2006).

Bioactive compounds can be protected by nanoencapsulation from absorption and ensures controlled release of beneficial live probiotic species to promote healthy and targeted gut function. Thus the viability of probiotic organisms including *Lactobacillus acidophilus, Lactobacillus casei, Lactobacillus rhamnosus*, and *Bifidobacterium* spp. within freeze dried yogurt can be improved by encapsulation of this bioactive component with calcium alginate (Kailasapathy and Rybka 1997). The bioavailability of antioxidant component from mainly from tomato; lycopene can be increased by synthesizing nanoparticles of lycopene and incorporating in tomato juice, pasta sauce, and jam (Auweter et al. 1999). Casein a bulky found milk protein, was used to make nanosized micelles and has been employed as a vehicle for delivering sensitive health promoting ingredients like various vitamin (Semo et al. 2007). Biopolymer nanofibers prepared by electrospinning technique by zein for encapsulating beta-carotene show the potential of nanotechnology in food and nutraceutical formulation and catings, bioactive food packaging, and food processing industries (Fernandez et al. 2009). New naturally derived carrier nanotubes for nanoencapsulation of nutrients, supplements, and pharmaceuticals are assembled from hydrolysed milk protein α-lactalbumin (Graveland-Bikker and de Kruif 2006).

3.5.2 Interactive Foods

Nanotechnology is helping to develop interactive foods which can allow consumers to choose and modify the food depending on their own nutritional needs or choice of tastes and flavors. The nanocapsules containing flavor or color enhancers or added nutritional elements would remain in dormant phase in the food and will only be released when triggered by the consumer (Dunn 2004). Efforts are made to develop foods which are capable of changing their color, flavor, or nutritional properties according to a consumer nutritional needs, allergies, or taste preference. Nanotechnology can facilitate techniques to make foods such as soft drinks, ice cream, chocolate, or chips to be commercially marketed as 'health' foods by reducing fat, carbohydrate or calorie content or by increasing protein, fiber or vitamin content. In addition, nanotechnology can help in the production of stronger flavorings, colorings, and nutritional additives, and processing aids to increase the speed of manufacturing and lowers costs of ingredients and processing (Burdo 2005). Utilization of nanofilters and membranes successfully can screen out or pass through certain molecules based on the shape and or size to remove toxins or adjust flavors, Nestle and Unilever are reported to be developing a nanoemulsion based ice cream with a lower fat content that retains a fatty texture and flavor (Renton 2006).

3.5.3 Texture

Commonly Texture Described as What Things Are Made of and how they Feel on Contact. Textures can Be Illustrated as Rough, Smooth, Hard, Soft, Liquid, Solid, Lumpy, Gritty Etc. Consumer like or Dislike Food because of its Taste, but the Texture of the Food Also Plays a Part in whether Consumer like it or Not. Texturing Is Big Business and the Science of Food Structure Even Has its Own Logy that Is Food Rheology.

Reduction of the size of food molecules to nanosized crystals creates more particles for providing greater surface area. Smaller particles improve food's spreadability and stability, and can assist in developing healthier minimal fat food products. Multiple emulsions such as water-in-oil-in-water can distribute the lipids more evenly to reduce extra stabilizers and thickeners to achieve a desirable food texture (Garti and Benichou 2004). Food texturing researchers, prepared nanoscale assays can activate the taste receptors of human tongue and can reduce the bitterness naturally inherent in some foods. (Wenner 2008). In beverages industry, photocatalytic process developed using gold nanoparticles by Lin et al. (2008) for decreasing the aging period and enhancing the sensory quality of sorghum spirits. Contreras et al. (2009) showed that zinc nanoparticles can be used to optimize conditions for surface enhancement of infrared absorption of food components. This technique able to demonstrate that butter treated with zinc naoparticles exerts trans fat spectral information along with the degree and the unsaturation of the acyl groups. These results clearly indicate the potential of nanomaterials in real time imaging sector to reveal useful information concerning food allergens, bioactive compounds, and microbial pathogens.

3.6 Safety and Societal Implications

Recently, interest has extensively grown in safety issues regarding the use of nanoparticles, nanocomposites, nanoconjugates in food packaging. Awareness about food safety and quality as well as its potential impact on consumers are key issues related to food processing and packaging which are developed by nanotechnology (Jain et al. 2016). Researchers are particularly more concerned with the possible ways of nanoparticles migration from the process to packaging material into the packed food and whether this migration would have a negative impact on the safety of food and consumer or quality of the packaged product (Bradley et al. 2011). Nanoparticles have much larger surface area to volume ratios, thus they may exhibit substantially different physicochemical and biological properties compared to conventionally larger sized particles (Ingale and Chaudhari 2013). There are three key factors majorly concerned of nanoparticle toxicity test strategies includes, physicochemical characterization, in vitro assays and in vivo studies (Oberdörster et al. 2005).

The existing functional safety laws, safety testing protocols, and the workplace health procedures are apparently inadequate to measure the exposure and assess the risks posed by nanofoods, nanofood packaging material and nanobased chemicals. Still the industries following established guidelines in the safety assessments of nanomaterials used for manufacturing nanofoods and nanopackaging materials and are not assessed as new chemicals. Novel experimental protocols and research tests should be performed to generate hazard and exposure information leading to risk assessments and to reliably answer concerns about the possible toxicological effects of exposure to nanoparticles, nanocomposites in the food product. A study shows that toxicities of nanoparticles and large particles were similar when the dose was expressed in surface area (Monteiller et al. 2007).

Toxicological evaluation of nanomaterials in food applications has done by high content screening technique and utilizing Zebrafish model can provide valuable developmental toxicity information in terms of endpoint identification and mechanism elucidation (Donofrio 2006). Nevertheless, currently there is a huge demand for low cost in vitro assays without reducing the efficiency and reliability of the risk assessment, since in vivo experiments are expensive, slow and ethically questionable (Siripireddy et al. 2017; Maddinedi et al. 2015, 2017; Tammina et al. 2017; Sannapaneni et al. 2016; Ranjan et al. 2016).

3.7 Conclusion and Perspective

Numerous varied opportunities for nanotechnology exist to play an important role in techniques of agriculture, food processing and packaging. The uses and benefits of nanotechnology are countless, from productivity enhancement through nanotechnology driven advanced precision farming and maximization of output in terms of yield with profit and minimization of inputs of fertilizers through better monitoring and targeted action only where required is desirable. Precision farming enables plants to use water, pesticides, and fertilizers more efficiently and reduces its excessive use. Use of nanotechnology may bring great benefits to farmers through ease in food production and to the food industry through development of new products through food processing, preservation, and packaging.

Expected applications of agricultural food nanotechnology include nanosensors or nanobiosensors for detecting contaminants and for soil quality and for plant health monitoring, for steady release and efficient dosage of water and fertilizers for plants, nanocapsules for agrochemical delivery, creating biofuels, nanocomposites for bioplastic film coatings used in food packaging, antimicrobial nanocomposites used for applications in decontamination of food, nanobiosensors for identification of pathogen contamination, and improving plant breeding (Dasgupta et al. 2015a, b). Additionally existing efforts are extra oriented to effectively reduce the negative impact of developed agrochemical products in the environment and human health, rather than its direct applications to improve the properties for food production. Agro formulations with higher bioavailability and efficacy and better selectivity will be seen actively functioning the near future of agriculture.

The use of biopolymers in the food industry has solved feasibility problems related to their relatively high cost and overall performance when compared to those of synthetic polymers. Several nanocomposites can provide active and or smart properties to food packaging materials, such as antimicrobial properties, oxygen scavenging ability, enzyme immobilization, or indication of the degree of exposure to some degradation related factor. Nanocomposites can not only protect the food against environmental factors, but also incorporate properties to the packaging material so it may actually enhance quality of foods to be consumed. Nanotechnological developed Membrane filters and filtration processes are gaining more attention and focus in food industry due to its exponential advantages (environmental friendliness, cost saving, and product improvement) when compared with other traditional methods. Use of nanotechnology in food industry, has provided sensors and diagnostic devices with improved sensitivity and selectivity to monitor food processes and assure food quality measurements along the real time production lines. Moreover antimicrobial edible films and coatings are utilized for improving the shelf life of food products without impairing consumer acceptability.

In addition to antimicrobial properties of antimicrobial edible films and coatings, zinc nanocomposites has presented modifications in the structure and properties of packaging materials like mechanical and thermal resistance. The design of a food packaging is significant since each industrial trait influences the physical integrity of the developed packaging and, therefore, ensures the protection of the packaged food. Understanding gap in addressing and framing the authority regulations of nanotechnology usage for foods, food additives, and food packaging materials is in progress through various regional and international agencies. Majorly the potential benefits of nanotechnology in agriculture and next generation food industry need to be balanced concerning for the soil, water, environment, and the occupational health of workers.

References

Afonso MD, Ferrer J, Bórquez R (2004) An economic assessment of proteins recovery from fish meal effluents by ultrafiltration. Trends Food Sci Technol 15(10):506–512

Aider M (2010) Chitosan application for active bio-based films production and potential in the food industry: review. LWT Food Sci Technol 43:837–842

Ajayan PM, Schadler LS, Braun PV (2003) Nanocomposite science and technology. Wiley-VCH, Weinheim. ISBN: 978-3-527-30359-5

Alexandre M, Dubois P (2000) Polymer-layered silicate nanocomposites: preparation, properties and uses of a new class of materials. Mater Sci Eng 28:1–63

Anand P, Nair HB, Sung B, Kunnumakkara AB, Yadav VR, Tekmal RR, Aggarwal BB (2010) Design of curcuminloaded PLGA nanoparticles formulation with enhanced cellular uptake, and increased bioactivity in vitro and superior bioavailability in vivo. Biochem Pharmacol 79 (3):330–338

Anarjan N, Mirhosseini H, Baharin BS, Tan CP (2011) Effect of processing conditions on physicochemical properties of sodium caseinate-stabilized astaxanthin nanodispersions. LWT Food Sci Technol 44(7):1658–1665

Anker M, Berntsen J, Hermansson AM, Stading M (2001) Improved water vapour barrier of whey protein films by addition of an acetylated monoglyceride. Innovative Food Sci Emerg Technol 3:81–92

Applerot G, Lipovsky A, Dror R, Perkas N, Nitzan Y, Lubart R, Gedanken A (2009) Enhanced antibacterial activity of nanocrystalline ZnO due to increased ROS-mediated cell injury. Adv Funct Mater 19(6):842–852

Arora A, Padua GW (2010) Review: nanocomposites in food packaging. J Food Sci 75(1):43–49

Arvanitoyannis IS (2008) The use of chitin and chitosan for food packaging applications. In: Chiellini E (ed) Environmentally compatible food packaging. Cambridge, CRC Wood head Publishing, pp 137–158

Auweter H, Bohn H, Haberkorn H, Horn D, Luddecke E, Rauschenberger V (1999) Production of carotenoid preparations in the form of coldwater-dispersible powders and the use of the novel carotenoid preparations. US Patent 5968251 (in English)

Avella M, De Vlieger JJ, Errico ME, Fischer S, Vacca P, Volpe MG (2005) Biodegradable starch or clay nanocomposite films for food packaging applications. Food Chem 93(3):467–474

Ayranci E, Tunc S (2003) A method for the measurement of the oxygen permeability and the development of edible films to reduce the rate of oxidative reactions in fresh foods. Food Chem 80:423–431

Bejrapha P, Min SG, Surassmo S, Choi MJ (2010) Physicothermal properties of freeze-dried fish oil nanocapsules frozen under different conditions. Dry Technol 28(4):481–489

Bejrapha P, Surassmo S, Choi M, Nakagawa K, Min S (2011) Studies on the role of gelatin as a cryo- and lyo-protectant in the stability of capsicum oleoresin nanocapsules in gelatin matrix. J Food Eng 105(2):320–331

Belhaj N, Arab-Tehrany E, Linder M (2010) Oxidative kinetics of salmon oil in bulk and in nanoemulsion stabilized by marine lecithin. Process Biochem 45(2):187–195

Benichou A, Aserin A, Garti N, (2004) Double emulsions stabilized with hybrids of natural polymers for entrapment and slow release of active matters. Adv Colloid Interf Sci 108–109:29–41. ISSN 0001-8686, https://doi.org/10.1016/j.cis.2003.10.013. (http://www.sciencedirect.com/science/article/pii/S0001868603001350) Keywords: Double emulsions; Slow, Control and sustained release; Biopolymeric emulsifiers; Steric stabilization

Beverlya RL, Janesa ME, Prinyawiwatkulaa W, Nob HK (2008) Edible chitosan films on ready-to-eat roast beef for the control of Listeria monocytogenes. Food Microbiol 25:534–537

Bharadwaj RK, Mehrabia AR, Hamiltona C, Trujilloa C, Murgaa M, Fana R, Chavira A, Thompsonb AK (2002) Structure-property relationships in crosslinked polyester-clay nanocomposites. Polymer (Guildf) 43:3699–3705

Bin Hussein MZ, Yahaya AH, Zainal Z, Kian LH (2005) Nanocomposite-based controlled release formulation of an herbicide, 2,4-dichlorophenoxyacetate incapsulated in zinc-aluminium-layered double hydroxide. Sci Technol Adv Mater 6(8):956–962

Bogue R (2005) Developments in biosensors – where are tomorrow's markets? Sens Rev 25(3):180–184. https://doi.org/10.1108/02602280510606426

Bogue R (2008) Nanosensors: a review of recent progress. Sens Rev 28(1):12–17. https://doi.org/10.1108/02602280810849965

Bourtoom T (2008) Edible films and coatings: characteristics and properties. Int Food Res J 15(3):237–248

Bourtoom T (2009) Review article. Edible protein films: properties enhancement. Int Food Res J 16:1–9

Boxall A, Chaudhry Q, Sinclair C, Jones A, Aitken R, Jefferson B, Watts C (2007) Current and predicted environmental exposure to engineered nanoparticles. Central Science Laboratory, York

Bradley EL, Castle L, Chaudhry Q (2011) Applications of nanomaterials in food packaging with a consideration of opportunities for developing countries. Trends Food Sci Technol 22(11):604–610

Brock DA, Douglas TE, Queller DC, Strassmann JE (2011) Primitive agriculture in a social amoeba. Nature 469(7330):393–396

Brody AL, Bugusu B, Han JH, Koelsch Sand C, McHugh TH (2008) Innovative food packaging. J Food Sci 73(8):107e117

Buonocore G, Conte A, Corbo MR, Sinigaglia M, Del Nobile MA (2005) Mono- and multilayer active films containing lysozyme as antimicrobial agent. Innovative Food Sci Emerg Technol 6:459–464

Burdo O (2005) Nanoscale effects in food production technologies. J Eng Phys Thermophys 78(1):90–97

Cagri A, Ustunol Z, Ryser ET (2001) Antimicrobial, mechanical, and moisture barrier properties of low pH whey protein-based edible films containing p-amino benzoic or sorbic acids. J Food Sci 66(6):865–870

Cassano A, Donato L, Conidi C, Drioli E (2008) Recovery of bioactive compounds in kiwifruit juice by ultrafiltration. Innovative Food Sci Emerg Technol 9(4):556–562

Chabeaud A, Vandanjon L, Bourseau P, Jaouen P, Chaplain- Derouiniot M, Guerard F (2009) Performances of ultrafiltration membranes for fractionating a fish protein hydrolysate: application to the refining of bioactive peptidic fractions. Sep Purif Technol 66(3):463–471

Chaudhry Q, Scotter M, Blackburn J, Ross B, Boxall A, Castle L, Aitken R, Watkins R (2008a) Applications and implications of nanotechnologies for the food sector. Food Addit Contam 25 (3):241–258

Chaudhry Q, Scotter M, Blackburn J, Ross B, Boxall A, Castle L, Aitken R, Watkins R (2008b) Applications and implications of nanotechnologies for the food sector. Food Addit Contam: Part A 25(3):241–258

Chen H, Weiss J, Shahidi F (2006) Nanotechnology in nutraceuticals and functional foods. Food Technol 60(3):30–36

Cheng Y, Liua Y, Huanga J, Lia K, Zhang W, Xiana Y, Jin L (2009) Combining biofunctional magnetic nanoparticles and ATP bioluminescence for rapid detection of *Escherichia coli*. Talanta 77(4):1332–1336

Cheong JN, Tan CP, Yaakob B, Che M, Misran M (2008) α-Tocopherol nanodispersions: preparation, characterization and stability evaluation. Int J Food Eng 89(2):204–209

Cheryan M (1998) Ultrafiltration and microfiltration handbook, 2nd edn. CRC Press, Boca Raton

Choi MJ, Ruktanonchai U, Min SG, Chun JY, Soottitantawat A (2010) Physical characteristics of fish oil encapsulated by ß-cyclodextrin using an aggregation method or polycaprolactone using an emulsion–diffusion method. Food Chem 119(4):1694–1703

Chung YL, Ansari S, Estevez L, Hayrapetyan S, Giannelis EP, Lai HM (2010) Preparation and properties of biodegradable starch-clay nanocomposites. Carbohydr Polym 79(2):391e396

Coma V (2008) Bioactive packaging technologies for extended shelf life of meat-based products. Meat Sci 78(2):90–103

Coma V, Deschamps A, Martial-Gros A (2003) Bioactive packaging materials from edible chitosan polymer-antimicrobial assessment on dairy related contaminants. J Food Sci 68 (9):2788–2792

Connolly (2008) Nanosensor developments in some European universities. Sens Rev 28(1):18–21. https://doi.org/10.1108/02602280810849974

Contreras MP, Avula RY, Singh RK (2009) Evaluation of nano zinc (ZnO) for surface enhancement of ATR–FTIR spectra of butter and spread. Food Bioprocess Technol 3(4):629–635. https://doi.org/10.1007/s11947-009-0237-4

CSIRO (2006) Farm factories: harvesting bioplastics. http://www.solve.csiro.au/0806/article6.htm. Accessed 20 June 2015

Cushen M, Kerry J, Morris M, Cruz-Romero M, Cummins E (2012) Nanotechnologies in the food industry – recent developments, risks and regulation. Trends Food Sci Technol 24(1):30–46. ISSN 0924-2244, https://doi.org/10.1016/j.tifs.2011.10.006. (http://www.sciencedirect.com/science/article/pii/S0924224411002378)

Dandekar PP, Jain R, Patil S, Dhumal R, Tiwari D, Sharma S, Vanage G, Patravale V (2010) Curcumin-loaded hydrogel nanoparticles: application in anti-malarial therapy and toxicological evaluation. J Pharm Sci 99(12):4992–5010

Dasgupta N, Ranjan S, Mundekkad D, Ramalingam C, Shanker R, Kumar A (2015a) Nanotechnology in agro-food: from field to plate. Food Res Int 69:381–400

Dasgupta N, Ranjan S, Mundra S, Ramalingam C, Kumar A (2015b) Fabrication of food grade vitamin E nanoemulsion by low energy approach, characterization and its application. Int J Food Prop 19(3):700–708

Dasgupta N, Ranjan S, Chakraborty AR, Ramalingam C, Shanker R, Kumar A (2016) Nanoagriculture and water quality management. In: Eric L (ed) Sustainable agriculture reviews: nanoscience in food and agriculture 1. Springer-Nature, Heidelberg. https://doi.org/10.1007/978-3-319-39303-2_1

Dasgupta N, Ranjan S, Ramalingam C (2017) Applications of nanotechnology in agriculture and water quality management. Environ Chem Lett 15(4):591–605

daSilva AC, Deda DK, da Róz AL, Prado RA, Carvalho CC, Viviani V, Leite FL (2013) Nanobiosensors based on chemically modified AFM probes: a useful tool for metsulfuron-methyl detection. Sensors (Basel) 13(2):1477–1489

Daufin G, Escudier JP, Carrère H, Bérot S, Fillaudeau L, Decloux M (2001) Recent and emerging applications of membrane processes in the food and dairy industry. Food Bioprod Process 79(2):89–102

deAzeredo HMC (2009) Nanocomposites for food packaging applications. Food Res Int 42:1240–1253

deBarros STD, Andrade CMG, Mendes ES, Peres L (2003) Study of fouling mechanism in pineapple juice clarification by ultrafiltration. J Membr Sci 215(1–2):213–224

dePaz E, Martin A, Estrella A, Rodriguez-Rojo S, Matias AA, Duarte CMM, Cocero MJ (2012) Formulation of β- carotene by precipitation from pressurized ethyl acetate-on water emulsions for application as natural colorant. Food Hydrocoll 26(1):17–27

Devlieghere F, Vermeiren L, Debevere J (2004) New preservation technologies: possibilities and limitations. Int Dairy J 14:273–285

Donofrio R (2006) Rapid safety testing of food nanomaterials using high content screening and zebrafish model. In: Nano and micro technologies in the food and health food industries conference, 25–26 October 2006 Amsterdam, The Netherlands

Dube A, Ken N, Nicolazzo JA, Ian L (2010) Effective use of reducing agents and nanoparticle encapsulation in stabilizing catechins in alkaline solution. Food Chem 122(3):662–667

Dunn J (2004) A mini revolution. Food Manufacture, London. http://www.foodmanufacture.co.uk/news/fullstory.php/aid/472/A%20mini%20revolution.htm. Accessed 23 Apr 2015

Durango AM, Soares NFF, Andrade NJ (2006) Microbiological evaluation of an edible antimicrobial coating on minimally processed carrots. Food Control 17:336–341

Echavarria AP, Torras C, Pagan J, Ibarz A (2011) Fruit juice processing and membrane technology application. Food Eng Rev 3(3–4):136–158

Emamifar A, Kadivar M, Shahedi M, Soleimanian-Zad S (2010) Evaluation of nanocomposite packaging containing Ag and ZnO on shelf life of fresh orange juice. Innovative Food Sci Emerg Technol 11(4):742–748. ISSN 1466-8564, https://doi.org/10.1016/j.ifset.2010.06.003. (http://www.sciencedirect.com/science/article/pii/S1466856410000664) Keywords: Orange juice; Nanocomposite; Nano-ZnO; Nanosilver

Espitia PJP, Soares FFN, dosReis CJS, deAndrade NJ, Cruz SR, Medeiros EAA (2012) Zinc oxide nanoparticles: synthesis, antimicrobial activity and food packaging applications. Food Bioprocess Technol 5:1447–1464. https://doi.org/10.1007/s11947-012-0797-6

Eswaranandam S, Hettiarachhy NS, Johnson MG (2004) Antimicrobial activity of citric, lactic, malic, or tartaric acids and nisin incorporated soy protein film against *Listeria monocytogenes*, *Eschericchia coli* O157:H7, and *Salmonella gaminara*. J Food Sci 69(3):FMS79–FMS84

Ezhilarasi PN, Karthik P, Chhanwal N, Anandharamakrishnan C (2013) Nanoencapsulation techniques for food bioactive components: a review. Food Bioprocess Technol 6:628–647. https://doi.org/10.1007/s11947-012-0944-0

Fan W, Sun J, Chen Y, Qiu J, Zhang Y, Chi Y (2009) Effects of chitosan coating on quality and shelf life of silver carp during frozen storage. Food Chem 115:66–70

Fernandez A, Torres-Giner S, Lagaron JM (2009) Novel route to stabilization of bioactive antioxidants by encapsulation in electrospun fibers of zein prolamine. Food Hydrocoll 23 (5):1427–1432

Fernandez-Saiz P, Lagaron JM, Ocio MJ (2009) Optimization of the biocide properties of chitosan for its application in the design of active films of interest in the food area. Food Hydrocoll 23:913–921

Ferreira I, Rocha S, Coelho M (2007) Encapsulation of antioxidants by spray-drying. Chem Eng Trans 11(9):713–717

Flores SK, Costa D, Yamashita F, Gerschenson LN, Grossmann MV (2010) Mixture design for evaluation of potassium sorbate and xanthan gum effect on properties of tapioca starch films obtained by extrusion. Mater Sci Eng C 30:196–202

Fox PF, McSweeney P, Cogan TM, Guinee TP (2004) Application of membrane technology to cheese production. Cheese: major cheese groups, 3rd edn, vol 2, pp 261–286

Frederiksen HK, Kristensen HG, Pedersen M (2003) Solid lipid microparticle formulations of the pyrethroid gamma-cyhalothrin-incompatibility of the lipid and the pyrethroid and biological properties of the formulations. J Control Release 86(2–3):243–252

Fu J, Park B, Siragusa G, Jones L, Tripp R, Zhao Y, Cho YJ (2008) An Au/Si hetero-nanorod based biosensor for Salmonella detection. Nanotechnology 19:1–7

Fujii M, Xing Z, Huaqing X, Hiroki A, Koji T, Tatsuya I, Hidekazu A, Tetsuo S (2005) Measuring the thermal conductivity of a single carbon nanotube. Phys Rev Lett 95:065502

Gadang VP, Hettiarachchy NS, Johnson MG, Owens C (2008) Evaluation of antibacterial activity of whey protein isolate coating incorporated with nisin, grape seed extract, malic acid, and EDTA on a turkey frankfurter system. J Food Sci 73(8):389–394

Galaverna G, DiSilvestro G, Cassano A, Sforza S, Dossena A, Drioli E, Marchelli R (2008) A new integrated membrane process for the production of concentrated blood orange juice: effect on bioactive compounds and antioxidant activity. Food Chem 106(3):1021–1030

García M, Martino M, Zaritizky N (2000) Lipid addition to improve barrier properties of edible starch-based films and coatings. J Food Sci 65(6):941–947

Garti N, Benichou A (2004) Recent developments in double emulsions for food applications. In: Friberg S, Larsson K, Sjoblom J (eds) Food emulsions, 4th edn. Marcel Dekker, New York, pp 353–412

Ghule K, Ghule AV, Chen BJ, Ling YC (2006) Preparation and characterization of ZnO nanoparticles coated paper and its antibacterial activity study. Green Chem 8(12):1034–1041

Golmohamadi A, Möller G, Powers J, Nindo C (2013) Effect of ultrasound frequency on antioxidant activity, total phenolic and anthocyanin content of red raspberry puree. Ultrason Sonochem 20(5):1316–1323

Gou M, Men K, Shi H, Xiang M, Zhang J, Song J, Long J, Wan Y, Luo F, Zhao X, Qian Z (2011) Curcumin-loaded biodegradable polymeric micelles for colon cancer therapy in vitro and in vivo. Nanoscale 3(4):1558–1567

Graveland-Bikker JF, deKruif CG (2006) Unique milk protein based nanotubes: food and nanotechnology meet. Trends Food Sci Technol 17(5):196–203

Hadaruga NG, Hadaruga DI, Paunescu V, Tatu C, Ordodi VL, Bandur G, Lupea AX (2006) Thermal stability of the linoleic acid/α- and β-cyclodextrin complexes. Food Chem 99 (3):500–508

Han C, Zhao Y, Leonard SW, Traber MG (2004) Edible coating to improve storability and enhance nutritional value of fresh and frozen strawberries (Fragaria x Ananassa) and raspberries (*Rubus ideaus*). Postharvest Biol Technol 33:67–78

Han C, Lederer C, Mcdaniel M, Zhao Y (2005) Sensory evaluation of fresh strawberries (Fragaria ananassa) coated with chitosan-based edible coatings. J Food Sci 70(3):S172–S178

Heyang J, Fei X, Cuilan J, Yaping Z, Lin H (2009) Nanoencapsulation of lutein with hydroxypropylmethyl cellulose phthalate by supercritical antisolvent. Chin J Chem Eng 17(4):672–677

Horner SR, Mace CR, Rothberg LJ, Miller BL (2006) A proteomic biosensor for enteropathogenic *E. coli*. Biosens Bioelectron 21(8):1659–1663

Hosseini MH, Razavi SH, Mousavi MA (2009) Antimicrobial, physical and mechanical properties of chitosanbased films incorporated with thyme, clove and cinnamon essential oils. J Food Process Preserv 33:727–743

Ingale AG, Chaudhari AN (2013) Biogenic synthesis of nanoparticles and potential applications: an eco-friendly approach. J Nanomed Nanotechnol 4:165. https://doi.org/10.4172/2157-7439.1000165

Ingale AG, Goto S (2013) Identification of antigenic epitopes, homology modelling and structural characterization of capsule biosynthesis protein (CapA) from *Compylobacter jejunni*. Gene Ther Mol Biol 15:74–84

Ingale AG (2014) Nanocomposite for surface coating. In: Soni S, Salhotra A, Suar M (eds) Handbook of research on diverse applications of nanotechnology in biomedicine, chemistry, and engineering. IGI Global, Hershey, pp 612–623

Jafari SM, He Y, Bhandari B (2007a) Encapsulation of nanopartricles of D-limonene by spray drying: role of emulsifiers and emulsifying agent. Dry Technol 25(6):1079–1089

Jafari SM, He Y, Bhandari B (2007b) Production of submicron emulsions by ultrasound and microfluidization techniques. J Food Eng 82(4):478–488

Jafari SM, Assadpoor E, Bhandari B, He Y (2008) Nanoparticle encapsulation of fish oil by spray drying. Food Res Int 41(2):172–183

Jain A, Ranjan S, Dasgupta N, Ramalingam C (2016) Nanomaterials in food and agriculture: an overview on their safety concerns and regulatory issues. Crit Rev Food Sci Nutr. https://doi.org/10.1080/10408398.2016.1160363

Janardan S, Pothini S, Ragul G, Anjaneyulu U, Shivendu R, Dasgupta N, Ramalingam C, Swamiappan S, Vijayakrishna K, Sivaramakrishna A (2016) Assessment on the antibacterial activity of nanosized silica derived from hypercoordinated silicon(iv) precursors. RSC Adv 6 (71):66394–66406

Jasińska M, Dmytrów I, Mituniewicz-Małek A, Wąsik K (2010) Cow feeding system versus milk utility for yoghurt manufacture. Acta Sci Pol Technol Aliment 9(2):189–199

Jin T, Sun D, JY S, Zhang H, Sue HJ (2009) Antimicrobial efficacy of zinc oxide quantum dots against Listeria monocytogenes, Salmonella enteritidis, and Escherichia coli O157:H7. J Food Sci 74(1):M46–M52

Jin T, Gurtler JB (2011) Inactivation of Salmonella in liquid egg albumen by antimicrobial bottle coatings infused with allyl isothiocyanate, nisin and zinc oxide nanoparticles. J Appl Microbiol 110:704–712. https://doi.org/10.1111/j.1365-2672.2011.04938.x

Jincheng W, Xiaoyu Z, Siahao C (2010) Preparation and properties of nanoencapsulated capsaicin by complex coacervation method. Chem Eng Commun 197(7):919–933

Jones PBC (2006) A nanotech revolution in agriculture and the food industry. Information Systems for Biotechnology, Blacksburg. Available from http://www.isb.vt.edu/articles/jun0605.htm. Accessed 19 Apr 2014

Jones N, Ray B, Ranjit KT, Manna AC (2008) Antibacterial activity of ZnO nanoparticle suspensions on a broad spectrum of microorganisms. FEMS Microbiol Lett 279:71–76

Kailasapathy K, Rybka S (1997) *L. acidophilus* and *Bifidobacterium spp.*—their therapeutic potential and survival in yoghurt. Aust J Dairy Technol 52:28–35

Kaiser Helmut Consultancy (2004) Study: nanotechnology in food and food processing industry worldwide 2003–2006–2010–2015. Available at www.hkc22.com/Nanofood.html

Karbowiak T, Debeaufort D, Voilley A, Trystram G (2010) From macroscopic to molecular scale investigations of mass transfer of small molecules through edible packaging applied at interfaces of multiphase food products. Innovative Food Sci Emerg Technol 11:352–360

Kentish S, Wooster TJ, Ashokkumar M, Balachandran S, Mawson R, Simons L (2008) The use of ultrasonics for nanoemulsion preparation. Innovative Food Sci Emerg Technol 9(2):170–175

Kriegel C, Kit KM, McClements DJ, Weiss J (2009) Influence of surfactant type and concentration on electrospinning of chitosan–poly(ethylene oxide) blend nanofibers. Food Biophys 4:213–228. https://doi.org/10.1007/s11483-009-9119-6

Kris-Etherton PM, Hecker KD, Bonanome A, Coval SM, Binkoski AE, Hilpert KF, Griel AE, Etherton TD (2002) Bioactive compounds in foods: their role in the prevention of cardiovascular disease and cancer. Am J Med 11(3-9S2):71–88

Kristo E, Koutsoumanis KP, Biliaderis CG (2008) Thermal, mechanical and water vapor barrier properties of sodium caseinate films containing antimicrobials and their inhibitory action on *Listeria monocytogenes*. Food Hydrocoll 22:373–386

Kumari A, Yadav SK, Pakade YB, Singh B, Yadav SC (2010) Development of biodegradable nanoparticles for delivery of quercetin. Colloids Surf B:Biointerf 80(2):184–192

Kwon SS, Nam YS, Lee JS, BS K, Han SH, Lee JY, Chang IS (2002) Preparation and characterization of coenzyme Q10-loaded PMMA nanoparticles by a new emulsification process based on microfluidization. Colloids Surf A Physicochem Eng Asp 210:95–104

Leong TSH, Wooster TJ, Kentish SE, Ashokkumar M (2009) Minimising oil droplet size using ultrasonic emulsification. Ultrason Sonochem 16(6):721–727

Leong WF, Lai OM, Long K, Yaakob B, Mana C, Misran M, Tan CP (2011) Preparation and characterisation of water-soluble phytosterol nanodispersions. Food Chem 129(1):77–83

Li Y, YT C, Luo D (2005) Multiplexed detection of pathogen DNA with DNA-based fluorescence nanobarcodes. Nat Biotechnol 23(7):885–889

Li JH, Hong RY, Li MY, Li HZ, Zheng Y, Ding J (2009) Effects of ZnO nanoparticles on the mechanical and antibacterial properties of polyurethane coatings. Prog Org Coat 64 (4):504–509

Lin D, Zhao Y (2007) Innovations in the development and application of edible coatings for fresh and minimally processed fruits and vegetables. Compr Rev Food Sci Food Saf 6(3):60–75

Lin LY, Peng C, Wang H, Chuang C, Yu T, Chen K, Chiu W, Peng RY (2008) Acceleration of maturity of young sorghum (kaoliang) spirits by linking nanogold photocatalyzed process to conventional biological aging—a kinetic approach. Food Bioprocess Technol 1:234–245

Liu F, Wen LX, Li ZZ, Yu W, Sun HY, Chen JF (2006) Porous hollow silica nanoparticles as controlled delivery system for water-soluble pesticide. Mater Res Bull 41(12):2268–2275

Luo Y, Zhang B, Whent M, Yu L, Wang Q (2011) Preparation and characterization of zein/chitosan complex for encapsulation of α-tocopherol, and its in vitro controlled release study. Colloids Surf B: Biointerf 85(2):145–152

Maddinedi SB, Mandal BK, Ranjan S, Dasgupta N (2015) Diastase assisted green synthesis of size-controllable gold nanoparticles. RSC Adv 5(34):26727–26733

Maddinedi SB, Mandal BK, Patil SH, Andhalkar VV, Ranjan S, Dasgupta N (2017) Diastase induced green synthesis of bilayered reduced graphene oxide and its decoration with gold nanoparticles. J Photochem Photobiol B Biol 166:252–258

Maftoonazad N, Ramaswamy HS, Marcotte M (2007) Moisture sorption behavior, and effect of moisture content and sorbitol on thermo-mechanical and barrier properties of pectin based edible films. Int J Food Eng 3(4):Article 10. Available at: http://www.bepress.com/ijfe/vol3/iss4/art10

Marques PT, Lima AMF, Bianco G, Laurindo JB, Borsali R, Le Meins JF, Soldi V (2006) Thermal properties and stability of cassava starch films crosslinked with tetraethylene glycol diacrylate. Polym Degrad Stab 91(4):726–732

Maynard AD, Aitken RJ, Butz T, Colvin V, Donaldson K, Oberdörster G, Philbert MA, Ryan J, Seaton A, Stone V, Tinkle SS, Tran L, Walker NJ, Warheit B (2006) Safe handling of nanotechnology. Nature 444:267–269

McLamore ES, Diggs A, Marzal PC, Shi J, Blakeslee JJ, Peer WA, Murphy AS, Porterfield DM (2010) Non-invasive quantification of endogenous root auxin transport using an integrated flux microsensor technique. Plant J 63(6):1004–1016

Mitrakas GE, Koutsoumanis KP, Lazarides HN (2008) Impact of edible coating with or without anti-microbial agent on microbial growth during osmotic dehydration and refrigerated storage of a model plant material. Innovative Food Sci Emerg Technol 9:550–555

Mohammad AW, Ching YN, Ying PL, Gen HN (2012) Ultrafiltration in food processing industry: review on application, membrane fouling, and fouling control. Food Bioprocess Technol 5:1143–1156

Monteiller C, Tran L, MacNee W, Faux S, Jones A, Miller B, Donaldson K (2007) The pro-inflammatory effects of low-toxicity lowsolubility particles, nanoparticles and fine particles, on epithelial cells in vitro: the role of surface area. Occup Environ Med 64:609–615

Moresi M, Lo Presti S (2003) Present and potential applications of membrane processing in the food industry. Ital J Food Sci 15:3–34
Mousavi SR, Rezaei M (2011) Nanotechnology in agriculture and food production. J Appl Environ Biol Sci 1(10):414–419
Mozafari RM, Flanagan J, Matia-Merino L, Awati A, Omri A, Suntres EZ, Singh H (2006) Recent trends in the lipid-based nanoencapsulation of antioxidants and their role in foods. J Sci Food Agric 86:2038e2045
Mukerjee A, Vishwanatha JK (2009) Formulation, characterization and evaluation of curcumin loaded PLGA nanosphere for cancer therapy. J Anticancer Res 29(10):3867–3875
Nakagawa K, Surassmo S, Min SG, Choi MJ (2011) Dispersibility of freeze-dried poly(epsilon-caprolactone) nanocapsules stabilized by gelatin and the effect of freezing. J Food Eng 102(2):177–188
Neethirajan S, Jayas DS (2011) Nanotechnology for the food and bioprocessing industries. Food Bioprocess Technol 4:39. https://doi.org/10.1007/s11947-010-0328-2
No HK, Meyres SP, Prinyawiwatkull W, Xu Z (2007) Applications of chitosan for improvement of quality and shelf life of foods: a review. J Food Sci 72:R87–R100
Oberdörster G, Maynard A, Donaldson K, Castranova V, Fitzpatrick J, Ausman K, Carter J, Karn B, Kreyling W, Lai D, Olin S, Monteiro-Riviere N, Warheit D, Yang H (2005) Principles for characterizing the potential human health effects from exposure to nanomaterials: elements of a screening strategy. Part Fibre Toxicol 2(8):1–35
Ojagh SM, Rezaei M, Razavi SH, Hosseini MH (2010) Effect of chitosan coatings enriched with cinnamon oil on the quality of refrigerated rainbow trout. Food Chem 120:193–198
Otles S, Yalcin B (2010) Nano-biosensors as new tool for detection of food quality and safety. LogForum 6(4):67–70
Ozimek L, Pospiech E, Narine S (2010) Nanotechnologie in food and meat processing. Acta Sci Pol Technol Aliment 9(4):401–412
Palchetti I, Mascini M (2008) Electroanalytical biosensors and their potential for food pathogen and toxin detection. Anal Bioanal Chem 391:455–471
Park SI, Daeschel MA, Zhao Y (2004) Functional properties of antimicrobial lysozyme-chitosan composite films. J Food Sci 69(8):215–221
Pintado CM, Ferreira MASS, Sousa I (2009) Properties of whey protein-based films containing organic acids and nisin to control *Listeria monocytogenes*. J Food Prot 72(9):1891–1896
Plexus Institute (2006) New nanotechnology food research—if it glows don't eat it. http://www.plexusinstitute.org/news-events/show_news.cfm?id=164. Accessed 20 Oct 2016
Pouliot Y (2008) Membrane processes in dairy technology—from a simple idea to worldwide panacea. Int Dairy J 18(7):735–740
Pranoto Y, Rakshit SK, Salokhe VM (2005) Enhancing antimicrobial activity of chitosan films by incorporating garlic oil, potassium sorbate and nisin. LWT Food Sci Technol 38:859–865
Preetz C, Rube A, Reiche I, Hause G, Mader K (2008) Preparation and characterization of biocompatible oil-loaded polyelectrolyte nanocapsules. Nanomedicine: nanotechnology. Biol Med 4(2):106–114
Rai V, Acharya S, Dey N (2012) Implications of nanobiosensors in agriculture. J Biomater Nanobiotechnol 3:315–324
Ranjan S, Dasgupta N, Chakraborty AR, Samuel SM, Ramalingam C, Shanker R, Kumar A (2014) Nanoscience and nanotechnologies in food industries: opportunities and research trends. J Nanopart Res 16:2464. https://doi.org/10.1007/s11051-014-2464-5
Ranjan S, Dasgupta N, Srivastava P, Ramalingam C (2016) A spectroscopic study on interaction between bovine serum albumin and titanium dioxide nanoparticle synthesized from microwave-assisted hybrid chemical approach. J Photochem Photobiol B Biol 161:472–481
Ravichandran R (2010) Nanotechnology applications in food and food processing: innovative green approaches, opportunities and uncertainties for global market. Int J Green Nanotechnol Phys Chem 1(2):72–96. https://doi.org/10.1080/19430871003684440
Renton A (2006) Welcome to the world of nanofoods. The Observer, London. http://observer.guardian.co.uk/foodmonthly/futureoffood/story/0,,1971266,00.html. Accessed 17 June 2015

Rhim J, Ng PKW (2007) Natural biopolymer-based nanocomposite films for packaging applications. Crit Rev Food Sci Nutr 47(4):411–433

Ribeiro C, Vicente AA, Teixeira JA, Miranda C (2007) Optimization of edible coating composition to retard strawberry fruit senescence. Postharvest Biol Technol 44:63–70

Ribeiro HS, Chua BS, Ichikawab S, Nakajima M (2008) Preparation of nanodispersions containing β-carotene by solvent displacement method. Food Hydrocoll 22(1):12–17

Rickman D, Luvall JC, Shaw J, Mask P, Kissel D, Sullivan D (1999) Precision Agriculture: Changing the Face of Farming. www.ghccmsfcnasagove/precisionag/

Rivett J, Speer DV (2009) Oxygen scavenging film with good interplay adhesion. US Patent 75141512 (in English)

Rodriguez A, Nerin C, Batlle R (2008) New cinnamon-based active paper packaging against Rhizopusstolonifer food spoilage. J Agric Food Chem 56(15):6364–6369

Rojas-Grau MA, Bustillos ARD, Friedman M, Henika PR, Martin-Belloso O, Mc Hugh TH (2006) Mechanical, barrier and antimicrobial properties of apple puree edible films containing plant essential oils. J Agric Food Chem 54:9262–9267

Roller S (2003) Introduction. In: Roller (ed) Natural antimicrobials for the minimal processing of foods. CRC Press, Boca Raton, pp 1–10

Rosenberg M (1995) Current and future applications for membrane processes in the dairy industry. Trends Food Sci Technol 6(1):12–19

Ruiz-Altisenta M, Ruiz-Garciaa L, Moredaa GP, Lub R, Hernandez-Sancheza N, Correaa EC, Diezmaa B, Nicolaïc B, García-Ramosd J (2010) Sensors for product characterization and quality of specialty crops—a review. Comput Electron Agric 74:176–194

Samelis J, Sofos JN (2003) Organic acids. In: Roller (ed) Natural antimicrobials for the minimal processing of foods. CRC Press, Boca Raton

Saxena A, Tripathi BP, Kumar M, Shahi VK (2009) Membrane-based techniques for the separation and purification of proteins: an overview. Adv Colloid Interf Sci 145(1–2):1–22

Sebti I, Martial-Gros A, Carnet-Pantiez A, Grelier S, Coma V (2005) Chitosan polymer as bioactive coating and film against *Aspergillus niger* contamination. J Food Sci 70(2):100–104

Semo E, Kesselman W, Danino D, Livney YD (2007) Casein micelle as a natural nano-capsular vehicle for nutraceuticals. Food Hydrocoll 21(6):936–942

Shen XL, JM W, Chen Y, Zhao G (2010) Antimicrobial and physical properties of sweet potato starch films incorporated with potassium sorbate or chitosan. Food Hydrocoll 24:285–290

Shukla A, Dasgupta N, Ranjan S, Singh S, Chidambram R (2017) Nanotechnology towards prevention of anaemia and osteoporosis: from concept to market. Biotechnol Biotechnol Equip 31(5):863–879

Silva HD, Cerqueira MA, Souza BWS, Ribeiro C, Avides MC, Quintas MAC, Coimbra JSR, Cunha MGC, Vicente AA (2011) Nanoemulsions of β-carotene using a highenergy emulsification–evaporation technique. J Food Eng 102(2):130–135

Silvestre C, Duraccio D, Cimmino S (2011) Food packaging based on polymer nanomaterials. Prog Polym Sci 36(12):1766–1782

Simões ADN, Tudela JA, Allende A, Puschmann R, Gil MI (2009) Edible coatings containing chitosan and moderate modified atmospheres maintain quality and enhance phytochemicals of carrot sticks. Postharvest Biol Technol 51:364–370

Siripireddy B, Mandal BK, Ranjan S, Dasgupta N, Ramalingam C (2017) Nano-zirconia – evaluation of its antioxidant and anticancer activity. J Photochem Photobiol B Biol 170:125–133. https://doi.org/10.1016/j.jphotobiol.2017.04.004

Sondi I, Salopek-Sondi B (2004) Silver nanoparticles as antimicrobial agent: a case study on E. coli as a model for Gram-negative bacteria. J Colloid Interface Sci 275:177–182

Sorrentino A, Gorrasi G, Vittoria V (2007) Potential perspectives of bionanocomposites for food packaging applications. Trends Food Sci Technol 18:84–95

Sowasod N, Charinpanitkul ST, Tanthapanichakoon W (2008) Nanoencapsulation of curcumin in biodegradable chitosan via multiple emulsion/solvent evaporation. Int J Pharm 347:93–101

Sozer N, Kokini JL (2009) Nanotechnology and its applications in the food sector. Trends Biotechnol 27(2):82e89

Srinivas PR, Philbert M, Vu TQ, Huang Q, Kokini JL, Saos E, Chen H, Peterson CM, Friedl KE, McDade-Ngutter C, Hubbard V, Starke-Reed P, Miller N, Betz JM, Dwyer J, Milner J, Ross SA (2010) Nanotechnology research: applications in nutritional sciences. J Nutr 140 (1):119e124

Stutzenberger FJ, Latour RA, Sun Y, Tzeng T (2007) Adhesin-specific nanoparticles and process for using same. US Patent No 20070184120 (in English)

Surassamo S, Bejrapha P, Min SG, Choi MJ (2010) Effect of surfactants on capsicum oleoresin loaded nanocapsules formulated through emulsion diffusion method. Food Res Int 43(1):8–17

Suyatma NE, Copinet A, Tighzert L, Coma V (2004) Mechanical and barrier properties of biodegradable films made from chitosan and poly (lactic acid) blends. J Polym Environ 12:1–6

Tachaprutinun A, Udomsup T, Luadthong C, Wanichwecharungruang S (2009) Preventing the thermal degradation of astaxanthin through nanoencapsulation. Int J Pharm 374(1–2):119–124

Tammina SK, Mandal BK, Ranjan S, Dasgupta N (2017) Cytotoxicity study of Piper nigrum seed mediated synthesized SnO2 nanoparticles towards colorectal (HCT116) and lung cancer (A549) cell lines. J Photochem Photobiol B Biol 166:158–168

Tharanathan RN (2003) Biodegradable films and composite coatings: past, present and future. Trends Food Sci Technol 14:71–78

The U.S. Department of Agriculture (2015) Nanoscale Science and Engineering for Agriculture and Food Systems. http://nifa.usda.gov/program/food-science-technology-programs. Accessed on 15th July 2015

Tiyaboonchai W, Tungpradit W, Plianbangchang P (2007) Formulation and characterization of curcuminoids loaded solid lipid nanoparticles. Int J Pharm 337(1–2):299–306

U.S. Food and Drug Administration (2014) Food safety for moms-to-be: medical professionals – foodborne pathogens. http://www.fda.gov/food/resourcesforyou/healtheducators/ucm091681.htm

Vargas M, Albors A, Chiralt A, González-Maríınez C (2006) Quality of cold-stored strawberries as affected by chitosan–oleic acid edible coatings. Postharvest Biol Technol 41:164–171

Vargas M, Pastor C, Chiralt A, McClements DJ, González- Martínez C (2008) Recent advances in edible coatings for fresh and minimally processed fruits. Crit Rev Food Sci Nutr 48 (6):496–511

Vásconez MB, Flores SK, Campos CA, Alvarado J, Gerschenson LN (2009) Antimicrobial activity and physical properties of chitosan–tapioca starch based edible films and coatings. Food Res Int 42(7):762–769

Vashist SK (2007) A review of microcantilevers for sensing applications. J Nanotechnol online 3. https://doi.org/10.2240/azojono0115

Vicentini DS, AJr S, Laranjeira MCM (2010) Chitosan/poly (vinyl alcohol) films containing ZnO nanoparticles and plasticizers. Mater Sci Eng C 30(4):503–508

Walia N, Dasgupta N, Ranjan S, Chen L, Ramalingam C (2017) Fish oil based vitamin D nanoencapsulation by ultrasonication and bioaccessibility analysis in simulated gastro-intestinal tract. Ultrason Sonochem 39:623–635

Wang L, Li X, Zhang G, Dong J, Eastoe J (2007a) Oil-in-water nanoemulsions for pesticide formulations. J Colloid Interface Sci 314(1):230–235

Wang J, Zhou G, Chen C, Yu H, Wang T, Ma Y, Jia G, Gao Y, Li B, Sun J, Li Y, Jiao F, Zhao Y, Chai Z (2007b) Acute toxicity and biodistribution of different sized titanium dioxide particles in mice after oral administration. Toxicol Lett 168(2):176–185

Wang JC, Chen SH, ZC X (2008a) Synthesis and properties research on the nanocapsulated capsaicin by simple coacervation method. J Dispers Sci Technol 29(5):687–695

Wang X, Jiang Y, Wang YW, Huang MT, Hoa CT, Huang Q (2008b) Enhancing anti-inflammation activity of curcumin through O/W nanoemulsions. Food Chem 108(2):419–424

Weiss J, Takhistov P, McClements DJ (2006) Functional materials in food nanotechnology. J Food Sci 71(9):107–116

Wenner M (2008) Magnifying taste: new chemicals trick the brain into eating less. Scientific American Magazine:96–99

Xing F, Cheng G, Yi K, Ma L (2004) Nanoencapsulation of capsaicin by complex coacervation of gelatin, acacia, and tannins. J Appl Polym Sci 96(6):2225–2229

Xu YX, Kimb KM, Hanna MA, Nag D (2005) Chitosan–starch composite film: preparation and characterization. Ind Crop Prod 21:185–192

Yuan Y, Gao Y, Mao L, Zhao J (2008a) Optimisation of conditions for the preparation of β-carotene nanoemulsions using response surface methodology. Food Chem 107 (3):1300–1306

Yuan Y, Yanxiang G, Zhao J, Mao L (2008b) Characterization and stability evaluation of β-carotene nanoemulsions prepared by high pressure homogenization under various emulsifying conditions. Food Res Int 41(1):61–68

Zhao X, Hilliard LR, Mechrey SJ, Wang Y, Bagwe RP, Jin S, Tan W (2004) A rapid bioassay for single bacterial cell quantitation using bioconjugated nanoparticles. Proc Natl Acad Sci 101:15027–15032

Zhao R, Torley P, Halley PJ (2008) Emerging biodegradable materials: starch- and protein-based bio-nanocomposites. J Mater Sci 43:3058–3071

Zhao L, Xiong H, Peng H, Wang Q, Han D, Bai C, Liu Y, Shi S, Deng B (2011) PEG-coated lyophilized pro-liposomes: preparation, characterizations and in vitro release evaluation of vitamin E. Eur Food Res Technol 232(4):647–654

Ziani K, Fernandez Pan I, Royo M, Maté J (2009) Antifungal activity of films and solutions based on chitosan. Food Hydrocoll 23:2309–2314

Zimet P, Livney YD (2009) Beta-lactoglobulin and its nanocomplexes with pectin as vehicles for ω-3 polyunsaturated fatty acids. Food Hydrocoll 23(4):1120–1126

Zivanovic S, Chi S, Draughon AF (2005) Antimicrobial activity of chitosan films enriched with essential oils. J Food Sci 70:M45–M51. https://doi.org/10.1111/j.1365-2621.2005.tb09045.x

Chapter 4
Plant Nano-nutrition: Perspectives and Challenges

Hassan El-Ramady, Neama Abdalla, Tarek Alshaal, Ahmed El-Henawy, Mohammed Elmahrouk, Yousry Bayoumi, Tarek Shalaby, Megahed Amer, Said Shehata, Miklós Fári, Éva Domokos-Szabolcsy, Attila Sztrik, József Prokisch, Elizabeth A.H. Pilon-Smits, Marinus Pilon, Dirk Selmar, Silvia Haneklaus, and Ewald Schnug

Abstract The global agriculture is facing many challenges including sustainable use and conservation of natural resources, climate change, urbanization, and pollution resulting from agrochemicals (e.g., fertilizers and pesticides). So, the sustainable agriculture is an urgent issue and hence the suitable agro-technological interventions are essential (e.g., nano- and bio-technology) for ensuring the safety and sustainability of relevant production system. Biotechnology and nanotechnology also can be considered emerging solutions to resolve the global food crisis.

H. El-Ramady (✉) • T. Alshaal • A. El-Henawy
Soil and Water Deparment, Faculty of Agriculture, Kafrelsheikh Uni., Kafr El-Sheikh, Egypt
e-mail: ramady2000@gmail.com; alshaaltarek@gmail.com; aelhenawy@yahoo.com

N. Abdalla
Plant Biotechnology Department, Genetic Engineering Division, National Research Center, Dokki, Giza, Egypt
e-mail: neama_ncr@yahoo.com

M. Elmahrouk • Y. Bayoumi
Horticulture Department, Faculty of Agriculture, Kafrelsheikh Uni., Kafr El-Sheikh, Egypt
e-mail: threemelmahrouk@yahoo.com; ybayoumi2002@yahoo.com.sg

T. Shalaby
Horticulture Department, Faculty of Agriculture, Kafrelsheikh Uni., Kafr El-Sheikh, Egypt

College of Agricultural and Food Sciences, King Faisal University, Al-Hassa, Saudi Arabia
e-mail: tashalaby@yahoo.com

M. Amer
Soils, Water and Environment Research Institute (SWERI), Agricultural Research Center, Giza, Egypt
e-mail: dr_megahed120@yahoo.com

S. Shehata
Vegetable crops Department, Faculty of Agriculture, Cairo University, Giza, Egypt
e-mail: said_shehata2@yahoo.com

K M Gothandam et al. (eds.), *Nanotechnology, Food Security and Water Treatment*, Environmental Chemistry for a Sustainable World 11,
https://doi.org/10.1007/978-3-319-70166-0_4

Nanoparticles or nanomaterials can be used in delivering different nutrients for plant growth. These nanoparticles as nanofertilizers have positive and negative effects on soils, soil-biota and plants. These effects mainly depend on multiple factors including nanofertilizer properties, plant species, soil fate and dynamics as well as soil microbial communities. Nanofertilizers could improve the nutrient use efficiencies through releasing of nutrients slowly and steadily for more than 30 days as well as reducing the loss of nutrients in agroecosystems and sustaining farm productivity. Here we review the plant nano-nutrition including the response of plants and soils to nanonutrients and their fate, dynamic, bioavailability, phytotoxicity, etc. Concerning the effects of nanonutrients on terrestrial environments are still an ongoing processes and it demands further researches as well as a knowledge gap towards different changes in shape, texture, color, taste and nutritional aspects on nanonutrients exposed plants as a major component in the food chain. Moreover, the interaction between nanonutrients and plants, soils, soil biota and the entire agroecosystem will be also highlighted.

Keywords Agri-nanotechnology • Plant nano-nutrition • Nanonutrients • Nanoparticles

Contents

M. Fári • É. Domokos-Szabolcsy
Plant Biotechnology Department, Debrecen Uni., Debrecen, Hungary
e-mail: fari@agr.unideb.hu; domokosszabolcsy@gmail.com

A. Sztrik • J. Prokisch
Institute of Bio- and Environmental Enegetics, Debrecen Uni., Debrecen, Hungary
e-mail: atyesz@gmail.com; jprokisch@agr.unideb.hu

E.A.H. Pilon-Smits • M. Pilon
Department of Biology, Colorado State University, Fort Collins, CO, USA
e-mail: epsmits@lamar.colostate.edu; pilon@lamar.colostate.edu

D. Selmar
Applied Plant Science Department, Institute for Plant Biology, TU Braunschweig, Braunschweig, Germany
e-mail: d.selmar@tu-bs.de

S. Haneklaus • E. Schnug
Institute of Crop and Soil Science (JKI), Federal Research Centre for Cultivated Plants, Braunschweig, Germany
e-mail: silvia.haneklaus@jki.bund.de; ewald.schnug@jki.bund.de

4.1 Introduction

A great challenge faces all countries all over the world concerning the feeding population from 6 to 9 billion by 2050 (Ditta et al. 2015). Due to about 40% of the population depends on agriculture in the developing countries; hence the agriculture can be considered a backbone in these countries. The global agriculture is facing several challenges including climate changes (Chalise and Naranpanawa 2016; Chen et al. 2016a, b; Ma et al. 2016; Villoria et al. 2016; Brown 2017), conservation of natural resources and their sustainable use (Ditta et al. 2015; Ma et al. 2016; Duran-Encalada et al. 2017), and urbanization (Maheshwari and Bristow 2016; Henderson et al. 2017) as well as environmental issues (e.g., pollution resulting from runoff and accumulation of fertilizers and pesticides) (Jacobsen and Hansen 2016; Qureshi et al. 2016; Ma et al. 2016; Kuppusamy et al. 2016a, b). So, the sustainable agriculture is an urgent issue and hence the suitable agro-technological interventions are essential like nano- and bio-technology for ensuring the safety and sustainability of relevant production system (Dasgupta et al. 2015a, b; Abhilash et al. 2016; Magalhães et al. 2017; Dasgupta et al. 2017; Ranjan et al. 2014; Shukla et al. 2017).

A great progress has been achieved in the nanotechnology sector in the last two decades (Belal and El-Ramady 2016; Liu et al. 2016; Reddy et al. 2016; Shalaby et al. 2016; Magalhães et al. 2017). The agricultural sector was one of the most important fields, which nanotechnology science involves leading a revolution in many applications such as the agri-food industries (Dasgupta et al. 2015a, b; Handford et al. 2015; He and Hwang 2016; Sarkar et al. 2016; Vélez et al. 2017), remediation of soils and waters from pollutants or nanoremediation (Belal and El-Ramady 2016; Jain et al. 2016a, b; Gillies et al. 2016; Gil-Díaz et al. 2016a; Kuppusamy et al. 2016a, b; Shao et al. 2016; Pulimi and Subramanian 2016), fertilizers and pesticides production (Mastronardi et al. 2015; Solanki et al. 2015; Subramanian et al. 2015; Chhipa and Joshi 2016; Dubey and Mailapalli 2016; Dwivedi et al. 2016; Panpatte et al. 2016), the precision farming (Gemtos et al. 2013; Ditta et al. 2015; Chhipa and Joshi 2016; Li et al. 2016a, b, c; Dubey and Mailapalli 2016; Shaw et al. 2016; Shalaby et al. 2016), post harvest and storage of crops (Flores-López et al. 2016; Luo et al. 2016; Mohammadi et al. 2016; Sogvar et al. 2016; Song et al. 2016; Li et al. 2017; Ray et al. 2017), etc.

Nanotechnology may have a hidden face in soils. The apparent face not only include the direct effects on soil microbial communities, and remediate of polluted soils, but also using natural nanoparticles like zeolites and nano-clays as soil amendments. Therefore, several applications of nanoparticles or nanomaterials in soils including (1) use of nanoparticles like zeolites and nano-clays in soil improvement (Xiong et al. 2015; Lateef et al. 2016; Danish et al. 2016), (2) soil application of nanofertilizers (Mastronardi et al. 2015; Subramanian et al. 2015; Dwivedi et al. 2016), (3) remediation of polluted soils (Araújo et al. 2015; Jain et al. 2015; Fajardo et al. 2015; Ibrahim et al. 2016; Jain et al. 2016a, b; Gillies et al. 2016; Gil-Díaz et al. 2016a), (4) using nano zero valent iron technique in the degradation of pollutants (Raman and Kanmani 2016; Yang et al. 2016; Vítková et al. 2016;

Yirsaw et al. 2016; Zhao et al. 2016; Xie et al. 2017) etc. Concerning the hidden face of nanotechnology in soils, it may be include the interaction between different nanoparticles and different environments. These different environmental compartments include plants, microbes, air and soil, which have been extensively studied (e.g., Abhilash et al. 2016; Du et al. 2016; Gil-Díaz et al. 2016b; Gillies et al. 2016; Song and Lee 2016; Yausheva et al. 2016; Wang et al. 2016a, b). So, the fate and behavior of nanomaterials in soils including transport, bioavailability and bio-toxicity of these nanoparticles should be addressed (Watson et al. 2015; Gogos et al. 2016; Li et al. 2016a, b, c; Wang et al. 2016a; Yirsaw et al. 2016, Shukla et al. 2017; Walia et al. 2017; Siripireddy et al. 2017; Maddinedi et al. 2015, 2017; Tammina et al. 2017; Sannapaneni et al. 2016; Ranjan et al. 2014, 2015, 2016; Nandita et al. 2016; Jain et al. 2016a, b). On the other hand, this behavior of nanoparticles in soils is mainly controlled by soil characterization particularly soil pH (Conway and Keller 2016), soil clay content (Zhang et al. 2016a; Chen et al. 2016a, b), soil organic matter (Majumdar et al. 2016), and soil cation exchange capacity (Watson et al. 2015; Gogos et al. 2016).

Plant nano-nutrition as a science is dealing with nanonutrients from different aspects including the uptake, translocation, metabolism, bioavailability of nanonutrients in rhizosphere for plant growth and development or it is nanotechnology application for the provision of nano-nutrients for the production of crops (Ditta et al. 2015). This branch of plant nutrition also includes nanoparticles phytotoxicity, the interaction between nanonutrients and plants (de la Rosa et al. 2016; Reddy et al. 2016; Zuverza-Mena et al. 2016; Wang et al. 2016b), soils (He et al. 2016; Dwivedi et al. 2016; Pachapur et al. 2016; Lateef et al. 2016), soil biota (Ibrahim et al. 2016; Maliszewska 2016) and the entire agroecosystem (Costa and Fadeel 2016; Fraceto et al. 2016; Servin and White 2016). Therefore, this review will focus on the plant nano-nutrition including the response of plants and soils to nanonutrients and their fate, dynamic, bioavailability, phytotoxicity, etc.

4.2 Nanotechnology in Agriculture and Its Challenges

More than 30,000 articles have been published since 2013 concerning nanotechnology with about 8,000 of those studies occurring in this year (June 21, 2016 through Science Direct or Springer Link). This reflects the significance of this science in our life including all fields or sectors. The agricultural field is an important one, which nanotechnology strongly invasives it. More than 100 books till now (June 21, 2016) have been published by Springer including some handbooks or encyclopedia (e.g. Bhushan 2016; Aliofkhazraei 2016; Aliofkhazraei and Makhlouf 2016; Egorova et al. 2016). Concerning nanotechnology definition, it is defined as "*the science, engineering and technology of controlling, building, and restricting materials and devices at the nanoscale*" according to Wang et al. (2016a, b). Nanotechnology penetrates all fields including medicine (Olivo and Dinish 2016; Khanna 2016; Ahmed and Jackson 2016; Steinhoff 2016; Zhang et al. 2016), pharmacology (Garvie-Cook 2016), industry (Bindal and Hamedi-Hagh

2016; Meguid 2016; Andrievski and Khatchoyan 2016), engineering (Jorio 2016; Singh and Gaharwar 2016), energy (Zhang 2016) and agriculture (Parisi et al. 2015; Dwivedi et al. 2016; Egorova et al. 2016; Ibrahim et al. 2016; Mehlhorn 2016; Piperigkou et al. 2016; Peters et al. 2016; Servin and White 2016; Wang et al. 2016a, b).

It is well known that, nanotechnology deals with the manufacturing, manipulation and characterization of different materials having a size range at the nanometer scale. Furthermore, reduction of the material size into the nano scale changing the physico-chemical properties comparing with the same material at larger-size scales (Peters et al. 2016). Therefore, it could be classified the agri-nanotechnology into three categories including plant- (phytonanotechnology), microbes (microbial or bio-nanotechnology) and animal-nanotechnology (zoo-nanotechnology). Phytonanotechnology refers to different applications of nanotechnology in both plant sciences and plant production systems (Wang et al. 2016a, b). Regarding the benefits and potential uses of nanotechnology in agriculture, there are significant applications including producing greater quantities of food with lower cost as well as energy sector and waste remediation (Fraceto et al. 2016; Servin and White 2016). However, many questions regarding different previous approaches and their risk in agricultural production remain unanswered. Numerous applications in agri-nanotechnology have developed engineering nanoparticles as smart delivery systems (e.g., Piperigkou et al. 2016; Peters et al. 2016; Servin and White 2016; Wang et al. 2016a).

Regarding the potential applications of nanotechnology in agricultural field, it is reported about these impacts including (1) increasing the crop productivity through using nano-agrochemicals (e.g. nanopesticides and nanofertilizers), (2) improving food security and its productivity, (3) improving soil quality *via* enhancing the water-holding capacity of soil (e.g. using nanoclays, nano-zeolites and hydrogels), (4) stimulating plant growth using nanomaterials by enhancing elemental uptake and use of nutrients (e.g. nano-SiO_2, TiO_2, ZnO and carbon nanotubes), (5) providing smart monitoring using nanosensors by wireless communication devices help farmers in maintaining farm with precise control and report timely needs of plants (Fraceto et al. 2016). These applications of nanotechnology in agricultural researches have been intensively used in both academic and industrial levels (Dasgupta et al. 2015a, b; Parisi et al. 2015; Fraceto et al. 2016) due to the unique properties of nanomaterials as well as its suitability candidates in designing and developing such novel nano-tools supporting the sustainability of agriculture (Ditta et al. 2015; Dwivedi et al. 2016).

It should be considered some important information regarding nanotechnology and its effects on plants including size, composition, concentration, surface charge and physical chemical properties of used nanoparticles/nanomaterials as well as their susceptibility of the plant species (Fraceto et al. 2016). Concerning the challenges of using nanotechnology in agriculture, some urgent issues should be kept in mind remain to be resolved in the near future including (1) the ecotoxicology of nanomaterials in agroecosystem (Bhatt and Tripathi 2011; Ma et al. 2013; Judy and Bertsch 2014; Anjum et al. 2015; Bour et al. 2015; Costa and Fadeel 2016;

Hu et al. 2016), (2) the pollution resulted from nanomaterials and its advanced regulations (Amenta et al. 2015; Du et al. 2016), (3) the sustainability and its biosafety of nanomaterials, (4) the development of carrier of nanomaterials (De Oliveira et al. 2014), and (5) the investigation of nanomaterial on the applied scale like the industrial level (Fraceto et al. 2016). Therefore, more developed techniques should be saved to monitor the fate of nanoparticles in different environments under different concentrations (Sadik et al. 2014). Moreover, more advanced carriers should be used for nanoparticles achieving high delivering the active agents (e.g. pesticides, nutrients, and fertilizers) enhancing the maximization of their efficiency seeking the sustainability (De Oliveira et al. 2014; Van Koetsem et al. 2016). Concerning the applied side, it should be investigated and then evaluated the nanomaterials on the commercial scale by comparing these nanoparticles with the commercial products as well as the interaction between different kinds of nanoparticles (Dimkpa et al. 2015). Regarding the ecotoxicology of nanoparticles, it should be studied the phytotoxic effects of nanoparticles on agroecosystems including plants, soil microbial communities, soil biota, water, air as well as human health (Bouguerra et al. 2016; Costa and Fadeel 2016; Fraceto et al. 2016; Servin and White 2016). Therefore, it could be concluded that, nanotechnology has a very strong link with the agriculture and penetrates several agricultural fields including fertilization, irrigation and water saving, nanoremediation of soils and water from pollutants as well as the agri-food sector. Many challenges are still needed for more researches and investigations for safe and sustainable using of nanomaterials in agriculture.

4.3 Nanotechnology in Soils: The Hidden Face

More than 110,000 articles have been published since 2013 concerning nanoparticles with about 25,000 of those studies occurring in this year (June 22, 2016 through Science Direct), whereas more than 37,000 articles belong nanomaterials from the same and previous period (about 10,000 only for 2016 through Science Direct). Several issues have been published concerning the nanotechnology in soils including nanoremediation of polluted soils (Araújo et al. 2015; Jain et al. 2015; Fajardo et al. 2015; Ibrahim et al. 2016; Jain et al. 2016a, b; Gillies et al. 2016; Gil-Díaz et al. 2016a), using nanoparticles/nanomaterial (e.g. zeolites and nano-clays) in soil improvements (e.g. Xiong et al. 2015; Lateef et al. 2016; Danish et al. 2016), soil application of nanofertilizers (Mastronardi et al. 2015; Subramanian et al. 2015; Dwivedi et al. 2016), using nano zero valent iron (nZVI) technique in the degradation of pollutants (e.g. Raman and Kanmani 2016; Yang et al. 2016; Vítková et al. 2016; Yirsaw et al. 2016; Zhao et al. 2016) etc.

The interaction between nanoparticles/nanomaterials and different environmental compartments including plants, microbes, air and soil was and still one of the most important issues in environmental nanotechnology. This previous interaction has been extensively studied by several researchers (e.g., Adams and Kanaroglou

2016; Abhilash et al. 2016; Du et al. 2016; Gil-Díaz et al. 2016a, b; Gillies et al. 2016; Song and Lee 2016; Yausheva et al. 2016; Wang et al. 2016a, b). Regarding the fate and behavior of nanoparticles in soils, once the nanoparticles enter the soil, different soil physico-chemical properties are largely controlled the transport, bioavailability and bio-toxicity of these nanoparticles (Watson et al. 2015; Gogos et al. 2016; Li et al. 2016a, b, c; Wang et al. 2016a, b; Yirsaw et al. 2016). Like other metals, the transportation, bioavailability, and sorption of metal nanoparticles in soils is totally governed by soil properties including soil pH (Waalewijn-Kool et al. 2014; Conway and Keller 2016), clay content (Zhang et al. 2016b), soil organic matter (Majumdar et al. 2016) and cation exchange capacity (Benoit et al. 2013; Dimkpa et al. 2015; Watson et al. 2015; Gogos et al. 2016).

On the other hand, the interaction between different types of nanoparticles in soils is an emerging issue and needs further studies to evaluate the behavior of nanoparticles in long-term field condition as well as their interactions in soils and the role of these interactions on plant nutrition (Dimkpa et al. 2015). Definitely, the interaction between these nanoparticles themselves in soils is governed by different soil properties as mentioned before and the properties of nanoparticles themselves (e.g. the size, shape and surface charge of nanoparticles). These previous properties of both soil and nanoparticles are the main factors control the dissolution, solubilization, agglomeration and aggregation of nanoparticles in soils (Dwivedi et al. 2016; Pachapur et al. 2016). It is reported that, the high organic content in soil enhances a strong binding between nanoparticles and soil decreasing the mobility, bioavailability and uptake of these nanoparticles and then the bio-toxicity by plants (Shoults-Wilson et al. 2011). Concerning silver nanoparticles in soils, it is also stated that, (1) the aggregation and its retention in soils of these particles is enhanced by both ionic strength and divalent cations (Thio et al. 2012), (2) the bioavailability of Ag ions is decreased by increasing soil pH due to a greater CEC raising the adsorption of Ag ions onto the soil surface and (3) as well as the hetero-aggregation of Ag-nanoparticles with natural colloids in soils reduces their mobility (Cornelis et al. 2013; Dwivedi et al. 2016; Pachapur et al. 2016; Troester et al. 2016).

What about the hidden face concerning the nanoparticles in soils? It could be drawn the complete portrait for nanoparticles in soils through the following questions: what is the interaction between nanoparticles and the different agroecosystem compartments including soil matrix (Dwivedi et al. 2016; Floris et al. 2016; Pachapur et al. 2016; Yang et al. 2016), plants (Zhang et al. 2016a; Yang et al. 2016), soil water (Zhao et al. 2016), soil air (Polis et al. 2013), soil biota (Shen et al. 2015; Xu et al. 2015; Durenkamp et al. 2016; He et al. 2016), aquatic environments like drinking water (Troester et al. 2016; Zhao et al. 2016) as well as the interaction between nanoparticles and pollutants (Xie et al. 2016) or themselves (Dimkpa et al. 2015; Li et al. 2016a, c; Pachapur et al. 2016). Therefore, the hidden face of nanoparticles in soils can be considered still not well understood and further studies should be suggested in monitoring nanoparticles in soils. It is reported that, the production of metal nanoparticles is expected to reach 58,000 tons by year 2020 according to United Nations Environment Programme (UNEP 2007). Concerning the environmental dynamics of metal oxide nanoparticles in homogeneous or heterogeneous systems, there are some factors controlling the fate, transport, transformation and toxicity of these metal nanoparticles. One of the most important factors controlling the dynamic of metal oxide nanoparticles is presence of pollutants in different environmental conditions. Hence, the interaction between these pollutants and metal nanoparticles will govern the transformation of metal oxides and their transport kinetics as well as the effects of pollutants on the toxicity of metal nanoparticles in both homogeneous and heterogeneous systems (Joo and Zhao 2016). In more details, it is resulted from the presence of contaminants decreasing in the bioavailability of these nanoparticles through sorption, hetero-aggregation, and/or complexation. Furthermore, the pollutants also control the fate and transport of these nanoparticles exhibiting their synergistic toxic effects (Joo and Zhao 2016).

One of the most important issues concerning nanoparticles in soils is the pollution resulted from these nanomaterials, which have great impacts on human health and soil ecosystems (Yang et al. 2016). It is found that, nano zero-valent iron (nZVI) represents about 70% from the metal-based nanoparticles widely used in environmental nanotechnologies, which have been extensively applied for *in situ*remediation across the world (Yang et al. 2016; Xie et al. 2016). The large amounts of these nanomaterials (nZVI) have been caused many ecotoxicity impacts on soil ecosystems including toxicity (Lefevre et al. 2016), cytotoxicity (Dong et al. 2016) and phytotoxicity (Xie et al. 2016) as well as increasing in oxidative stress (Chaithawiwat et al. 2016) and disruption of microbial community (Pawlett et al. 2013; Saccà et al. 2014) in both aquatic and terrestrial ecosystems (Yang et al. 2016).

Therefore, it could be concluded that, the nanotechnology has many applications in soil system including nanoremediation polluted lands, nanofertilizers, nanopesticides, precision farming using nanosensors (nano-farm), using nanomaterials in soil improvement, etc. There are further studies should be performed concerning the fate and behavior of nanomaterials in soils and their effects on agroecosystem. So, it could be called this interaction between nanomaterials and soil system by "the hidden face" because a lot of information in this area is still unknown.

4.4 Nanotechnology in Plant Nutrition

More than 110,000 articles have been published several years ago regarding plant nutrition, whereas in general these publications were nearly more than 6,000 annually as recorded by the Science Direct (accessed by June 22, 2016). Like other plant sciences, plant nutrition is a branch dealing with the physiology, metabolism, uptake and translocation of different nutrients by plants as well as the molecular plant nutrition. This science can be also involved the following topics: (1) agronomic or physiological efficiency of nutrients (Chaudhary et al.

2015; Loeppmann et al. 2016), (2) rhizosphere and its architecture (Szoboszlay et al. 2015; Oburger and Schmidt 2016), (3) the mineral nutrition and stress physiology (Matraszek et al. 2016; Kolenc et al. 2016), (4) molecular nutrient uptake and plant sensing (Mitra 2015; Rizwan et al. 2016a), and (5) nutrient cycles (Hobbie 2015; Ford et al. 2016). Many publications have been published several years ago concerning plant nutrition started 5000 BC when the ancient Egyptians recorded their paintings about plant nutrition (El-Ramady et al. 2014a), till Justus von Liebig (1803–1873; El-Ramady et al. 2014b) and the handbook of plant nutrition for Barker and Pilbeam (2015).

Nanoscience and nanotechnology have a great applications in agricultural sciences, although the applications in plant sciences as well as plant production systems have been received comparatively little interest (Wang et al. 2016a, b). Several nanomaterials have been synthesized through the physical, chemical and biological methods *via* bacteria, fungi, plants etc. These nanoparticles or nanomaterials have been investigated including many studies such as studying different effects of nanoparticles on the cellular morphology, functions, behavior and the effectiveness of nanoparticles on plants from the agricultural and the horticultural significance (Patra et al. 2013; Ditta et al. 2015; Rizwan et al. 2016b; Panpatte et al. 2016; Wang et al. 2016a, b). It is reported that, nanoparticles have been used in enhancing the crop productivity through the high efficiency of nutrients in the form of nanofertilizers, nanopesticides, or nanoherbicides by the plants (e.g. Tarafdar et al. 2013; Ditta et al. 2015; Panpatte et al. 2016; Rizwan et al. 2016b). These nanoparticles have several advantages in promoting the agricultural productivity including (1) enhancement plant seed germination and growth against stress (Table 4.1; Khan et al. 2016), (2) increasing the efficiency of water and fertilizer dosage, (3) sustainable management using nanosensors in pest detection, (4) using of nanocapsules for pesticides and herbicides in control pests (Ditta et al. 2015) and these nanoparticles can be considered as a next generation technology for sustainablc agriculture (Panpatte et al. 2016; Tolaymat et al. 2016) (Fig. 4.1).

Concerning the nanotechnology and plant nutrition, it is reported that plant nano-nutrition includes nanofertilizers as a source for nutrients (Ditta et al. 2015; Liu et al. 2016; Panpatte et al. 2016), nano-capsules as a nanoscale carriers (Meredith et al. 2016), nano-smart delivery systems (Ditta et al. 2015), nano-oligo cellulosic materials (Mohamed et al. 2016), clay nanotubes or halloysite (Peixoto et al. 2016; Donaldson 2016), micro-fabricated xylem vessels (Ditta et al. 2015), and nanopesticides (Ditta et al. 2015; Subramanian et al. 2016). It could notice that, many agricultural practices can be achieved seeking for the improvement of crop production through using many applications of nanoagrochemicals such as nanopesticides (e.g. nanofungicides, nanoherbicides, etc), nanofertilizers, nanosensors (Subramanian et al. 2016).

Therefore, it could be concluded that, there is a strong link between plant nutrition and nanotechnology through many applications including different nano-agrochemicals such as nanofertilizers, nanopesticides, nanosensors, nano-capsules as a nanoscale carriers etc. Definitely, the using of nanotechnology applications in plant nutrition can help in improving crop production, saving the time and costs as well as

Table 4.1 A comparison between nano copper oxide and nano selenium role in impacting on plants including oxidative stress and antioxidative defense system

Plant species	Stressor level (Exp. medium)	Form & dosage (exposure period)	Different effects of Se on stressed plants and associated potential mechanisms (Reference)
Nano-CuO stress			
Oryza sativa L.	Nano-CuO: 2.5, 10, 50, 100, and 1000 mg L^{-1} (hydroponic)	30 days and size <50 nm nano-CuO	nano-CuO increased MDA and proline contents and APX and SOD activities (Da Costa and Sharma 2016)
Zea mays L.	Nano-CuO: 0.01, 0.02 ppm for root expo.; 8 mg kg^{-1} foliar spray (hydrop.)	Exposure root and leaf 3 weeks and size <50 nm nano-CuO	0.02 ppm root exposure and 8 ppm foliar spray decreased GPX, CAT and succinate dehydrogenase activities but increased SOD and glucose-6-phosphate dehydrogenase activities in leaves (Adhikari et al. 2016)
Transgenic cotton and conventional cotton	10, 200 and 1,000 mg L^{-1} nano-CuO (hydroponics)	10 days and nano CuO diameter were 30 ± 10 nm	Plant hormones IAA and ABA were significantly at high level inhibited and also reduced the uptake of B, Mo, Mn, Mg, Zn and Fe; plants biomass and height not affected at 10 mg kg^{-1} nano-CuO exposure (Le Van et al. 2016)
Halimione portulacoides	Nano-CuO: 10 mg kg^{-1} (hydroponic)	8 days and size of nano CuO <50 nm	No metal translocation Cu in roots of *Halimione*, less Cu accumulation in roots when Cu available in the form of nanoparticles, Cu can form aggregates, reducing SA and thus decreasing Cu availability (Andreotti et al. 2015)
Brassica juncea L.	Nano-CuO: 20, 50, 100, 200, 400, 500 mg L^{-1} (*in vitro* exp.)	14 days and size <50 nm nano-CuO	Nano-CuO increased H_2O_2 and MDA formation; induced POD and SOD but inhibited APX activity in roots and shoots (Nair and Chung 2015)
Medicago sativa L. and *Lactuca sativa* L.	5, 10 and 20 mg L^{-1} nano-CuO (hydroponic)	15 days and nanoparticles 10–100 nm	Nano-CuO reduced CAT activity in alfalfa and increased APX activity in roots of both lettuce and alfalfa. (Hong et al. 2015)

(continued)

Table 4.1 (continued)

Plant species	Stressor level (Exp. medium)	Form & dosage (exposure period)	Different effects of Se on stressed plants and associated potential mechanisms (Reference)
Glycine max L.	Nano-CuO: 50, 100, 200, 400 and 500 mg L^{-1} (*in vitro* ex.)	14 days and size <50 nm nano-CuO	Nano-CuO at 100–500 mg L $^{-1}$ significantly increased the H_2O_2 level, POD activity in roots (Nair and Chung 2014a)
Arabidopsis thaliana L.	Nano-CuO: 0.5, 1, 2, 5, 10, 20, 50, and 100 mg L^{-1} (*in vitro* ex.)	21 days and size 30 nm nano-CuO	Nano-CuO concentration-dependently increased $O_2^{\bullet}$ and H_2O_2 formation in leaves and roots; induced antioxidant, sulfur assimilation, GSH biosynthesis genes (Nair and Chung 2014b)
Nano-Se stress			
Lycopersicum esculentum Mill. cv. Halil	Temperature stress: 10, 25, 40 °C for 24 h (hydroponic)	2.5, 5, 8 μM Se as Na_2SeO_4 & nano-Se 1, 4, 8, 12 μM (3 d)	Se and nano-Se can improve fresh and dry weight of shoot, diameter, root fresh and dry weight and root of tomato plants under high and/or low temperature stress; Se and nano-Se increases relative water content and root volume significantly after a short-term of high and/or low temperature stress (Haghighi et al. 2014)
Arundo donax L. 2 ecotypes Blossom and 20SZ	Nano-Se: 100 mg L^{-1} Se as *Lactobacillus casei* (*in virto* experiment)	0.1, 1, 10, 50, 100 mg L^{-1} Se as Na_2SeO_4 (8–16 d)	Both *Arundo* ecotypes could uptake and accumulate nano-Se however in lower concentration comparing to the selenate; the toxic level of selenate was 20 and 50 mg L^{-1} for Blossom and 20SZ accumulating 920 and 896 mg kg^{-1} Se in clusters resp. (Domokos-Szabolcsy et al. 2014)

(continued)

Table 4.1 (continued)

Plant species	Stressor level (Exp. medium)	Form & dosage (exposure period)	Different effects of Se on stressed plants and associated potential mechanisms (Reference)
Nicotinia tabacum L. cv. Ottawa, Petit, Havana	Nano-Se: 1, 10, 20, 50, 100 mg L^{-1} Se as *Lactobacillus acidophilus* (*in virto* exper.)	1, 10, 20, 50, 100 mg L^{-1} Se as Na_2SeO_4 (8 and 16 d)	Nano-Se (50–100 mg kg^{-1}) stimulated callus initiation, micro-shoot formation on callus surface and root regeneration; SeO_4^{-2} (50–100 mg kg^{-1}) inhibited both callus and root formation; SeO_4^{-2} can get into plant tissue and in excess as a pro-oxidant can damage directly and/or indirectly root generation growth and regeneration of explants; nano-Se generated roots 80% of plantlets on 8th day and >90% on 16th day (Domokos-Szabolcsy et al. 2012)

Abbreviations: *AsA* Reduced ascorbate, *GST* glutathione-*S*-transferase, *MTs* metallothioneins, *Pn* net photosynthesis, *Gs* stomatal conductance, *Tr* transpiration rate, *APX* ascorbate peroxidase activity, *CAT* catalase activity, *MDA* malondialdehyde, *SOD* superoxide dismutase, *CAT* catalase, *POD* peroxidase, *GPX* guaiacol peroxidase, *IAA* indole-3-aceticacid, *ABA* abscisic acid

alleviating the abiotic stress-induced damage through the activation of plant defense system. This reflects the role of nanomaterials in (1) giving the plant protection against reactive oxygen species, (2) protecting plants against the oxidative stress through enhancing the plant antioxidative enzymes including superoxide dismutase, catalase, peroxidase and (3) the acting of nanomaterials itself as inducers of oxidative stress. The small size of these nanomaterials can help plants easily in penetrating and regulating water channels and then assisting the germination of seeds and plant growth as well as the big surface area of nanomaterials improves the adsorption and delivery substances (Khan et al. 2016).

4.5 Nanofertilizers and Nanonutrients

As mentioned before, a real revolution in nanoparticles or nanomaterials has been achieved to penetrate all fields including industrial, pharmaceutical, medicinal, and agricultural sectors. These nanoparticles represent a magic solution in many sectors such as nanoremediation of polluted water and lands (Gil-Díaz et al. 2016a, b; Gillies et al. 2016; Peters et al. 2016; Zhao et al. 2016). Several studies regarding nanoparticles have been published focusing on many fields including energy sector (Le Croy et al. 2016), food industries (Khan and Oh 2016; Rizwan et al. 2016b; Souza and Fernando 2016), synthesis of nanoparticles (Ahmed et al. 2016a; Bennur

Fig. 4.1 Some experiments using nanoparticles in plant nutrition including *in vitro* trials (*photo 1*: nano-Se and in a very hard rooting moth orchids plant – *Phalaenopsis spp.* – unpublished data; whereas *photo 2*: nano-Se and tobacco plant growth), microfarm (*photo 3*: nano-Se and some sprouts production) and field experiments (*photo 4*: using of nitrogen nanofertilizer in lettuce growth). All photos by El-Ramady except no.1 by Elmahrouk; a kind permission from Dr. Mohamed Sharaf for *photo 4*; the bar in *photo 1* represents 2 cm

et al. 2016; Kuppusamy et al. 2016a, b; Sodipo and Abdul Aziz 2016; Singh et al. 2016), nanosensors (Priyadarshini and Pradhan 2016), oil lubrication (Dai et al. 2016), the dynamics and effects of metal oxide nanoparticles (Joo and Zhao 2016; Mustafa and Komatsu 2016; Rizwan et al. 2016b), biomedical applications (Durán et al. 2016; Karimzadeh et al. 2016; Jones et al. 2016; Zarschler et al. 2016), and antimicrobial purposes (Ahmed et al. 2016b; Samiei et al. 2016; Ambika and Sundrarajan 2016).

It is well known that, nanoparticles can (in homogeneous or heterogeneous) interact with soil, air, sediment, and plants and release into different previous environmental compartments. Several studies concerning the beneficial and negative effects of nanoparticles on plants have been reported (Rizwan et al. 2016b; Wang et al. 2016a, b). Definitely, these effects were controlled by plant and nanoparticles characterization including the source, type, the duration of exposure to plants and size of these nanoparticles as well as the plant species (Rico et al. 2014; Bandyopadhyay et al. 2015; Lalau et al. 2015; Cox et al. 2016; Du et al. 2016; Mustafa and Komatsu 2016). On the other hand, the factor of environmental media is also significant in orientation the interaction between nanoparticles and plants including *in vitro* (Domokos-Szabolcsy et al. 2012, 2014; El-Ramady et al. 2014c; Gomez-Garay et al. 2014; Homaee and Ehsanpour 2015; Castiglione et al. 2016; Kumari et al. 2016; Sarmast and Salehi 2016), micro-farm (El-Ramady et al. 2014a, b, c; El-Ramady et al. 2015a, b; El-Ramady et al. 2016a, b), pots (Alidoust and Isoda 2013; Sri Sindhura et al. 2014; Rico et al. 2015a; Gogos et al. 2016; Moll et al. 2016), hydroponic (Schwabe et al. 2013; Haghighi et al. 2014; Zhang et al. 2015a, b; López-Moreno et al. 2016; Tripathi et al. 2016) and field experimentation (Suriyaprabha et al. 2012).

In the early 2000s, the *nano-era* began when more than 35 countries have been initiated research programs in nanotechnology, resulting in a steady increase in engineered nanomaterials production (Zuverza-Mena et al. 2016). The release of engineered nanomaterials in the environment is considered an important issue (Nowack et al. 2013, 2014, 2015; Keller et al. 2013, 2014; Wigger et al. 2015; Conway and Keller 2016; Caballero-Guzman and Nowack 2016; Nowack et al. 2016; Park et al. 2016; Tolaymat et al. 2016). It has been resulted from the rapid proliferation of nanoproducts use increasing in the exposure to humans through different environmental systems (e.g. water, air, sediments and soils). Regarding the exposure to nanomaterials, it could be happened directly through the unintentional release during both use and consumption of nanoproducts and through remediation purposes for polluted sites (Nowack et al. 2013). The indirect pathway for nanomaterials can be occurred *via* the sewage treatment plants and landfills as well as incineration plants (Park et al. 2016; Caballero-Guzman and Nowack 2016).

There are several metal/metalloid oxide nanoparticles (e.g., TiO_2, ZnO, CeO_2) can be release into the aquatic environments through both wastewater discharge and runoff (Osmond and McCall 2010). It is reported that, ZnO and TiO_2 nanoparticles have been extensively used in the production of skincare products, with more than 33,000 tons of sunscreens produced containing up to 25% of ZnO nanoparticles as well as at least 25% of the sunscreen used (~4000–6000 tons/year) was released in reef areas (Joo and Zhao 2016; Caballero-Guzman and Nowack 2016). Concerning the uptake pathway by plants and translocation of metal nanoparticles, two confirmed pathways in plant system have been reported including the uptake from the soil application (root to leaves or fruits) or from the foliar application (leaves to roots) (Ma et al. 2015; Du et al. 2016). Due to the existence of metal nanoparticles and their interaction with plant system, several physiological, agronomical, photosynthetic parameters and antioxidant activities have been reported (Du et al. 2016; Reddy et al. 2016; Rizwan et al. 2016b; Zuverza-Mena et al. 2016).

It is reported that, fertilizers are an important outputs in enhancing the agricultural production but its problem includes the environmental pollution from overuse and the low use efficiency (only 30–40% from the total applied fertilizers, whereas the rest is lost form the agroecosystem through leaching or evaporation or degradation) (Panpatte et al. 2016). Therefore, using the nanofertilizers or nanonutrients will increase the efficiency use of nutrients as well as reducing the possibility for environment pollution if nanofertilizers will be used in proper amounts. On the other hand, nanofertilizers can be mainly produced from the encapsulation of fertilizers within a nanoparticle. Concerning the main techniques for encapsulating fertilizers within nanoparticles, it is reported according to Rai et al. (2012) that, these techniques include (1) encapsulating the nutrients inside nano-porous materials, (2) coating with thin polymer film and (3) delivering nutrients as particle or emulsions in nanoscale dimensions (Panpatte et al. 2016).

Nanofertilizers can be defined as fertilizers contained within nano-structured formulations which can be delivered to targeted sites to allow release of active ingredients keeping the plant nutrient demands (Wang et al. 2016a, b). Therefore, among different agricultural inputs, nanofertilizers can be quite considered promising in enhancing the growth, nutrition and then productivity of crops as well as regulating the release of nutrients and improving efficiency use of nutrients under controlled environmental conditions (Yuvaraj and Subramanian 2015; Subramanian et al. 2015, 2016). Different crops have been fertilized using nanonutrients or nanofertilizers as presented in Table 4.2. These studies include fate and behavior of these nanonutrients in different agroecosystem compartments (soil, water and soil micro-biota), and evaluation of the potential mechanism for toxicity of nanonutrients and their tolerance in both soil–micro-biota and plants (Anjum et al. 2015). It is found that, the manufactured nanoparticles are not always more toxic than other chemical species containing the same elements. For example, CuO nanoparticles are slightly more toxic comparing with other Cu ions, whereas ZnO nanoparticles have a similar toxicity of Zn ions but in case of manganese and iron oxide nanoparticles the toxicity were less than their ionic counterparts as well as significant enhancement the growth of lettuce seedlings by 12–54%, respectively (Liu et al. 2016). That means nanoparticles of Mn and Fe can be considered effective nanofertilizers in increasing the agronomic productivity (Liu et al. 2016).

Therefore, it could be summarized that, the global food production suffer from several challenges including climate changes, pollution, poverty, the wrong agricultural practices, tenure, and low soil fertility. To save the enough and safe food for all people, new strategies in the agricultural production should be performed particularly using the applications of technology in agriculture. These applications include the precision farming and nanofertilizers or nanonutrients. A part from the great advantages of nanofertilizers including the high use efficiency, saving in costs

Table 4.2 A list of some important articles published recently on plant nano-nutrition

Focus area of the study	References
The effects of uncoated and coated cerium oxide nanoparticles with citric acid, bulk cerium oxide, cerium acetate on the nutritional quality of tomato fruits	Barrios et al. (2016)
Physiological and biochemical response of plants to engineered nanomaterials and their levels in the environment including the interactions with the plants, soil microorganisms and potential accumulation in the food chain	de la Rosa et al. (2016)
The interaction between metal oxide nanoparticles with higher terrestrial plants including the physiological and biochemical aspects	Du et al. (2016)
Selenium and nano-selenium in plant nutrition including different nano-fertilizers of selenium	El-Ramady et al. (2016a)
Synthesis, using and characterization of nano-zeolite and nano–composite as an environment friendly slow release fertilizer	Lateef et al. (2016)
Study of the phytotoxicity of ionic ($FeCl_3$), micro- and nano-sized zerovalent iron in three macrophyte plants (*Lepidium sativum*, *Sinapis alba* and *Sorghum saccharatum*)	Libralato et al. (2016)
Study the effects of stabilized nanoparticles of oxide Cu, Zn, Mg, and Fe in low concentrations on lettuce seed germination to establish and know these nano-toxicants or nanonutrients	Liu et al. (2016)
Effects of soil organic matter contents on cerium translocation and different physiological processes in kidney bean plants exposed to cerium oxide nanoparticles increasing the antioxidant enzyme activities in the aerial tissues of plants	Majumdar et al. (2016)
Toxicity of heavy metals and metal-containing nanoparticles on plants including different strategies for plant tolerance mechanisms and transport through cell wall and plasma membrane and the vacuole	Mustafa and Komatsu (2016)
The next generation technology for nanoparticles and its application in nanofertilizers in frame of the sustainable agriculture	Panpatte et al. (2016)
Different lessons learned from nanotechnology including the toxicity of engineered nanomaterials to terrestrial plants	Reddy et al. (2016)
The next steps for understanding engineered nanoparticle exposure and risk in frame of nanotechnology in agriculture	Servin and White (2016)
Using of nano-TiO_2 particles in removal and bioaccumulation of cadmium in soybean plants cultivated in contaminated soils	Singh and Lee (2016)
Selenium nanoparticles as a nutrition supplement including different methods for synthesis of Se-nanoparticles	Skalickova et al. (2016)
Nano zerovalent iron from synthesis to environmental applications including the impacts on living organisms, microorganisms, aquatic and soil, plants and its toxicity assessment	Stefaniuk et al. (2016)
The role of nano-silicon and its foliar application on heavy metals accumulation in different rice cultivars	Wang et al. (2016c)
Exposure of different engineered nanoparticles (metal oxide of iron, copper, zinc, etc) to plants including the physiological and biochemical responses	Zuverza-Mena et al. (2016)
Different copper nano-scales and its fate in the soil–plant system including the toxicity and its potential mechanisms	Anjum et al. (2015)

(continued)

Table 4.2 (continued)

Focus area of the study	References
The applications of nanotechnology in agriculture sector including the controlled release of different agrochemicals	Aouada and de Moura (2015)
The using, applications and perspectives of nanoparticles in sustainable agricultural crop production	Ditta et al. (2015)
The potentials of engineered nanoparticles as fertilizers in increasing the agronomic productions	Liu and Lal (2015)
The emerging contaminants and opportunities for risk mitigation from both nanopesticides and nanofertilizers	Kah (2015)
The strategic role of nanotechnology in fertilizers sector including their potential and limitations	Mastronardi et al. (2015)
Nano-fertilizers and their smart delivery system including the comparison between conventional and nano-fertilizers	Solanki et al. (2015)
Nano-fertilizers and their using for balanced crop nutrition including synthesis of nanofertilizers, characterization, uptake by plants	Subramanian et al. (2015)
Implications of nanotechnology on plant productivity including nanofertilizers and its rhizospheric environment	Thul and Sarangi (2015)
Using silicon nanoparticles in alleviating chromium (VI) phytotoxicity in pea (*Pisum sativum* L.) seedlings	Tripathi et al. (2015)
The role of foliar application with nano-silicon in alleviating cadmium toxicity in rice seedlings	Wang et al. (2015)

and time, a lot of problems are still need solution and further researches. These problems include the nanotoxicity, ecotoxicity, genotoxicity for all environmental compartments.

4.6 Plant Nano-nutrition

The using of nanoparticles or nanomaterials in plant nutrition could be defined as the plant nano-nutrition. It includes using these nanofertilizers or nanonutrients in supplying cultivated plants with the essential nutrients for their growth and productivity (e.g., ZnO, SiO_2, iron oxide, CuO, Mn oxide, phosphorus, nitrogen nanoparticles, etc.). That means simply the source of different nutrients for plants will be through the applied nanonutrients or nanofertilizers as well as the release of nutrients will be also slowly and steadily for a long time (more than 30 days) creating a new approach in improving the nutrient use efficiency (Kah 2015; Mastronardi et al. 2015; Subramanian et al. 2015). Globally, the use of nanofertilizer in a large scale is still limited, although the customized fertilizers have a significant role in sustaining the farm productivity (Subramanian et al. 2015).

As mentioned before, nanonutrients are nutrients in nano-dimensions ranging from 30 to 40 nm having the capability in holding of many nutrient ions because of their high surface area; the release of nutrients is slowly and steadily in commensuration with crop demands (Subramanian et al. 2015). It is also reported about the effects of nanonutrients on the nutritional quality of crops by a few researchers (e.g., Peralta-Videa et al. 2014; Barrios et al. 2016; Reddy et al. 2016; Servin and White 2016). These effects include in general beside different positive effects on plant defense and growth aspects improving the nutritional quality of edible crops. The previous effects definitely depend on the type, concentration and size of metal/metalloid nanoparticles or nanonutrients, plant species, soil and its characterization and climatic conditions. The nutritional quality of edible crops has been become one of the most important issues in plant nano-nutrition, which many studies focused on it (e.g., Rico et al. 2013, 2014, 2015a; Reddy et al. 2016; Servin and White 2016). Many previous studied have been shown effects of nanoparticles on the nutritional quality of some crop plants exposure to CeO_2 nanoparticles such as wheat (Rico et al. 2014), rice (Rico et al. 2013), barley (Rico et al. 2015a), soybean (Lopez-Moreno et al. 2010; Hernandez-Viezcas et al. 2013; Peralta-Videa et al. 2014), tomato (Wang et al. 2013; Barrios et al. 2016) and cucumber (Zhao et al. 2014).

Therefore, plant nano-nutrition is a branch of plant nutrition dealing with the nutrition and development of plants using nanonutrients or nanomaterials. These nanonutrients have a positive effects in plant growth as well as the nutritional quality of crop plants. Definitely, there are some gaps in plant nano-nutrition particularly in the use of these nanonutrients on the large scale and the global level. The different interactions between nanonutrients or nanomaterials with different agroecosystem compartments is still not fully understood and needs further more researches.

4.7 Conclusion

No doubt that, metal/metalloid nanoparticles have a vital role in the emerging nanotechnological sectors but the progress in researches related to the impact of these nanoparticles on the terrestrial environments is still not fully understood. In general, the concentration of nanoparticles less than 50 mg kg^{-1} thought to be beneficial for plant growth, whereas the higher concentrations will be inhibited as well as toxic. Therefore, more in depth studies should be carried out on the plant cell level including gene expression, production of reactive oxygen species, signal transduction etc. (Reddy et al. 2016). Studies of omics in the nanotoxicology should be also considered (Costa and Fadeel 2016; Hu et al. 2016) as well as the plant physiology and biochemistry resulted from nanonutrients exposure (Barrios et al. 2016; de la Rosa et al. 2016; Zuverza-Mena et al. 2016). More researches are also needed in investigating the response of the terrestrial ecosystems to the combined metal nanoparticles and the interaction between nanoparticles and pollutants under different climatic conditions including the arid, flooding, etc. (Du et al. 2016). More investigations in the rhizospheric region are required in order to evaluate the interaction among nanonutrients besides their properties and different soil characterization as well as plant species (Majumdar et al. 2016; Zuverza-Mena et al. 2016). It is worth to mention that, an urgent need is also wanted concerning development of a framework in measuring the availability and total soil contents of nanoparticles or nanonutrients as well as their extraction (Rodrigues et al. 2016).

The safe and enough food production were and will still the global emerging issue. To produce these enough amounts of food, it should focus on the plant nutrition science and how develop its applications to achieve this global target. Plant nano-nutrition can be considered one of the most important tools that help us in performing a high use efficiency of nutrients saving their amount and time. A global regulation for use these nanonutrients or nanofertilizers should be established as soon as possible.

Acknowledgements Authors thank the outstanding contribution of STDF research teams (Science and Technology Development Fund, Egypt) and MBMF/DLR (the Federal Ministry of Education and Research of the Federal Republic of Germany), (Project ID 5310) for their help. Great support from this German-Egyptian Research Fund (GERF) is gratefully acknowledged.

References

Abhilash PC, Tripathi V, Adil Edrisi S, Kant Dubey R, Bakshi M, Dubey PK, Singh HB, Ebbs SD (2016) Sustainability of crop production from polluted lands. Energ Ecol Environ 1(1):54–65. https://doi.org/10.1007/s40974-016-0007-x

Adams MD, Kanaroglou PS (2016) Mapping real-time air pollution health risk for environmental management: combining mobile and stationary air pollution monitoring with neural network models. J Environ Manag 168:133–141. https://doi.org/10.1016/j.jenvman.2015.12.012

Adhikari T, Sarkar D, Mashayekhi H, Xing BS (2016) Growth and enzymatic activity of maize (*Zea mays* L.) plant: solution culture test for copper dioxide nano particles. J Plant Nutr 39 (1):99–115

Ahmed W, Jackson MJ (2016) Surgical tools and medical devices. Springer, Cham. https://doi.org/10.1007/978-3-319-33489-9

Ahmed S, Ahmad M, Swami BL, Ikram S (2016a) A review on plants extract mediated synthesis of silver nanoparticles for antimicrobial applications: a green expertise. J Adv Res 7(1):17–28

Ahmed S, Ikram AS, Yudha SS (2016b) Biosynthesis of gold nanoparticles: a green approach. J Photochem Photobiol B Biol 161:141–153

Alidoust D, Isoda A (2013) Effect of γFe_2O_3 nanoparticles on photosynthetic characteristic of soybean (*Glycine max* L. Merr.): foliar spray versus soil amendment. Acta Physiol Plant 35:3365–3375. https://doi.org/10.1007/s11738-013-1369-8

Aliofkhazraei M (2016) Handbook of nanoparticles. Springer, Cham. https://doi.org/10.1007/978-3-319-15338-4

Aliofkhazraei M, Makhlouf ASH (2016) Handbook of nanoelectrochemistry: electrochemical synthesis methods, properties, and characterization techniques. Springer, Cham. https://doi.org/10.1007/978-3-319-15266-0

Ambika S, Sundrarajan M (2016) [EMIM] BF_4 ionic liquid-mediated synthesis of TiO_2 nanoparticles using *Vitex negundo Linn* extract and its antibacterial activity. J Mol Liq. https://doi.org/10.1016/j.molliq.2016.06.079

Amenta V, Aschberger K, Arena M, Bouwmeester H, Moniz FB, Brandhoff P et al (2015) Regulatory aspects of nanotechnology in the agri/feed/food sector in EU and non-EU countries. Regul Toxicol Pharmacol 73:463–476. https://doi.org/10.1016/j.yrtph.2015.06.016

Andreotti F, Mucha AP, Caetano C, Rodrigues P, Gomes CR, Almeida CMR (2015) Interactions between salt marsh plants and Cu nanoparticles: effects on metal uptake and phytoremediation processes. Ecotoxicol Environ Saf 120:303–309

Andrievski RA, Khatchoyan AV (2016) Nanomaterials in extreme environments: fundamentals and applications, Springer series in materials science series no. 230. Springer, Cham. https://doi.org/10.1007/978-3-319-25331-2

Anjum NA, Adam V, Kizek R, Duarte AC, Pereira E, Iqbal M, Lukatkin AS, Ahmad I (2015) Nanoscale copper in the soil–plant system – toxicity and underlying potential mechanisms. Environ Res 138:306–325

Aouada FA, de Moura MR (2015) Nanotechnology applied in agriculture: controlled release of agrochemicals. In: Rai M et al (eds) Nanotechnologies in food and agriculture. Springer, Cham, pp 103–118. https://doi.org/10.1007/978-3-319-14024-7_5

Araújo R, Castro ACM, Fiúza A (2015) The use of nanoparticles in soil and water remediation processes. Mater Today Proc 2(1):315–320

Bandyopadhyay S, Plascencia-Villa G, Mukherjee A, Rico CM, José-Yacamán M, Peralta-Videa JR, Gardea-Torresdey JL (2015) Comparative phytotoxicity of ZnO NPs bulk ZnO, and ionic zinc onto the alfalfa plants symbiotically associated with Sinorhizobium meliloti in soil. Sci Total Environ 515:60–69

Barker AV, Pilbeam DJ (2015) Handbook of plant nutrition, Books in soils, plants, and the environment series, 2nd edn. CRC Press, Boca Raton

Barrios AC, Medina-Velo IA, Zuverza-Mena N, Dominguez OE, Peralta-Videa JR, Gardea-Torresdey JL (2016) Nutritional quality assessment of tomato fruits after exposure to uncoated and citric acid coated cerium oxide nanoparticles, bulk cerium oxide, cerium acetate and citric acid. Plant Physiol Biochem. https://doi.org/10.1016/j.plaphy.2016.04.017

Belal E, El-Ramady H (2016) Nanoparticles in water, soils and agriculture. In: Ranjan S et al (eds) Nanoscience in food and agriculture 2, Sustainable agriculture reviews 21. Springer, Cham. https://doi.org/10.1007/978-3-319-39306-3_10

Bennur T, Khan Z, Kshirsagar R, Javdekar V, Zinjarde S (2016) Biogenic gold nanoparticles from the Actinomycete *Gordonia amarae*: application in rapid sensing of copper ions. Sensors Actuators B Chem 233:684–690

Benoit R, Wilkinson KJ, Sauve S (2013) Partitioning of silver and chemical speciation of free Ag in soils amended with nanoparticles. Chem Cent J 7:75

Bhatt I, Tripathi BN (2011) Interaction of engineered nanoparticles with various components of the environment and possible strategies for their risk assessment. Chemosphere 82(3):308–317

Bhushan B (2016) Encyclopedia of nanotechnology. Springer, Dordrecht. https://doi.org/10.1007/978-94-007-6178-0

Bindal A, Hamedi-Hagh S (2016) Silicon nanowire transistors. Springer, Cham. https://doi.org/10.1007/978-3-319-27177-4

Bouguerra S, Gavina A, Ksibi M, da Graça Rasteiro M, Rocha-Santos T, Pereira R (2016) Ecotoxicity of titanium silicon oxide ($TiSiO_4$) nanomaterial for terrestrial plants and soil invertebrate species. Ecotoxicol Environ Saf 129:291–301

Bour A, Mouchet F, Silvestre J, Gauthier L, Pinelli E (2015) Environmentally relevant approaches to assess nanoparticles ecotoxicity: a review. J Hazard Mater 283:764–777

Brown I (2017) Climate change and soil wetness limitations for agriculture: spatial risk assessment framework with application to Scotland. Geoderma 285:173–184

Caballero-Guzman A, Nowack B (2016) A critical review of engineered nanomaterial release data: are current data useful for material flow modeling? Environ Pollut 213:502–517

Castiglione MR, Giorgetti L, Bellani L, Muccifora S, Bottega S, Spanò C (2016) Root responses to different types of TiO_2 nanoparticles and bulk counterpart in plant model system *Vicia faba* L. Environ Exp Bot 130:11–21

Chaithawiwat K, Vangnai A, McEvoy JM, Pruess B, Krajangpan S, Khan E (2016) Role of oxidative stress in inactivation of *Escherichia coli* BW25113 by nanoscale zero-valent iron. Sci Total Environ 565:857–862

Chalise S, Naranpanawa A (2016) Climate change adaptation in agriculture: a computable general equilibrium analysis of land-use change in Nepal. Land Use Policy 59:241–250
Chaudhary DR, Gautam RK, Yousuf B, Mishra A, Jha B (2015) Nutrients, microbial community structure and functional gene abundance of rhizosphere and bulk soils of halophytes. Appl Soil Ecol 91:16–26
Chen S, Chen X, Xu J (2016a) Impacts of climate change on agriculture: evidence from China. J Environ Econ Manag 76:105–124
Chen S, Han Y, Fu C, Zhang H, Zhu Y, Zuo Z (2016b) Micro and nano-size pores of clay minerals in shale reservoirs: implication for the accumulation of shale gas. Sediment Geol 342:180–190
Chhipa H, Joshi P (2016) Nanofertilisers, nanopesticides and nanosensors in agriculture. In: Ranjan S et al (eds) Nanoscience in food and agriculture 1, Sustainable agriculture reviews 20. Springer, Cham, pp 247–282. https://doi.org/10.1007/978-3-319-39303-2_9
Conway JR, Keller A (2016) Gravity-driven transport of three engineered nanomaterials in unsaturated soils and their effects on soil pH and nutrient release. Water Res 98:250–260
Cornelis G, Pang L, Doolette C, Kirby JK, McLaughlin MJ (2013) Transport of silver nanoparticles in saturated columns of natural soils. Sci Total Environ 463–464:120–130
Costa PM, Fadeel B (2016) Emerging systems biology approaches in nanotoxicology: towards a mechanism-based understanding of nanomaterial hazard and risk. Toxicol Appl Pharmacol 299:101–111
Cox A, Venkatachalam P, Sahi S, Sharma N (2016) Silver and titanium dioxide nanoparticle toxicity in plants: a review of current research. Plant Physiol Biochem 107:147–163
Da Costa MVJ, Sharma PK (2016) Effect of copper oxide nanoparticles on growth, morphology, photosynthesis, and antioxidant response in *Oryza sativa*. Photosynthetica 54(1):110–119. https://doi.org/10.1007/s11099-015-0167-5
Dai W, Kheireddin B, Gao H, Liang H (2016) Roles of nanoparticles in oil lubrication. Tribol Int 102:88–98
Danish M, Gu X, Lu S, Naqvi M (2016) Degradation of chlorinated organic solvents in aqueous percarbonate system using zeolite supported nanozero valent iron (Z-nZVI) composite. Environ Sci Pollut Res Int 23(13):13298–13307. https://doi.org/10.1007/s11356-016-6488-5.
Dasgupta N, Ranjan S, Mundra S, Ramalingam C, Kumar A (2015a) Fabrication of food grade vitamin E nanoemulsion by low energy approach, characterization and its application. Int J Food Prop 19(3):700–708
Dasgupta N, Ranjan S, Mundekkad D, Ramalingam C, Shanker R, Kumar A (2015b) Nanotechnology in agro-food: from field to plate. Food Res Int 69:381–400. https://doi.org/10.1016/j.foodres.2015.01.005
Dasgupta N, Ranjan S, Ramalingam C (2017) Applications of nanotechnology in agriculture and water quality management. Environ Chem Lett 15(4):591–605
de la Rosa G, García-Castaneda C, Vazquez-Núnez E, Alonso-Castro AJ, Basurto-Islas G, Mendoza A, Cruz-Jimenez G, Molina C (2016) Physiological and biochemical response of plants to engineered NMs: implications on future design. Plant Physiol Biochem. https://doi.org/10.1016/j.plaphy.2016.06.014
De Oliveira JL, Campos EVR, Bakshi M, Abhilash PC, Fraceto LF (2014) Application of nanotechnology for the encapsulation of botanical insecticides for sustainable agriculture: prospects and promises. Biotechnol Adv 32:1550–1561. https://doi.org/10.1016/j.biotechadv.2014.10.010
Dietzel A (2016) Microsystems for pharmatechnology: manipulation of fluids, particles, droplets, and cells. Springer, Cham. https://doi.org/10.1007/978-3-319-26920-7
Dimkpa CO, McLean JE, Britt DW, Anderson AJ (2015) Nano-CuO and interaction with nano-ZnO or soil bacterium provide evidence for the interference of nanoparticles in metal nutrition of plants. Ecotoxicology 24:119–129. https://doi.org/10.1007/s10646-014-1364-x
Ditta A, Arshad M, Ibrahim M (2015) Nanoparticles in sustainable agricultural crop production: applications and perspectives. In: Siddiqui MH et al (eds) Nanotechnology and plant sciences. Springer, Cham, pp 55–75. https://doi.org/10.1007/978-3-319-14502-0_4

Domokos-Szabolcsy E, Marton L, Sztrik A, Babka B, Prokisch J, Fari M (2012) Accumulation of red elemental selenium nanoparticles and their biological effects in *Nicotinia tabacum*. Plant Growth Regul 68:525–531

Domokos-Szabolcsy E, Alladalla NA, Alshaal T, Sztrik A, Márton L, El-Ramady H (2014) *In vitro* comparative study of two *Arundo donax* L. ecotypes' selenium tolerance. Int J Hortic Sci 20 (3–4):119–122. ISSN 1585-0404

Donaldson L (2016) Halloysite clay nanotubes hold promise. Mater Today 19(1):5–6

Dong H, Xie Y, Zeng G, Tang L, Liang J, He Q, Zhao F, Zeng Y, Wu Y (2016) The dual effects of carboxymethyl cellulose on the colloidal stability and toxicity of nanoscale zero-valent iron. Chemosphere 144:1682–1689

Du W, Tan W, Peralta-Videa JR, Gardea-Torresdey JL, Ji R, Yin Y, Guo H (2016) Interaction of metal oxide nanoparticles with higher terrestrial plants: physiological and biochemical aspects. Plant Physiol Biochem. https://doi.org/10.1016/j.plaphy.2016.04.024

Dubey A, Mailapalli DR (2016) Nanofertilisers, nanopesticides, nanosensors of pest and nanotoxicity in agriculture. In: Lichtfouse E (ed) Sustainable agriculture reviews, Sustainable agriculture reviews 19. Springer, Cham, pp 307–330. https://doi.org/10.1007/978-3-319-26777-7_7

Durán N, Nakazato G, Seabra AB (2016) Antimicrobial activity of biogenic silver nanoparticles, and silver chloride nanoparticles: an overview and comments. Appl Microbiol Biotechnol. https://doi.org/10.1007/s00253-016-7657-7

Duran-Encalada JA, Paucar-Caceres A, Bandala ER, Wright GH (2017) The impact of global climate change on water quantity and quality: a system dynamics approach to the US–Mexican transborder region. Eur J Oper Res 256(2):567–581

Durenkamp M, Pawlett M, Ritz K, Harris JA, Neal AL, McGrath SP (2016) Nanoparticles within WWTP sludges have minimal impact on leachate quality and soil microbial community structure and function. Environ Pollut 211:399–405

Dwivedi S, Saquib Q, Al-Khedhairy AA, Musarrat J (2016) Understanding the role of nanomaterials in agriculture. In: Singh DP et al (eds) Microbial inoculants in sustainable agricultural productivity. Springer, Delhi, pp 271–288. https://doi.org/10.1007/978-81-322-2644-4_17

Egorova EM, Kubatiev AA, Schvets VI (2016) Biological effects of metal nanoparticles. Springer, Cham. https://doi.org/10.1007/978-3-319-30906-4

El-Ramady HR, Abdalla NA, Alshaal TA, Elhawat N, Domokos-Szabolcsy É, Prokisch J, Sztrik A, Fári M (2014a). Nano-selenium: from *in vitro* to micro farm experiments. The international conference "*Biogeochemical Processes at Air-Soil-Water Interfaces and Environmental Protection*" for the European society for soil conservation, Imola–Ravenna, Italy 23–26 June 2014. Doi:10.13140/2.1.2260.4481

El-Ramady HR, Alshaal TA, Amer M, Domokos-Szabolcsy É, Elhawat N, Prokisch J, Fári M (2014b) Soil quality and plant nutrition. In: Ozier-Lafontaine H, Lesueur-Jannoyer M (eds) Sustainable agriculture reviews 14: agroecology and global change, Sustainable agriculture reviews 14. Springer, Cham, pp 345–447. https://doi.org/10.1007/978-3-319-06016-3_11

El-Ramady HR, Alshaal TA, Shehata SA, Domokos-Szabolcsy É, Elhawat N, Prokisch J, Fári M, Marton L (2014c) Plant nutrition: from liquid medium to micro-farm. In: Ozier-Lafontaine H, Lesueur-Jannoyer M (eds) Sustainable agriculture reviews 14 (agroecology and global change), Sustainable agriculture reviews 14. Springer, Cham, pp 449–508. https://doi.org/10.1007/978-3-319-06016-3_12

El-Ramady H, Abdalla N, Alshaal T, El-Henawy A, Faizy SE-DA, Shams MS, Shalaby T, Bayoumi Y, Elhawat N, Shehata S, Sztrik A, Prokisch J, Fári M, Pilon-Smits EA, Domokos-Szabolcsy É (2015a) Selenium and its role in higher plants. In: Lichtfouse E, Schwarzbauer J, Robert D (eds) Environmental chemistry for a sustainable world, Pollutants in buildings, water and living organisms, vol 7. Springer, Cham, pp 235–296. https://doi.org/10.1007/978-3-319-19276-5_6

El-Ramady H, Domokos-Szabolcsy É, Shalaby TA, Prokisch J, Fári M (2015b) Selenium in agriculture: water, air, soil, plants, food, animals and nanoselenium. In: Lichtfouse E (ed) Environmental chemistry for a sustainable world, CO_2 sequestration, biofuels and depollution, vol 5, pp 153–232. https://doi.org/10.1007/978-3-319-11906-9_5. Springer Berlin.

El-Ramady H, Abdalla N, Taha HS, Alshaal T, El-Henawy A, Faizy SE-DA, Shams MS, Youssef SM, Shalaby T, Bayoumi Y, Elhawat N, Shehata S, Sztrik A, Prokisch J, Fári M, Domokos-Szabolcsy É, Pilon-Smits EA, Selmar D, Haneklaus S, Schnug E (2016a) Selenium and nano-selenium in plant nutrition. Environ Chem Lett 14(1):123–147. https://doi.org/10.1007/s10311-015-0535-1

El-Ramady H, Alshaal T, Abdalla N, Prokisch J, Sztrik A, Fári M, Domokos-Szabolcsy É (2016b). Selenium and nano-selenium biofortified sprouts using micro-farm systems. In: Bañuelos, GS, Lin, Z-Q, Guilherme, LRG, dos Reis, AR (eds) Proceedings of the 4th International Conference on Selenium in the Environment and human health, Sao Paulo, Brazil, 18–21 October 2015, CRC, Taylor & Francis Group, London, UK, pp 189–190, Doi:10.13140/RG.2.1.1065.9925

Fadel TR, Steevens JA, Thomas TA, Linkov I (2015) The challenges of nanotechnology risk management. Nano Today 10:6–10. https://doi.org/10.1016/j.nantod.2014.09.008

Fajardo C, Gil-Díaz M, Costa G, Alonso J, Guerrero AM, Nande M, Lobo MC, Martín M (2015) Residual impact of aged nZVI on heavy metal-polluted soils. Sci Total Environ 535:79–84. https://doi.org/10.1016/j.scitotenv.2015.03.067

Flores-López ML, Cerqueira MA, de Rodríguez DJ, Vicente AA (2016) Perspectives on utilization of edible coatings and nano-laminate coatings for extension of postharvest storage of fruits and vegetables. Food Eng Rev 8:292–305. https://doi.org/10.1007/s12393-015-9135-x

Floris B, Galloni P, Sabuzi F, Conte V (2016) Metal systems as tools for soil remediation. Inorg Chim Acta. https://doi.org/10.1016/j.ica.2016.04.003

Ford H, Roberts A, Jones L (2016) Nitrogen and phosphorus co-limitation and grazing moderate nitrogen impacts on plant growth and nutrient cycling in sand dune grassland. Sci Total Environ 542(Part A):203–209

Fraceto LF, Grillo R, de Medeiros GA, Scognamiglio V, Rea G, Bartolucci C (2016) Nanotechnology in agriculture: which innovation potential does it have? Front Environ Sci 4:20. https://doi.org/10.3389/fenvs.2016.00020

Gardea-Torresdey JL, Rico CM, White JC (2014) Trophic transfer, transformation and impact of engineered nanomaterials in terrestrial environments. Environ Sci Technol 48:2526–2540

Garvie-Cook H (2016) Novel (Trans)dermal drug delivery strategies: micro- and nano-scale assessments. Springer, Cham. https://doi.org/10.1007/978-3-319-28901-4

Gemtos T, Fountas S, Tagarakis A, Liakos V (2013) Precision agriculture application in fruit crops: experience in handpicked fruits. Proced Technol 8:324–332

Gil-Díaz M, Diez-Pascual S, Gonzalez A, Alonso J, Rodríguez-Valdes E, Gallego JR, Lobo MC (2016a) A nanoremediation strategy for the recovery of an As-polluted soil. Chemosphere 149:137–145. https://doi.org/10.1016/j.chemosphere.2016.01.106

Gil-Díaz M, Gonzalez A, Alonso J, Lobo MC (2016b) Evaluation of the stability of a nanoremediation strategy using barley plants. J Environ Manag 165:150–158. https://doi.org/10.1016/j.jenvman.2015.09.032

Gillies G, Mackenzie K, Kopinke F-D, Georgi A (2016) Fluorescence labelling as tool for zeolite particle tracking in nanoremediation approaches. Sci Total Environ 550:820–826. https://doi.org/10.1016/j.scitotenv.2016.01.009

Gogos A, Moll J, Klingenfuss F, van der Heijden M, Irin F, Green MJ, Zenobi R, Bucheli TD (2016) Vertical transport and plant uptake of nanoparticles in a soil mesocosm experiment. J Nanobiotechnol 14:40. https://doi.org/10.1186/s12951-016-0191-z

Gomez-Garay A, Pintos B, Manzanera JA, Lobo C, Villalobos N, Martín L (2014) Uptake of CeO_2 nanoparticles and its effect on growth of *Medicago arborea In Vitro* plantlets. Biol Trace Elem Res 161:143–150. https://doi.org/10.1007/s12011-014-0089-2

Haghighi M, Abolghasemi R, Teixeira da Silva JA (2014) Low and high temperature stress affect the growth characteristics of tomato in hydroponic culture with Se and nano-Se amendment. Sci Hortic 178:231–240. https://doi.org/10.1016/j.scienta.2014.09.006

Handford CE, Dean M, Spence M, Henchion M, Elliott CT, Campbell K (2015) Awareness and attitudes towards the emerging use of nanotechnology in the agri-food sector. Food Control 57:24–34

He X, Hwang H-M (2016) Nanotechnology in food science: functionality, applicability, and safety assessment. J Food Drug Anal 24(4):671–681

He S, Feng Y, Ni J, Sun Y, Xue L, Feng Y, Yu Y, Lin X, Yang L (2016) Different responses of soil microbial metabolic activity to silver and iron oxide nanoparticles. Chemosphere 147:195–202

Henderson J, Storeygard A, Deichmann U (2017) Has climate change driven urbanization in Africa? J Dev Econ 124:60–82

Hernandez-Viezcas JA, Castillo-Michel H, Andrews JC, Cotte M, Rico CM, Peralta-Videa JR, Priester JH, Holden PA, Gardea-Torresdey JL (2013) In situ synchrotron fluorescence mapping and coordination of CeO_2 and ZnO nanoparticles in soil cultivated soybean (*Glycine max*). ACS Nano 7:1415–1423

Hobbie SE (2015) Plant species effects on nutrient cycling: revisiting litter feedbacks. Trends Ecol Evol 30(6):357–363

Homaee MB, Ehsanpour AA (2015) Physiological and biochemical responses of potato (Solanum tuberosum) to silver nanoparticles and silver nitrate treatments under in vitro conditions. Ind J Plant Physiol. (October–December 2015) 20(4):353–359. Doi:https://doi.org/10.1007/s40502-015-0188-x

Hong J, Rico CM, Zhao L, Adeleye AS, Keller AA, Peralta-Videa JR, Gardea-Torresdey JL (2015) Toxic effects of copper-based nanoparticles or compounds to lettuce (*Lactuca sativa*) and alfalfa (*Medicago sativa*). Environ Sci Proc Impacts 17(1):177–185

Hu X, Li D, Gao Y, Mu L, Zhou Q (2016) Knowledge gaps between nanotoxicological research and nanomaterial safety. Environ Int 94:8–23

Ibrahim RK, Hayyan M, AlSaadi MA, Hayyan A, Ibrahim S (2016) Environmental application of nanotechnology: air, soil, and water. Environ Sci Pollut Res Int. https://doi.org/10.1007/s11356-016-6457-z

Jacobsen BH, Hansen AL (2016) Economic gains from targeted measures related to non point pollution in agriculture based on detailed nitrate reduction maps. Sci Total Environ 556:264–275

Jain R, Jordan N, Schild D, van Hullebusch ED, Weiss S, Franzen C, Farges F, Hübner R, Lens PNL (2015) Adsorption of zinc by biogenic elemental selenium nanoparticles. Chem Eng J 260:855–863. https://doi.org/10.1016/j.cej.2014.09.057

Jain A, Ranjan S, Dasgupta N, Ramalingam C (2016a) Nanomaterials in food and agriculture: an overview on their safety concerns and regulatory issues. Crit Rev Food Sci Nutr 15:1–21

Jain R, Dominic D, Jordan N, Rene ER, Weiss S, van Hullebusch ED, Hubner R, Lens PNL (2016b) Higher Cd adsorption on biogenic elemental selenium nanoparticles. Environ Chem Lett. https://doi.org/10.1007/s10311-016-0560-8

Janardan S, Suman P, Ragul G, Anjaneyulu U, Shivendu R, Dasgupta N, Ramalingam C, Swamiappan S, Vijayakrishna K, Sivaramakrishna A (2016) Assessment on the antibacterial activity of nanosized silica derived from hypercoordinated silicon(iv) precursors. RSC Adv 6 (71):66394–66406

Jang MH, Lim M, Hwang YS (2014) Potential environmental implications of nanoscale zero-valent iron particles for environmental remediation. Environ Health Toxicol 29:e2014022

Jones DE, Ghandehari H, Facelli JC (2016) A review of the applications of data mining and machine learning for the prediction of biomedical properties ofnanoparticles. Comput Methods Prog Biomed 132:93–103

Joo SH, Zhao D (2016) Environmental dynamics of metal oxide nanoparticles in heterogeneous systems: a review. J Hazard Mater. https://doi.org/10.1016/j.jhazmat.2016.02.068

Jorio A (2016) Bioengineering applications of carbon nanostructures, Nanomedicine and nanotoxicology series. Springer, Cham. https://doi.org/10.1007/978-3-319-25907-9

Judy JD, Bertsch PM (2014) Bioavailability, toxicity and fate of manufactured nanomaterials in terrestrial ecosystems. In: Sparks (ed) Advances in agronomy, vol 123. Elsevier Inc. https://doi.org/10.1016/B978-0-12-420225-2.00001-7, San Diego, pp 1–64

Kah M (2015) Nanopesticides and nanofertilizers: emerging contaminants or opportunities for risk mitigation? Front Chem 3:64. https://doi.org/10.3389/fchem.2015.00064.eCollection2015.

Karimzadeh I, Aghazadeh M, Ganjali MR, Norouzi P, Shirvani-Arani S, Doroudi T, Kolivand PH, Marashi SA, Gharailou D (2016) A novel method for preparation of bare and poly (vinylpyrrolidone) coated superparamagnetic iron oxide nanoparticles for biomedical applications. Mater Lett 179:5–8

Keller A, McFerran S, Lazareva A, Suh S (2013) Global life cycle releases of engineered nanomaterials. J Nanopart Res 15:1–17

Keller AA, Vosti W, Wang H, Lazareva A (2014) Release of engineered nanomaterials from personal care products throughout their life cycle. J Nanopart Res 16:2489. https://doi.org/10.1007/s11051-014-2489-9

Khan I, Oh D-H (2016) Integration of nisin into nanoparticles for application in foods. Innovative Food Sci Emerg Technol 34:376–384

Khan MN, Mobin M, Abbas ZK, AlMutairi KA, Siddiqui ZH (2016) Role of nanomaterials in plants under challenging environments. Plant Physiol Biochem. https://doi.org/10.1016/j.plaphy.2016.05.038

Khanna VK (2016) Implantable medical electronics: prosthetics, drug delivery, and health monitoring. Springer, Cham. https://doi.org/10.1007/978-3-319-25448-7

Kolenc Z, Vodnik D, Mandelc S, Javornik B, Kastelec D, Čerenak A (2016) Hop (*Humulus lupulus* L.) response mechanisms in drought stress: proteomic analysis with physiology. Plant Physiol Biochem 105:67–78

Kumari R, Singh JS, Singh DP (2016) Biogenic synthesis and spatial distribution of silver nanoparticles in the legume mungbean plant (*Vigna radiata* L). Plant Physiol Biochem. https://doi.org/10.1016/j.plaphy.2016.06.001

Kuppusamy S, Palanisami T, Megharaj M, Venkateswarlu K, Naidu R (2016a) *In-Situ* remediation approaches for the management of contaminated sites: a comprehensive overview. In: de Voogt P (ed) Reviews of environmental contamination and toxicology, vol 1. Reviews of environmental contamination and toxicology, vol 236. Springer, Switzerland, pp 1–115. Doi:10.1007/978-3-319-20013-2_1

Kuppusamy P, Yusoff MM, Maniam GP, Govindan N (2016b) Biosynthesis of metallic nanoparticles using plant derivatives and their new avenues in pharmacological applications: an updated report. Saudi Pharm J 24(4):473–484

Lalau CM, De Almeida Mohedano R, Schmidt EC et al (2015) Toxicological effects of copper oxide nanoparticles on the growth rate photosynthetic pigment content, and cell morphology of the duckweed Landoltia punctata. Protoplasma 252:221–229

Lateef A, Nazir R, Jamil N, Alam S, Shah R, Khan MN, Saleem M (2016) Synthesis and characterization of zeolite based nano–composite: an environment friendly slow release fertilizer. Microporous Mesoporous Mater 232:174–183

Le Croy GE, Yang S-T, Yang F, Liu Y, Fernando KAS, Bunker CE, Hu Y, Luo PG, Sun Y-P (2016) Functionalized carbon nanoparticles: syntheses and applications in optical bioimaging and energy conversion. Coord Chem Rev 320–321:66–81

Le Van N, Ma C, Shang J, Rui Y, Liu S, Xing B (2016) Effects of CuO nanoparticles on insecticidal activity and phytotoxicity in conventional and transgenic cotton. Chemosphere 144:661–670

Lefevre E, Bossa N, Wiesner MR, Gunsch CK (2016) A review of the environmental implications of *in situ* remediation by nanoscale zero valent iron (nZVI): behavior, transport and impacts on microbial communities. Sci Total Environ 565:889–901

Li Q, Chen X, Zhuang J, Chen X (2016a) Decontaminating soil organic pollutants with manufactured nanoparticles. Environ Sci Pollut Res 23:11533–11548. https://doi.org/10.1007/s11356-016-6255-7

Li Z, Lin Z, Wang N, Wang J, Liu W, Sun K, Qing Fu Y, Wang Z (2016b) High precision NH_3 sensing using network nano-sheet Co_3O_4 arrays based sensor at room temperature. Sensors Actuators B Chem 235:222–231

Li H, Ye X, Guo X, Geng Z, Wang G (2016c) Effects of surface ligands on the uptake and transport of gold nanoparticles in rice and tomato. J Hazard Mater 314:188–196

Li H, Li H, Liu J, Luo Z, Joyce D, He S (2017) Nano-silver treatments reduced bacterial colonization and biofilm formation at the stem-ends of cut gladiolus Eerde spikes. Postharvest Biol Technol 123:102–111

Libralato G, Costa Devoti A, Zanella M, Sabbioni E, Mičetić I, Manodori L, Pigozzo A, Manenti S, Groppi F, Volpi Ghirardini A (2016) Phytotoxicity of ionic, micro- and nano-sized iron in three plant species. Ecotoxicol Environ Saf 123:81–88

Liu R, Lal R (2015) Potentials of engineered nanoparticles as fertilizers for increasing agronomic productions. Sci Total Environ 514:131–139. https://doi.org/10.1016/j.scitotenv.2015.01.104

Liu R, Zhang H, Lal R (2016) Effects of stabilized nanoparticles of copper, zinc, manganese, and iron oxides in low concentrations on Lettuce (*Lactuca sativa*) seed germination: nanotoxicants or nanonutrients? Water Air Soil Pollut 227:42. https://doi.org/10.1007/s11270-015-2738-2

Loeppmann S, Blagodatskaya E, Pausch J, Kuzyakov Y (2016) Substrate quality affects kinetics and catalytic efficiency of exo-enzymes in rhizosphere and detritusphere. Soil Biol Biochem 92:111–118

Lopez-Moreno ML, de la Rosa G, Hernandez-Viezcas JA, Castillo-Michel H, Botez CE, Peralta-Videa JR, Gardea-Torresdey JL (2010) Evidence of the differential biotransformation and genotoxicity of ZnO and CeO_2 nanoparticles on soybean (*Glycine max*) plants. Environ Sci Technol 14:7315–7320

López-Moreno ML, Avilés LL, Pérez NG, Irizarry BÁ, Perales O, Cedeno-Mattei Y, Román F (2016) Effect of cobalt ferrite ($CoFe_2O_4$) nanoparticles on the growth and development of *Lycopersicon lycopersicum* (tomato plants). Sci Total Environ 550:45–52

Luo Z, Xu Y, Ye Q (2016). Effect of nano-SiO_2-LDPE packaging on biochemical, sensory, and microbiological quality of Pacific white shrimp Penaeus vannamei during chilled storage. Fish Sci 81:983–993 Doi:https://doi.org/10.1007/s12562-015-0914-3

Ma H, William PL, Diamond SA (2013) Ecotoxicity of manufactured ZnO nanoparticles: a review. Environ Pollut 172:76–85

Ma CX, White JC, Dhankher OP, Xing B (2015) Metal-based nanotoxicity and detoxification pathways in higher plants. Environ Sci Technol 49(12):7109–7122

Ma J, Hung H, Macdonald RW (2016) The influence of global climate change on the environmental fate of persistent organic pollutants: a review with emphasis on the Northern Hemisphere and the Arctic as a receptor. Glob Planet Chang 146:89–108

Maddinedi SB, Mandal BK, Ranjan S, Dasgupta N (2015) Diastase assisted green synthesis of size-controllable gold nanoparticles. RSC Adv 5(34):26727–26733

Maddinedi SB, Mandal BK, Patil SH, Andhalkar VV, Ranjan S, Dasgupta N (2017) Diastase induced green synthesis of bilayered reduced graphene oxide and its decoration with gold nanoparticles. J Photochem Photobiol B Biol 166:252–258

Magalhães WLE, Zanoni PRS, Helm CV, Lazzarotto M, Satyanarayana KG (2017) Nanotechnology applied to improve functionality in food. In: Grumezescu AM (ed) Nutrient delivery: a volume in nanotechnology in the agri-food industry. Academic Press, Amsterdam, pp 177–219

Maheshwari B, Bristow KL (2016) Peri-urban water, agriculture and urbanization. Agric Water Manag 176:263–265

Majumdar S, Peralta-Videa JR, Trujillo-Reyes J, Sun Y, Barrios AC, Niu G, Flores- Margez JP, Gardea-Torresdey JL (2016) Soil organic matter influences cerium translocation and physiological processes in kidney bean plants exposed to cerium oxide nanoparticles. Sci Total Environ 569–570:201–211

Maliszewska I (2016) Effects of the biogenic gold nanoparticles on microbial community structure and activities. Ann Microbiol 66:785–794. https://doi.org/10.1007/s13213-015-1158-2

Mastronardi E, Tsae P, Zhang X, Monreal C, DeRosa MC (2015) Strategic role of nanotechnology in fertilizers: potential and limitations. In: Rai M et al (eds) Nanotechnologies in food and agriculture. Springer, Cham, pp 25–67. https://doi.org/10.1007/978-3-319-14024-7_2

Matraszek R, Hawrylak-Nowak B, Chwil S, Chwil M (2016) Macroelemental composition of cadmium stressed lettuce plants grown under conditions of intensive sulphur nutrition. J Environ Manag 180:24–34

Meguid SA (2016) Advances in nanocomposites: modeling, characterization and applications. Springer, Cham. https://doi.org/10.1007/978-3-319-31662-8

Mehlhorn H (2016) Nanoparticles in the fight against parasites, Parasitology research monographs series, vol 8. Springer, Cham. https://doi.org/10.1007/978-3-319-25292-6

Meredith AN, Harper B, Harper SL (2016) The influence of size on the toxicity of an encapsulated pesticide: a comparison of micron- and nano-sized capsules. Environ Int 86:68–74

Miller MA, Weissleder R (2016) Imaging the pharmacology of nanomaterials by intravital microscopy: toward understanding their biological behavior. Adv Drug Deliv Rev. https://doi.org/10.1016/j.addr.2016.05.023

Mitra GN (2015) Regulation of nutrient uptake by plants: a biochemical and molecular approach. Springer, Delhi, India. https://doi.org/10.1007/978-81-322-2334-4

Mohamed AL, El-Naggar ME, Shaheen TI, Hassabo AG (2016) Novel nano polymeric system containing biosynthesized core shell silver/silica nanoparticles for functionalization of cellulosic based material. Microsyst Technol 22:979–992. https://doi.org/10.1007/s00542-015-2776-0

Mohammadi A, Hashemi M, Hosseini SM (2016) Postharvest treatment of nanochitosan-based coating loaded with *Zataria multiflora* essential oil improves antioxidant activity and extends shelf-life of cucumber. Innovative Food Sci Emerg Technol 33:580–588

Moll J, Gogos A, Bucheli TD, Widmer F, van der Heijden MGA (2016) Effect of nanoparticles on red clover and its symbiotic microorganisms. J Nanobiotechnol 14:36. https://doi.org/10.1186/s12951-016-0188-7.

Mustafa G, Komatsu S (2016) Toxicity of heavy metals and metal-containing nanoparticles on plants. Biochim Biophy Acta Proteins Proteomics 1864(8):932–944

Nair PMG, Chung IM (2014a) A mechanistic study on the toxic effect of copper oxide nanoparticles in soybean (*Glycine max* L.) root development and lignification of root cells. Biol Trace Elem Res 162(1-3):342–352

Nair PMG, Chung IM (2014b) Impact of copper oxide nanoparticles exposure on *Arabidopsis thaliana* growth, root system development, root lignification, and molecular level changes. Environ Sci Pollut Res 21(22):12709–12722

Nair PMG, Chung IM (2015) Study on the correlation between copper oxide nanoparticles induced growth suppression and enhanced lignification in Indian mustard (*Brassica juncea* L.) Ecotoxicol Environ Safe 113:302–313

Noguera-Oviedo K, Aga DS (2016) Lessons learned from more than two decades of research on emerging contaminants in the environment. J Hazard Mater 316:242–251

Nowack B, David RM, Fissan H, Morris H, Shatkin JA, Stintz M, Zepp R, Brouwer D (2013) Potential release scenarios for carbon nanotubes used in composites. Environ Int 59:1–11

Nowack B, Mueller NC, Krug HF, Wick P (2014) How to consider engineered nanomaterials in major accident regulations? Environ Sci Eur 26:2

Nowack B, Baalousha M, Bornhoft N, Chaudhry Q, Cornelis G, Cotterill J, Gondikas A, Hassellov M, Lead JR, Mitrano DM, von der Kammer F, Wontner-Smith T (2015) Progress towards the validation of modeled environmental concentrations of engineered nanomaterials by analytical measurements. Environ Sci Nano 2:421–428

Nowack B, Bornhoft N, Ding Y, Riediker M, Jimenez AS, Sun T, van Tongeren M, Wohlleben W (2016) The flows of engineered nanomaterials from production, use, and disposal to the environment. In: Viana M (ed) Indoor and outdoor nanoparticles: determinants of release and exposure scenarios, Hdb Env Chem, vol 48. Springer, Cham, pp 209–232. https://doi.org/10.1007/698_2015_402

Oburger E, Schmidt H (2016) New methods to unravel rhizosphere processes. Trends Plant Sci 21 (3). https://doi.org/10.1016/j.tplants.2015.12.005

Olivo M, Dinish US (2016) Frontiers in biophotonics for translational medicine: in the celebration of year of light (2015), Progress in optical science and photonics series, vol 3. Springer, Singapore. https://doi.org/10.1007/978-981-287-627-0

Osmond MJ, McCall MJ (2010) Zinc oxide nanoparticles in modern sunscreens: an analysis of potential exposure and hazard. Nanotoxicology 4:15–41

Pachapur VL, Larios AD, Cledón M, Brar SK, Verma M, Surampalli RY (2016) Behavior and characterization of titanium dioxide and silver nanoparticles in soils. Sci Total Environ 563–564:933–943

Panpatte DG, Jhala YK, Shelat HN, Vyas RV (2016) Nanoparticles: the next generation technology for sustainable agriculture. In: Singh DP et al (eds) Microbial inoculants in sustainable agricultural productivity. Springer, Delhi, India, pp 289–300. https://doi.org/10.1007/978-81-322-2644-4_18

Parisi C, Vigani M, Rodríguez-Cerezo E (2015) Agricultural nanotechnologies: what are the current possibilities? Nano Today 10:124–127. https://doi.org/10.1016/j.nantod.2014.09.009

Park CM, Chu KH, Heo J, Her N, Jang M, Son A, Yoon Y (2016) Environmental behavior of engineered nanomaterials in porous media: a review. J Hazard Mater 309:133–150

Patra P, Choudhury SR, Mandal S, Basu A, Goswami A, Gogoi R, Srivastava C, Kumar R, Gopal M (2013) Effect sulfur and ZnO nanoparticles on stress physiology and plant (*Vigna radiata*) nutrition. In: Giri PK, et al (eds) Advanced nanomaterials and nanotechnology. Springer proceedings in physics, vol 143. Springer-Verlag Berlin Heidelberg, pp 301–309. Doi:10.1007/978-3-642-34216-5_31

Pawlett M, Ritz K, Dorey RA, Rocks S, Ramsden J, Harris JA (2013) The impact of zero-valent iron nanoparticles upon soil microbial communities is context dependent. Environ Sci Pollut Res Int 20:1041–1049

Peixoto AF, Fernandes AC, Pereira C, Pires J, Freire C (2016) Physicochemical characterization of organosilylated halloysite clay nanotubes. Microporous Mesoporous Mater 219:145–154

Peralta-Videa JR, Hernandez-Viezcas JA, Zhao L, Diaz BC, Ge Y, Priester JH, Holden PA, Gardea-Torresdey JL (2014) Cerium dioxide and zinc oxide nanoparticles alter the nutritional value of soil cultivated soybean plants. Plant Physiol Biochem 80:128–135

Peters RJB, Bouwmeester H, Gottardo S, Amenta V, Arena M, Brandhoff P, Marvin HJP, Mech A, Botelho Moniz F, Pesudo LQ, Rauscher H, Schoonjans R, Undas AK, Vettori MV, Weigel S, Aschberger K (2016) Nanomaterials for products and application in agriculture, feed and food. Trends Food Sci Technol. https://doi.org/10.1016/j.tifs.2016.06.008

Piperigkou Z, Karamanou K, Engin AB, Gialeli C, Docea AO, Vynios DH, Pavão MSG, Golokhvast KS, Shtilman MI, Argiris A, Shishatskaya E, Tsatsakis AM (2016) Emerging aspects of nanotoxicology in health and disease: from agriculture and food sector to cancer therapeutics. Food Chem Toxicol 91:42–57

Polis P, Mosdorf P, Karwowska E, Jastrzebska A, Olszyna A, Kunicki A, Piramidowicz R, Anders K, Jusza A (2013) Influence of Al2O3/Pr nanoparticles on soil, air and water microorganisms. In: Öchsner A et al (eds) Characterization and development of biosystems and biomaterials, Advanced structured materials, vol 29. Springer, Berlin, pp 1–8. https://doi.org/10.1007/978-3-642-31470-4_1

Priyadarshini E, Pradhan N (2016) Gold nanoparticles as efficient sensors in colorimetric detection of toxic metal ions: a review. Sensors and Actuators B Chem. https://doi.org/10.1016/j.snb.2016.06.081

Pulimi M, Subramanian S (2016) Nanomaterials for soil fertilisation and contaminant removal. In: Ranjan S et al (eds) Nanoscience in food and agriculture 1, Sustainable agriculture reviews, vol 20. Springer, Cham, pp 229–246. https://doi.org/10.1007/978-3-319-39303-2_9

Qureshi MI, Awan U, Arshad Z, Rasli AM, Zaman K, Khan F (2016) Dynamic linkages among energy consumption, air pollution, greenhouse gas emissions and agricultural production in

Pakistan: sustainable agriculture key to policy success. Nat Hazards 84:367–381. https://doi.org/10.1007/s11069-016-2423-9

Rai V, Acharya S, Dey N (2012) Implications of nanobiosensors in agriculture. J Biomat Nanobiotechnol 3:315–324

Raman CD, Kanmani S (2016) Textile dye degradation using nano zero valent iron: a review. J Environ Manag 177:341–355. https://doi.org/10.1016/j.jenvman.2016.04.034

Ranjan S, Dasgupta N, Chakraborty AR, Melvin Samuel S, Ramalingam C, Shanker R, Kumar A (2014) Nanoscience and nanotechnologies in food industries: opportunities and research trends. J Nanopart Res 16(6):1–23

Ranjan S, Dasgupta N, Sudandiradoss C, Ramalingam C, Ashutosh K (2015) A novel approach to evaluate titanium dioxide nanoparticle-protein interaction through docking: an insight into the mechanism of action. Proc Nat Acad Sci India Sect B Biol Sci 87(3):937–943

Ranjan S, Dasgupta N, Srivastava P, Ramalingam C (2016) A spectroscopic study on interaction between bovine serum albumin and titanium dioxide nanoparticle synthesized from microwave-assisted hybrid chemical approach. J Photochem Photobiol B Biol 161:472–481

Ray M, Ray A, Dash S, Mishra A, Achary KG, Nayak S, Singh S (2017) Fungal disease detection in plants: traditional assays, novel diagnostic techniques and biosensors. Biosens Bioelectron 87:708–723

Reddy PVL, Hernandez-Viezcas JA, Peralta-Videa JR, Gardea-Torresdey JL (2016) Lessons learned: are engineered nanomaterials toxic to terrestrial plants? Sci Total Environ 568:470–479

Rico CM, Morales MI, Barrios AC, McCreary AC, Hong J, Lee W, Nunez J, Peralta-Videa JR, Gardea-Torresdey JL (2013) Effect of cerium oxide nanoparticles on the quality of rice (*Oryza sativa* L.) grains. J Agric Food Chem 61:11278–11285. https://doi.org/10.1021/jf404046v

Rico CM, Lee SC, Rubenecia R, Mukherjee A, Hong J, Peralta-Videa JR, Gardea-Torresdey JL (2014) Cerium oxide nanoparticles impact yield and modify nutritional parameters in wheat (*Triticum aestivum* L.) J Agric Food Chem 62:9669–9675

Rico CM, Barrios AC, Tan W, Rubenecia R, Lee SC, Varela-Ramirez A, Peralta-Videa JR, Gardea-Torresdey JL (2015a) Physiological and biochemical response of soil-grown barley (*Hordeum vulgare* L.) to cerium oxide nanoparticles. Environ Sci Pollut Res 22:10551–10558. https://doi.org/10.1007/s11356-015-4243-y

Rico CM, Peralta-Videa JR, Gardea-Torresdey JL (2015b) Chemistry, biochemistry of nanoparticles and their role in antioxidant defense system in plants. In: Siddiqui MH, Al-Whaibi MH, Mohammed F (eds) Nanotechnology and plant sciences. Nanoparticles and their impact on plants, pp 1–17. https://doi.org/10.1007/978-3-319-14502-0_1 Chapter 1

Rizwan M, Ali S, Adrees M, Rizvi H, Zia-ur-Rehman M, Hannan F, Qayyum MF, Hafeez F, Ok YS (2016a) Cadmium stress in rice: toxic effects, tolerance mechanisms, and management: a critical review. Environ Sci Pollut Res. https://doi.org/10.1007/s11356-016-6436-4

Rizwan M, Ali S, Qayyum MF, Ok YS, Adrees M, Ibrahim M, Zia-ur-Rehman M, Farid M, Abbas F (2016b) Effect of metal and metal oxide nanoparticles on growth and physiology of globally important food crops: a critical review. J Hazard Mater. https://doi.org/10.1016/j.jhazmat.2016.05.061

Rodrigues SM, Trindade T, Duarte AC, Pereira E, Koopmans GF, Römkens PFAM (2016) A framework to measure the availability of engineered nanoparticles in soils: trends in soil tests and analytical tools. Trends Anal Chem 75:129–140

Saccà ML, Fajardo C, Costa G, Lobo C, Nande M, Martin M (2014) Integrating classical and molecular approaches to evaluate the impact of nanosized zero-valent iron (nZVI) on soil organisms. Chemosphere 104:184–189

Sadik OA, Du N, Kariuki V, Okello V, Bushlyar V (2014) Current and emerging technologies for the characterization of nanomaterials. ACS Sustain Chem Eng 2:1707–1716. https://doi.org/10.1021/sc500175v

Samiei M, Farjami A, Dizaj SM, Lotfipour F (2016) Nanoparticles for antimicrobial purposes in Endodontics: a systematic review of in vitro studies. Mater Sci Eng C 58:1269–1278

Sarkar P, Irshaan S, Sivapratha S, Choudhary R (2016) Nanotechnology in food processing and packaging. In: Ranjan S et al (eds) Nanoscience in food and agriculture 1, Sustainable agriculture reviews, vol 20. Springer, Cham, pp 185–227. https://doi.org/10.1007/978-3-319-39303-2_9

Sarmast MK, Salehi H (2016) Silver nanoparticles: an influential element in plant nanobiotechnology. Mol Biotechnol 58:441–449. https://doi.org/10.1007/s12033-016-9943-0

Schwabe F, Schulin R, Limbach LK, Stark W, Bürge D, Nowack B (2013) Influence of two types of organic matter on interaction of CeO_2 nanoparticles with plants in hydroponic culture. Chemosphere 91(4):512–520

Servin AD, White JC (2016) Nanotechnology in agriculture: next steps for understanding engineered nanoparticle exposure and risk. NanoImpact 1:9–12. https://doi.org/10.1016/j.impact.2015.12.002

Shalaby T, Bayoumi Y, Abdalla N, Taha H, Alshaal T, Shehata S, Amer M, Domokos-Szabolcsy É, El-Ramady H (2016) Nanoparticles, soils, plants and sustainable agriculture. In: Ranjan S et al (eds) Nanoscience in food and agriculture 1, Sustainable agriculture reviews, vol 20. Springer, Cham. https://doi.org/10.1007/978-3-319-39303-2_10

Shao J, He Y, Zhang H, Chen A, Lei M, Chen J, Peng L, Gu J-D (2016) Silica fertilization and nano-MnO_2 amendment on bacterial community composition in high arsenic paddy soils. Appl Microbiol Biotechnol 100:2429–2437. https://doi.org/10.1007/s00253-015-7131-y

Shaw R, Lark RM, Williams AP, Chadwick DR, Jones DL (2016) Characterising the within-field scale spatial variation of nitrogen in a grassland soil to inform the efficient design of in-situ nitrogen sensor networks for precision agriculture. Agric Ecosyst Environ 230:294–306

Shen Z, Chen Z, Hou Z, Li T, Lu X (2015) Ecotoxicological effect of zinc oxide nanoparticles on soil microorganisms. Front Environ Sci Eng 9(5):912–918. https://doi.org/10.1007/s11783-015-0789-7

Shoults-Wilson W, Zhurbich O, McNear D, Tsyusko O, Bertsch P, Unrine J (2011) Evidence for avoidance of Ag nanoparticles by earthworms (*Eisenia fetida*). Ecotoxicology 20:385–396

Shukla A, Dasgupta N, Ranjan S, Singh S, Chidambram R (2017) Nanotechnology towards prevention of anaemia and osteoporosis: from concept to market. Biotechnol Biotechnol Equip 31(5):863–879

Singh A, Gaharwar AK (2016) Microscale technologies for cell engineering. Springer, Cham. https://doi.org/10.1007/978-3-319-20726-1

Singh J, Lee B-K (2016) Influence of nano-TiO_2 particles on the bioaccumulation of Cd in soybean plants (*Glycine max*): a possible mechanism for the removal of Cd from the contaminated soil. J Environ Manag 170:88–96

Singh P, Kim Y-J, Zhang D, Yang D-C (2016) Biological synthesis of nanoparticles from plants and microorganisms. Trends Biotechnol 34(7):588–599

Siripireddy B, Mandal BK, Ranjan S, Dasgupta N, Ramalingam C (2017) Nano-zirconia – Evaluation of its antioxidant and anticancer activity. J Photochem Photobiol B Biol 170:125–133

Skalickova S, Milosavljevic V, Cihalova K, Horky P, Richtera L, Adam V (2016) Perspective of selenium nanoparticles as a nutrition supplement. Nutrition. https://doi.org/10.1016/j.nut.2016.05.001

Sodipo BK, Abdul Aziz A (2016) Recent advances in synthesis and surface modification of superparamagnetic iron oxide nanoparticles with silica. J Magn Magn Mater 416:275–291

Sogvar OB, Saba MK, Emamifar A, Hallaj R (2016) Influence of nano-ZnO on microbial growth, bioactive content and postharvest quality of strawberries during storage. Innovative Food Sci Emerg Technol 35:168–176

Solanki P, Bhargava A, Chhipa H, Jain N, Panwar J (2015) Nano-fertilizers and their smart delivery system. In: Rai M et al (eds) Nanotechnologies in food and agriculture. Springer, Cham, pp 81–101. https://doi.org/10.1007/978-3-319-14024-7_4

Song U, Lee S (2016) Phytotoxicity and accumulation of zinc oxide nanoparticles on the aquatic plants *Hydrilla verticillata* and *Phragmites Australis*: leaf-type-dependent responses. Environ Sci Pollut Res 23:8539–8545. https://doi.org/10.1007/s11356-015-5982-5

Song H, Yuan W, Jin P, Wang W, Wang X, Yang L, Zhang Y (2016) Effects of chitosan/nano-silica on postharvest quality and antioxidant capacity of loquat fruit during cold storage. Postharvest Biol Technol 119:41–48

Souza VGL, Fernando AL (2016) Nanoparticles in food packaging: biodegradability and potential migration to food: a review. Food Packag Shelf Life 8:63–70. https://doi.org/10.1016/j.fpsl.2016.04.001

Sri Sindhura K, Prasad TNVKV, Panner Selvam P, Hussain OM (2014) Synthesis, characterization and evaluation of effect of phytogenic zinc nanoparticles on soil exo-enzymes. Appl Nanosci 4:819–827. https://doi.org/10.1007/s13204-013-0263-4

Stefaniuk M, Oleszczuk P, Ok YS (2016) Review on nano zerovalent iron (nZVI): from synthesis to environmental applications. Chem Eng J 287:618–632. https://doi.org/10.1016/j.cej.2015.11.046

Steinhoff G (2016) Regenerative medicine – from protocol to patient: 3. Tissue engineering, biomaterials and nanotechnology. Springer, Cham. https://doi.org/10.1007/978-3-319-28274-9

Subramanian KS, Manikandan A, Thirunavukkarasu M, Rahale CS (2015) Nano-fertilizers for balanced crop nutrition. In: Rai M et al (eds) Nanotechnologies in food and agriculture. Springer, Cham, pp 69–80. https://doi.org/10.1007/978-3-319-14024-7_3

Subramanian KS, Muniraj I, Uthandi S (2016) Role of actinomycete-mediated nanosystem in agriculture. In: Subramaniam G et al (eds) Plant growth promoting actinobacteria. Springer, Singapore, pp 233–247. https://doi.org/10.1007/978-981-10-0707-1_15

Suriyaprabha R, Karunakaran G, Yuvakkumar R, Prabu P, Rajendran V, Kannan N (2012) Growth and physiological responses of maize (*Zea mays* L.) to porous silica nanoparticles in soil. J Nanopart Res 14:1294. https://doi.org/10.1007/s11051-012-1294-6

Szoboszlay M, Lambers J, Chappell J, Kupper JV, Moe LA, McNear Jr. DH (2015) Comparison of root system architecture and rhizosphere microbial communities of Balsas teosinte and domesticated corn cultivars. Soil Biol Biochem 80:34–44

Tammina SK, Mandal BK, Ranjan S, Dasgupta N (2017) Cytotoxicity study of Piper nigrum seed mediated synthesized SnO2 nanoparticles towards colorectal (HCT116) and lung cancer (A549) cell lines. J Photochem Photobiol B Biol 166:158–168

Tarafdar JC, Sharma S, Raliya R (2013) Nanotechnology: interdisciplinary science of applications. Afr J Biotechnol 12(3):219–226

Thio BJR, Montes MO, Mahmoud MA, Lee DW, Zhou D, Keller AA (2012) Mobility of capped silver nanoparticles under environmentally relevant conditions. Environ Sci Technol 46:6985–6991

Thul ST, Sarangi BK (2015) Implications of nanotechnology on plant productivity and its rhizospheric environment. In: Siddiqui MH et al (eds) Nanotechnology and plant sciences. Springer, Cham, pp 37–53. https://doi.org/10.1007/978-3-319-14502-0_3

Tolaymat T, Abdelraheem W, El Badawy A, Dionysiou D, Genaidy A (2016) The path towards healthier societies, environments, and economies: a broader perspective for sustainable engineered nanomaterials. Clean Technol Environ Policy. https://doi.org/10.1007/s10098-016-1146-7

Tripathi DK, Singh VP, Prasad SM, Chauhan DK, Dubey NK (2015) Silicon nanoparticles (SiNp) alleviate chromium (VI) phytotoxicity in *Pisum sativum* (L.) seedlings. Plant Physiol Biochem 96:189–198

Tripathi DK, Singh S, Singh VP, Prasad SM, Dubey NK, Chauhan DK (2016) Silicon nanoparticles more effectively alleviated UV-B stress than silicon in wheat (*Triticum aestivum*) seedlings. Plant Physiol Biochem. https://doi.org/10.1016/j.plaphy.2016.06.026

Troester M, Brauch H-J, Hofmann T (2016) Vulnerability of drinking water supplies to engineered nanoparticles. Water Res 96:255–279

UNEP (2007) Emerging challenges-nanotechnology and the environment. In: GEO year book. United Nations Environment Programme Division of Early Warning and Assessment, Nairobi, pp 61–70

Van Koetsem F, Van Havere L, Du Laing G (2016) Impact of carboxymethyl cellulose coating on iron sulphide nanoparticles stability, transport, and mobilization potential of trace metals present in soils and sediment. J Environ Manag 168:210–218

Vélez MA, Perotti MC, Santiago L, Gennaro AM, Hynes E (2017) Bioactive compounds delivery using nanotechnology: design and applications in dairy food. In: Grumezescu AM (ed) Nutrient delivery: a volume in nanotechnology in the agri-food industry. Academic Press, Amsterdam, pp 221–250

Villoria NB, Elliott J, Müller C, Shin J, Zhao L, Song C (2016) Rapid aggregation of global gridded crop model outputs to facilitate cross-disciplinary analysis of climate change impacts in agriculture. Environ Model Softw 75:193–201

Vítková M, Rákosová S, Michálková Z, Komárek M (2016) Metal(loid)s behaviour in soils amended with nano zero-valent iron as a function of pH and time. J Environ Manag. https://doi.org/10.1016/j.jenvman.2016.06.003

Waalewijn-Kool PL, Rupp S, Lofts S, Svendsen C, van Gestel CAM (2014) Effect of soil organic matter content and pH on the toxicity of ZnO nanoparticles to *Folsomia candida*. Ecotoxicol Environ Saf 108:9–15

Walia N, Dasgupta N, Ranjan S, Chen L, Ramalingam C (2017) Fish oil based vitamin D nanoencapsulation by ultrasonication and bioaccessibility analysis in simulated gastro-intestinal tract. Ultrason Sonochem 39:623–635

Wang Q, Ebbs SD, Chen Y, Ma X (2013) Trans-generational impact of cerium oxide nanoparticles on tomato plants. Metallomics 5:753–759

Wang S, Wang F, Gao S (2015) Foliar application with nano-silicon alleviates Cd toxicity in rice seedlings. Environ Sci Pollut Res 22:2837–2845. https://doi.org/10.1007/s11356-014-3525-0

Wang J, Fang Z, Cheng W, Tsang PE, Zhao D (2016a) Ageing decreases the phytotoxicity of zero-valent iron nanoparticles in soil cultivated with *Oryza sativa*. Ecotoxicology. https://doi.org/10.1007/s10646-016-1674-2

Wang F, Liu X, Shi Z, Tong R, Adams CA, Shi X (2016b) Arbuscular mycorrhizae alleviate negative effects of zinc oxide nanoparticle and zinc accumulation in maize plants—a soil microcosm experiment. Chemosphere 147:88–97. https://doi.org/10.1016/j.chemosphere.2015.12.076

Wang S, Wang F, Gao S, Wang X (2016c) The role of nano-silicon and its foliar application on heavy metals accumulation in different rice cultivars. Water Air Soil Pollut 227:228. https://doi.org/10.1007/s11270-016-2928-6

Watson J-L, Fang T, Dimkpa CO, Britt DW, McLean JE, Jacobson A, Anderson AJ (2015) The phytotoxicity of ZnO nanoparticles on wheat varies with soil properties. Biometals 28:101–112. https://doi.org/10.1007/s10534-014-9806-8

Wigger H, Hackmann S, Zimmermann T, Köser J, Thöming J, Gleich A v (2015) Influences of use activities and waste management on environmental releases of engineered nanomaterials. Sci Total Environ 535:160–171

Xie Y, Cheng W, Tsang PE, Fang Z (2016) Remediation and phytotoxicity of decabromodiphenyl ether contaminated soil by zero valent iron nanoparticles immobilized in mesoporous silica microspheres. J Environ Manag 166:478–483

Xie Y, Dong H, Zeng G, Tang L, Jiang Z, Zhang C, Deng J, Zhang L, Zhang Y (2017) The interactions between nanoscale zero-valent iron and microbes in the subsurface environment: a review. J Hazard Mater 321:390–407

Xiong SJ, Xu WH, Xie WW, Chen R, Chen YQ, Chi SL, Chen X, Zhang JZ, Xiong ZT, Wang ZY, Xie DT (2015) Effect of nano zeolite on chemical fractions of Cd in soil and its uptake by cabbage. Huan Jing Ke Xue 36(12):4630–4641. (Article in Chinese)

Xu C, Peng C, Sun L, Zhang S, Huang H, Chen Y, Shi J (2015) Distinctive effects of TiO_2 and CuO nanoparticles on soil microbes and their community structures in flooded paddysoil. Soil Biol Biochem 86:24–23

Yang Y-F, Cheng Y-H, Liao C-M (2016) *In situ* remediation-released zero-valent iron nanoparticles impair soil ecosystems health: a *C. elegans* biomarker-based risk assessment. J Hazard Mater 317:210–220

Yausheva E, Sizova E, Lebedev S, Skalny A, Miroshnikov S, Plotnikov A, Khlopko Y, Gogoleva N, Cherkasov S (2016) Influence of zinc nanoparticles on survival of worms *Eisenia fetida* and taxonomic diversity of the gut microflora. Environ Sci Pollut Res. https://doi.org/10.1007/s11356-016-6474-y

Yirsaw BD, Megharaj M, Chen Z, Naidu R (2016) Environmental application and ecological significance of nano-zero valent iron. J Environ Sci 44:88–98

Yuvaraj M, Subramanian KS (2015) Controlled-release fertilizer of zinc encapsulated by a manganese hollow core shell. Soil Sci Plant Nutr 61(2):319–326. https://doi.org/10.1080/00380768.2014.979327

Zarschler K, Rocks L, Licciardello N, Boselli L, Polo E, Garcia KP, De Cola L, Stephan H, Dawson KA (2016) Ultrasmall inorganic nanoparticles: State-of-the-art and perspectives for biomedical applications. Nanomedicine 12(6):1663–1701

Zhang X-S (2016) Micro/nano integrated fabrication technology and its applications in microenergy harvesting. Springer, Berlin. https://doi.org/10.1007/978-3-662-48816-4

Zhang M, Naik RR, Dai L (2016) Carbon nanomaterials for biomedical applications, Springer series in biomaterials science and engineering series, vol 5. Springer, Cham. https://doi.org/10.1007/978-3-319-22861-7

Zhang P, Ma Y, Zhang Z (2016a) Interactions between engineered nanomaterials and plants: phytotoxicity, uptake, translocation, and biotransformation. In: Siddiqui MH et al (eds) Nanotechnology and plant sciences. Springer, Cham, pp 77–99. https://doi.org/10.1007/978-3-319-14502-0_5

Zhang W, Musante C, White JC, Schwab P, Wang Q, Ebbs SD, Ma X (2016b) Bioavailability of cerium oxide nanoparticles to *Raphanus sativus* L. in two soils. Plant Physiol Biochem. https://doi.org/10.1016/j.plaphy.2015.12.013

Zhao L, Peralta-Videa JR, Rico CM, Hernandez-Viezcas JA, Sun Y, Niu G, Servin A, Nunez JE, Duarte-Gardea M, Gardea-Torresdey JL (2014) CeO_2 and ZnO nanoparticles change the nutritional qualities of cucumber (*Cucumis sativus*). J Agric Food Chem 62:2752–2759. https://doi.org/10.1021/jf405476u

Zhao X, Liu W, Cai Z, Han B, Qian T, Zhao D (2016) An overview of preparation and applications of stabilized zero-valent iron nanoparticles for soil and groundwater remediation. Water Res 100:245–266. https://doi.org/10.1016/j.watres.2016.05.019

Zuverza-Mena N, Martínez-Fernandez D, Du W, Hernandez-Viezcas JA, Bonilla-Bird N, Lopez-Moreno ML, Komarek M, Peralta-Videa JR, Gardea-Torresdey JL (2016) Exposure of engineered nanomaterials to plants: insights into the physiological and biochemical responses-a review. Plant Physiol Biochem. https://doi.org/10.1016/j.plaphy.2016.05.037

Chapter 5
Toxicological Impact of Carbon Nanomaterials on Plants

Prakash M. Gopalakrishnan Nair

Abstract The fast growth of nanotechnology has resulted in the production and use of engineered nanoparticles with unique physical and chemical properties in various fields. The increased utilization of engineered nanoparticles enhances the risks associated with their release into the environment. The smaller size and modified physico-chemical properties raise concerns about their entry and adverse effects in plants. For instance, studies have shown that nanomaterials can be absorbed and translocated within plants. Since plants represent a major component of the ecosystem, the accumulation of engineered nanoparticles in plants is a threat to plants and the food chain.

This chapter reviews phytotoxic effect of carbon nanomaterials under *in vitro* and in vivo exposure conditions. Carbon nanomaterials are widely incorporated in commercial products used in agriculture. Recent studies have been conducted to test the toxic effects of carbon nanomaterials either alone or in combination with other chemicals in plants. Results reveal that the effect of carbon nanomaterials in plants are intricate and challenging and vary between different plant species, type of the nanomaterial and concentrations tested. Carbon nanomaterials were evidenced to penetrate through seed coats, enter into the plant cells and translocate into different plant parts. Exposure to carbon nanomaterials decreases seed germination, root growth and changes the roots architecture. Carbon nanomaterials inhibits seedling growth and changes morphological, physiological, biochemical, molecular, nutritional and genetic levels in plants. Modulation in the expression of genes related to cell division and plant development were also reported.

Keywords Plants • Engineered nanoparticles • Carbon nanomaterials • Phytotoxicity

---

P.M. Gopalakrishnan Nair (✉)
Department of Applied Bioscience, College of Life and Environmental Sciences, Konkuk University, Seoul, South Korea
e-mail: pmgnair@gmail.com

K M Gothandam et al. (eds.), *Nanotechnology, Food Security and Water Treatment*, Environmental Chemistry for a Sustainable World 11,
https://doi.org/10.1007/978-3-319-70166-0_5

Contents

5.1 Introduction

The beginning of the "nano-era" in early twentieth century revolutionized the nanotechnology industry with the production of a variety of engineered nanoparticles (Roco 2003). The engineered nanoparticles are broadly classified into four categories based on their composition (1) carbon-based materials (2) metal-based nanoparticles (3) dendrimers and (4) bio-inorganic complexes having one (nanolayers), two (nanowires and nanotubes) or three dimensions (quantum dots, metal nanoparticles and fullerenes) on the nanoscale (US EPA 2007). The unique physico-chemical characteristics of engineered nanoparticles vary based on their shape, surface composition and size and are used as free nanoparticles or incorporated in to different products. The engineered nanoparticles are used in several applications viz. biomedical imaging for diagnosis, drug and gene delivery, pharmaceuticals, cosmetics, fuel additives and electronics.

The engineered nanoparticles enter into the environment through emission from manufacturing processes or through various anthropogenic activities. Waste discharges from waste water treatment plants and the application of sewage sludge in agricultural fields results in the deposition of engineered nanoparticles in agricultural soils (Gottschalk et al. 2009; Westerhoff et al. 2011). According to the available datas 50, 55 and 100% of treated wastewater sludge or bio-solids are used for agriculture use in Australia, USA and Switzerland respectively (Gottschalk et al. 2009). As per the reports, soils, sediments and landfills are the sinks of approximately 80% of the carbon nanomaterials released to the environment (Keller et al. 2013). This has raised concerns regarding the high risk of carbon nanomaterials to agricultural regions (Keller and Lazareva 2014).

Nanotechnology offers many uses in agriculture and due to their ever increasing use, the potential risks remain unclear. In agriculture, the engineered nanoparticles are incorporated in nano-agrochemicals (nano-pesticides and nano-fertilizers), nano-biosensors and nano-biocomposites (Adhikari et al. 2012; Gopal et al. 2012; Sekhon 2014). Apart from the direct risks to plants, since plants belongs to the first trophic level of the terrestrial food chain, exposure to engineered nanoparticles will have significant implications on human health and environment (Klaine et al. 2008;

Ma et al. 2010; Rico et al. 2011). Therefore, it has become essential to conduct risk assessments which will be useful for the safer use of nanotechnology (Tolaymat et al. 2015). The initial studies on the effect of engineered nanoparticles on germination and seedling growth in plants was reported in soybean (Lin and Xing 2007). Later on, several morphological, physiological and genetical changes have been reported in plants exposed to different types of engineered nanoparticles (Reddy et al. 2016). The impact and efficacy of engineered nanoparticles depends on their composition, physico-chemical properties, size, concentration as well as plant species (Ma et al. 2010; Khodakovskaya et al. 2012; Dasgupta et al. 2015, 2016, 2017; Shukla et al. 2017; Walia et al. 2017; Siripireddy et al. 2017; Maddinedi et al. 2015, 2017; Tammina et al. 2017; Ranjan and Ramalingam 2016; Dasgupta and Ramalingam 2016; Sannapaneni et al. 2016; Ranjan et al. 2014, 2016; Jain et al. 2016). This review covers the effects of carbon nanomaterials on *in vitro* cultures, seed germination, growth, mixture toxicity with other chemicals and the possible mechanisms of toxicity.

5.2 Carbon Nanomaterials and Their Applications

The detection of carbon-based nanomaterials started with the identification of "buckminsterfullerene" or the "buckyball" (Kroto et al. 1985), followed by carbon nanotubes (fullerene derivatives) (Lijima 1991) and graphene (Klaine et al. 2008). Presently due to their special optical, mechanical, electrical and thermal properties, carbon nanomaterials are increasingly used in several applications (Hurt et al. 2006; Srivastava et al. 2015). The carbon nanomaterials contains materials of diverse structure and size such as fullerenes, nano-onions, nano-cones, nano-horns, carbon dots, carbon nanotubes, nano-beads, nano-fibers, nano-diamonds and graphene (Sharon et al. 2010; Chai et al. 2013).

Fullerenes are hollow spheres (cage like structure) with a hexagonal network of carbon atoms (Chichiriccò and Poma 2015). For example, the fullerene C_{60} consists of 60 carbon atoms located at the vertices of twenty hexagons and twelve pentagons (Yadav and Kumar 2008). The fullerenes are also produced as higher mass with different geometric structures, such as, C_{70}, C_{76}, C_{78} and C_{80} (Kikuchi et al. 1992) and are extensively used in lubricants, electronics, cosmetics, fuel cells and in dietary supplements (Loutfy et al. 2002).

The carbon nanotubes are cylindrical structures with open or closed ends and are mainly categorized into single walled carbon nanotubes (having an outer diameter of 0.8–2 nm), and multi-walled carbon nanotubes (having outer diameter of 5–20 nm) depending on the number of rolled graphene layers (De Volder et al. 2013). Carbon nanotubes are used in electronic devices, paper batteries, cables and wires, field emission devices, transistors, electrical circuits, composite materials, absorbent for pollutant removal from water and in biomedical imaging (De Heer et al. 1995; Yao et al. 1999; Fuhrer et al. 2000; Rueckes et al. 2000; Franklin et al. 2001; Kemp et al. 2003; Jerosz et al. 2011; Islam et al. 2015).

Graphene is one atom thick planar sheet of sp^2-bonded carbon atoms packed in a hexagonal honeycomb crystal lattice, having two dimensions (Georgakilas et al.

2015). Graphene has several applications in electronics, biochemical sensors and in solar cells (Choi et al. 2010). Grapheme oxide is obtained by oxidation of graphite and has several applications in biomedical fields, such as drug delivery, cancer photothermal therapy, tissue engineering, bio-sensing and biological imaging (Sun et al. 2008; Zhang et al. 2011a,b; Sheng et al. 2013).

The carbon nanomaterials are utilized in various environmental applications such as in solar cells, for the production of renewable energy, soil remediation, contaminant degradation and in the detection as sensors for pollutants (Mauter and Elimelech 2008; Rasool and Lee 2015). In agriculture, carbon based nanomaterials contributes to approximately 40% of the total engineered nanoparticles used and are mainly used either as additives or as active components (Gogos et al. 2012). For example, fungicides encapsulated in multiwalled carbon nanotubes were more toxic to *Alternaria alternate* compared to bulk pesticides which were not capsulated (Sarlak et al. 2014). In the case of fertilizer application, for slow and efficient release, encapsulation with graphene oxide films was found to be effective (Zhang et al. 2014). For example, Zhang et al. (2014) reported that encapsulation of potassium nitrate in graphene oxide prolonged the release into the soil thereby making the availability of potassium nitrate more efficiently to the plants. Carbon nanomaterials could be used as additives for the development of efficient fungicides due to antifungal properties (Wang et al. 2014). In nano-biotechnology areas, the ability of carbon nanomaterials to penetrate and enter into cells could be used for the purpose of delivery of DNA molecules (Liu et al. 2009a, b; Burlaka et al. 2015). The various applications of carbon nanomaterials in different areas are shown in Fig. 5.1.

5.3 Phytotoxic Effect of Carbon Nanomaterials on Plants

The effects of carbon nanomaterials in plants were mainly studied on their effect on seed germination, plant growth and development. Only few studies have been reported on the toxicity of carbon nanomaterials in combination with other chemicals and on the mechanism of toxicity in plants. According to the available reports, the responses of carbon nanomaterials in plants varied based on the plants species, concentration tested and the stage of development. The reported effects of carbon nanomarials in plants are summarized in Fig. 5.2.

5.3.1 Penetration of Carbon Nanomaterials in Plants

The ability of different types of carbon nanomaterials to penetrate in to the plants has been reported from several studies. Lin et al. (2009a, b) investigated the uptake and translocation of carbon nanomaterials in rice plants (*Oryza sativa* L.). They found that fullerene C_{70} could be easily taken up by roots and transported to shoots.

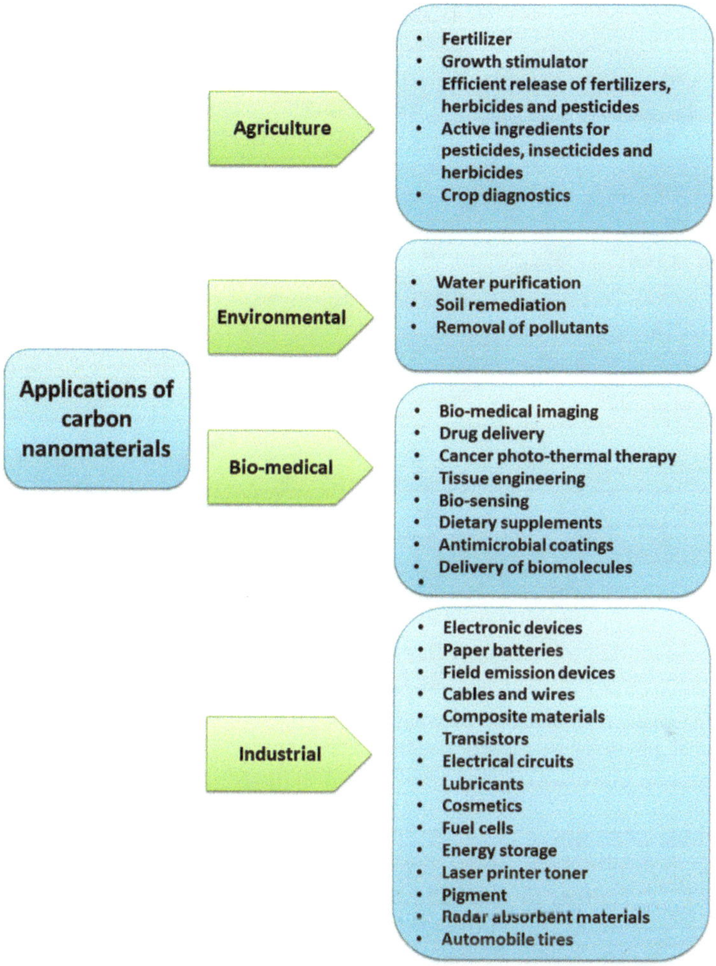

Fig. 5.1 The application of carbon nanomaterials for various uses in agriculture, environmental, bio-medical and industrial areas

It was also observed that C_{70} could be potentially transported downward from leaves to roots through phloem if C_{70} entered into plants through plant leaves. According to reports, the main pathway for the uptake of C_{60} fullerene to the plant from the soil is through the roots. Avanasi et al. (2014) reported that the highest accumulation of fullerene C_{60} (40–47%) occurred in the roots, followed by tuber (22–23%), stem (12–16%) and leaves (18–22%). Husen and Siddiqi (2014) observed that small sized carbon nanomaterials assimilated to the spaced areas after being transported through the capillary system and larger ones accumulated in the narrow passages thereby blocking the nutrient flow.

Wang et al. (2016) investigated the bio-accumulation of fullerenol (water-soluble derivative of fullerene carbon nanomaterial) nanoparticles in wheat using

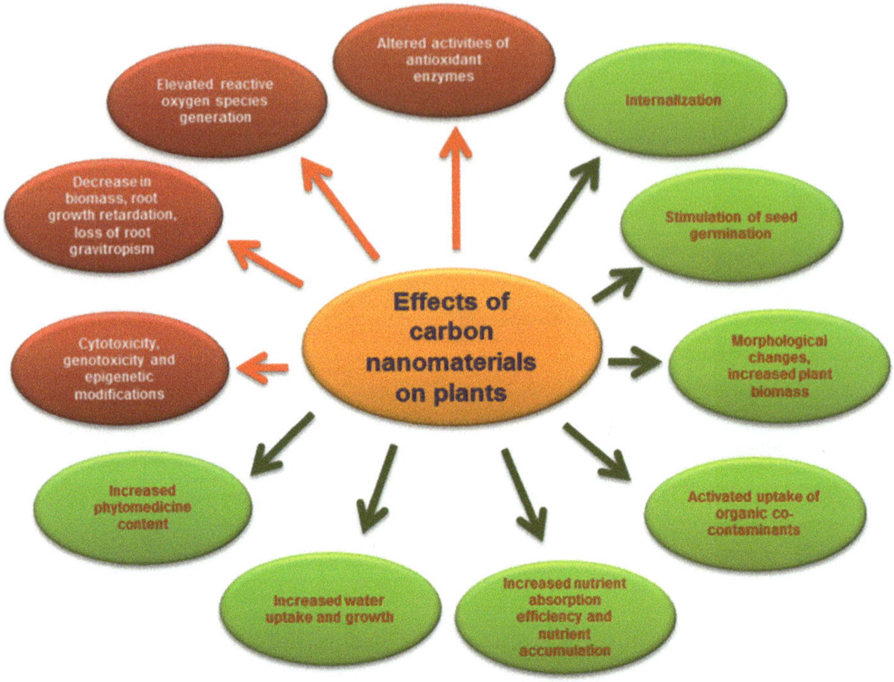

Fig. 5.2 Schematic representation showing various effects of carbon nanomaterials in plants at morphological, physiological, genetical and molecular levels

^{13}C-labelling techniques. The maximum bio-accumulation was observed in roots. Prolonged exposure with lower concentrations of fullerenol showed significant increase of ^{13}C content in roots and higher concentrations (10 μg/mL) suppressed the bio-accumulation. Only limited translocation of fullerenol from roots to stems and leaves was observed. Scanning electron microscopy analysis confirmed the presence of fullerenol nanoparticles in roots, with smaller particles being found in the vascular cylinder area of roots.

Samaj et al. (2004) demonstrated that single walled carbon nanotubes of length less than 500 nm labeled with fluorescein isothiocyanate penetrated the cell wall of the living plants by endocytosis. The results showed that though fluorescein isothiocyanate alone is not easily taken up by the plants, both of them jointly facilitated the absorption and penetration of carbon nanomaterials. In another study, Liu et al. (2009a, b) demonstrated the cellular uptake of both single walled carbon nanotubes and fluorescein isothiocyanate and single walled carbon nanotubes and DNA conjugates, demonstrating that single walled carbon nanotubes hold great promise as nano-transporters for walled plant cells. The penetration of chemically shortened single walled carbon nanotubes into the cell wall and cell membrane of tobacco (*Nicotiana tabacum*) and periwinkle (*Catharanthus roseus*) has also been reported (Liu et al. 2009a, b; Serag et al. 2011, 2012a, b, c).

Smirnova et al. (2011) demonstrated the presence of multi walled carbon nanotubes from Taunit inside the cells and tissues of seedling roots and leaves. The results showed the ability of multi walled carbon nanotubes to penetrate and accumulate inside the roots as well as their ability to be transported to the seedling leaves.

Ghosh et al. (2011) reported the internalization of multi walled carbon nanotubes inside the plant cells resulting in chromosomal aberrations, DNA fragmentation and apoptosis in Allium root cells. Wang et al. (2012a, b) revealed that wheat seedlings exposed to o- multi walled carbon nanotubes penetrated the cell wall and entered the cytoplasm through the roots. Larue et al. (2012) observed the uptake and translocation of ^{14}C-radio labeled multi walled carbon nanotubes through the roots to the leaves in wheat and rape seed. Lahiani et al. (2013) reported the penetration and presence of multi walled carbon nanotubes to the seed coats of barley, soybean and corn. Yan et al. (2016) reported that exposure to multi walled carbon nanotubes in rice resulted in the penetration to the cell walls of roots of rice seedlings.

Zhai et al. (2015) studied the vegetative uptake of differentially charged multi walled carbon nanotubes viz. neutral pristine multi walled carbon nanotubes, positively charged multi walled carbon nanotubes-NH_2 and negatively charged multi walled carbon nanotubes-COOH in model food chain plants viz. maize (*Zea mays*) and soybean (*Glycine max*). The aim of the study was to find the effect of cellular, charge and size selectivity on the uptake differentially charged multi walled carbon nanotubes on food crops. They observed that the multi walled carbon nanotubes were directly taken-up through the roots and translocated to stems and leaves of maize and soybean plants. Accumulation of multi walled carbon nanotubes was observed in the xylem and phloem cells and also in the cytoplasm, cell wall, cell membrane, chloroplast and mitochondria using transmission electron microscope studies. Overall the study showed that the uptake and translocation of different types of multi walled carbon nanotubes were based on the cellular, charge and size in maize and soybean.

The post uptake behavior of carbon nanotubes inside the plant cells was studied by Serag et al. (2012a, b, c). It was observed that the plant cells differentiating into tracheary elements incorporated the cup-stacked carbon nanotubes into cell structure via. oxidative cross-linking of monolignols to the nanotubes surface during lignin biosynthesis. In a later study, Serag et al. (2013) reported that the diameter and length of single walled carbon nanotubes are the major restraining features for their effective penetration into the plant cell wall. Zhao et al. (2015) reported the accumulation of grapheme oxide in root hair and root parenchyma cells of *A. thaliana*. However, no translocation of grapheme oxide from the roots to the stem or leaves was observed.

Due to their smaller size and altered physical, chemical and structural properties, the absorption and translocation of different types of carbon nanomaterials raises serious concerns about their toxic effects on plants and also on the environment since plants represent the interface between the environment and biosphere.

5.3.2 In vitro *Exposure Studies of Carbon Nanomaterials*

The vulnerability of carbon nanomaterials on plant cells under *in vitro* cell culture conditions has been investigated by few studies. Liu et al. (2013) studied the effects of water-soluble carboxy fullerenes [ws-C_{70}; $C_{70}(C(COOH)_2)2–4$] in tobacco BY-2 cells (*Nicotiana tobacum*, cv. Bright Yellow). The results showed cell boundary disruption and growth inhibition, possibly due to the adsorption of ws-C_{70} to the plant cell wall through hydrostatic interaction with the carboxylic groups of fullerenes. Shen et al. (2010) showed that exposure to single walled carbon nanotubes exerted adverse effects on Arabidopsis and rice leaf protoplasts through oxidative stress, leading to programmed cell death or apoptosis, DNA damage and chromatin condensation. Dang et al. (2012) reported that exposure to water-soluble single walled carbon nanotubes and fullerene $C_{70}(C(COOH)_2)4–8$ on *Nicotiana tabacum* BY-2 cells caused shrunken morphology and lower proliferation rates. An increase in reactive oxygen species generation, abnormal mitochondrial distribution, decreased mitochondrial activity and impaired actin cytoskeleton arrangement was observed. However, no cytotoxicity was observed after treatment with either single walled carbon nanotubes or $C_{70}(C(COOH)_2)4–8$. In comparison, exposure to $C_{70}(C(COOH)_2)4–8$ caused more serious adverse effects on BY-2 cells. Tan et al. (2009) also reported that rice (*Oryza sativa* L.) suspension cells treated with multi walled carbon nanotubes enhanced the reactive oxygen species generation and decreased the cell viability.

The toxic effects of graphene was studied in *A. thaliana* (Columbia ecotype) T_{87} cell suspensions using different toxicological end points viz. morphology, mitochondrial dysfunction, reactive oxygen species generation and also the translocation of graphene inside the cells (Begum and Fugetsu 2013). Transmission electron microscopy analysis revealed the uptake of graphene nanoparticles in to the cells probably through endocytosis. Graphene exposure also caused morphological changes, fragmented nuclei, membrane damage and mitochondrial dysfunction in exposed cells. The results showed that graphene induced cell death in *A. thaliana* T_{87} cells through mitochondrial damage mediated by reactive oxygen species generation.

5.3.3 *Phytotoxic Effects of Fullerenes*

Investigations on the inhibitory effect of water-soluble fullerene C_{70} $(C(COOH)_2)$ 4–8 at the cellular level has been studied by Liu et al. (2010) using the transgenic seedling lines expressing fluorescent makers. Root growth retardation and loss of root gravitropism was observed and fluorescence imaging revealed the abnormalities of root tips in hormone distribution, cell division, microtubule organization and mitochondrial activity. Jiang et al. (2014) observed that rice seeds germinated in the presence of 150 μg/mL of carbon nanotubes decreased the root length, root activity, and stem length of the germinated seedlings of rice. The results showed the toxicity of higher concentrations of carbon nanotubes on seed germination and root growth. Mushtaq (2011) investigated the inhibitory effects of carbon nanoparticles (size 30–50 nm; 0 to 5000 μg/mL) on seed germination and root elongation of cucumber (*Cucumis sativus*). Inhibitory effect was more obvious in root growth than the seed germination percentages.

Liu et al. (2013) provided the first direct evidence for the change of plant cell wall composition under fullerene treatment using atomic force microscopy ligand-receptor binding force measurement to the living plant cell. They studied the changes of tobacco plant cell wall (*Nicotiana tobacum* L. cv. Bright Yellow) after treatment with water-soluble carboxy fullerenes ($C_{70}(C(COOH)_2)(2–4)$). It was observed that, the $C_{70}(C(COOH)_2)(2–4)$ were adsorbed to the cell wall which led to the disruption of cell wall and membrane leading to cell growth inhibition. A time and dose dependent increase of glycosyl residue on the cell wall accompanied by enhanced reactive oxygen species generation was observed.

5.3.4 Effects of Single Walled Carbon Nanotubes

The species dependent toxicity of single walled carbon nanotubes was studied by Canas et al. (2008). It was observed that exposure to single walled carbon nanotubes significantly affected root elongation of tomato, cabbage, carrot and lettuce. Tomato (*Lycopersicon esculentum*) showed the highest degree of sensitivity to single walled carbon nanotubes. The results showed that functionalized single walled carbon nanotubes demonstrated different toxic behaviors and are less toxic than non-functionalized single walled carbon nanotubes. This work highlighted the importance of investigating the surface properties of carbon nanotubes in determining their phytotoxicity. Pourkhaloee et al. (2011) reported that in salvia (*S. macrosiphon*), pepper (*C. annuum*) and tall fescue (*F. arundinacea*) exposure to higher concentrations of single walled carbon nanotubes affected the development of seedlings.

Yan et al. (2013) reported that exposure to single walled carbon nanotubes in maize seedlings caused inhibition of root hair growth. Gene transcription analysis showed that exposure to single walled carbon nanotubes decreased the root hair associated gene (*RTH1*, *RTH3*) expression. Treatment with single walled carbon nanotubes up-regulated the expression of epigenetic modification enzyme genes resulting in the global deacetylation of histone H3. It was concluded that exposure to single walled carbon nanotubes increased the histone deacetylation as a result of accumulation of the nanoparticles in the root apex.

5.3.5 Toxic Effects of Multi Walled Carbon Nanotubes

The germination rate of maize and rye grass decreased after exposure to 2000 mg/L of multi walled carbon nanotubes (Lin and Xing 2007). Stampoulis et al. (2009) investigated the effect of multi walled carbon nanotubes exposure under hydroponic conditions on zucchini. Treatment with 1000 mg/L for 15 days resulted in a 60% biomass reduction compared to control and bulk carbon. Ghodake et al. (2010) reported that exposure to 40 μg/mL of multi walled carbon nanotubes reduced root length and number of root hairs in *Brassica juncea* and *Phaseolus mungo*. Phytotoxicity studies using soil and in vitro conditions in young seedlings of *Parodia ayopayana* exposed to single walled and multi walled carbon nanotubes were conducted by Basiuk et al. (2013). The plant height, width and total root lengths were affected after 22 weeks of growth in soil and 16 weeks under in vitro conditions. The stronger effect was observed under in vitro conditions. Exposure to single walled carbon nanotubes caused higher effects as compared to multi walled carbon nanotubes. Reduction in primary spine length and under developed micrometer-sized secondary spines was observed. This gave evidences for the strong and systemic phytotoxic effect of multi walled carbon nanotubes on plant growth and development.

Mondal et al. (2011) revealed the dose dependent toxicity of multi walled carbon nanotubes in mustard, where oxidized-multi walled carbon nanotubes exerted higher negative effects than pristine multi walled carbon nanotubes. At higher exposure concentrations, both pristine and oxidized-multi walled carbon nanotubes caused toxicity, reducing germination by 4.4% and 7.6% and dry biomass by 1.6 and 2.2-fold respectively. Zhai et al. (2015) reported that exposure to differently charged multi walled carbon nanotubes viz. neutral pristine multi walled carbon nanotubes, positively charged multi walled carbon nanotube-NH_2 and negatively charged multi walled carbon nanotube-COOH inhibited the growth of soybean.

The importance of cyto-genotoxic effects of multi walled carbon nanotubes in the plant system and the importance of epigenetic studies on nanoparticle toxicity was investigated by Ghosh et al. (2015). They investigated the cytotoxic, genotoxic and epigenetic effect of multi walled carbon nanotubes in *Allium cepa*. Uptake of multi walled carbon nanotubes into the root cells significantly altered the cellular morphology, compromised the membrane integrity and mitochondrial function. Induction of DNA damage, micronucleus formation and chromosome aberration was also observed. Apoptotic cell death was observed as indicated by formation of inter nucleosomal fragments. The cyto-genotoxic effects were also confirmed by the accumulation of cells in the sub-G0 phase of the cell cycle and an increase in CpG methylation using the isoschizomers MspI/HpaII. Significant increase in the levels of 5-methyl-deoxy-cytidine was revealed by High Performance Liquid Chromatography analysis of DNA samples.

Apart from studying the toxic effects of multi walled carbon nanotubes in plants, few studies have been conducted to understand their mixture toxicity along with other environmental pollutants. The implications of the application of multi walled carbon nanotubes alone or in combination with pesticides for their use in agriculture, especially in leafy vegetables were studied by Hamdi et al. (2015). They evaluated the effect of non-functionalized and amino-functionalized multi walled carbon nanotubes as well as the presence of multi walled carbon nanotubes on the coexistent pesticide accumulation in lettuce (*Lactuca sativa* L.). The phytotoxicity of multi walled carbon nanotubes during germination and growth of lettuce seeds was monitored after sowing into 1000 mg/L of multi walled carbon nanotubes-amended vermiculite. The seedlings were subsequently exposed to 100 ng/L of chlordane (cis-chlordane, trans-chlordane and trans-non achlor and p, p'-DDE in the irrigation solution for a 19-days during the growth period. No significant influence on seed germination and plant growth was found after multi walled carbon nanotubes and pesticide exposure. However, the presence and type of multi walled carbon nanotubes significantly influenced pesticide contents in the seedlings. In roots and shoots, exposure to the non-functionalized multi walled carbon nanotubes decreased the root and shoot pesticide content by 88% and 78%, respectively. Exposure to amino-functionalized multi walled carbon nanotubes decreased the pesticide content by 57% in the roots and 23% in the shoots. However, the presence of humic acid completely reversed the reduction in the accumulation of pesticides induced by amino-functionalized multi walled carbon nanotubes probably due to the strong competition over adsorption sites on the multi walled carbon nanotubes.

To determine the combined effect of multi walled carbon nanotubes and heavy metals on agricultural crops, Wang et al. (2014) studied the effects of carboxylated multi walled carbon nanotubes (2.5, 5, and 10 mg/L) and their combination with 20 μM lead and 5 μM cadmium (shortened as Pb + Cd) in *Vicia faba* L. seedlings. It was observed that exposure to multi walled carbon nanotubes disturbed the nutrient element homeostasis, induced oxidative stress and damaged the leaves. Exposure to combination of carboxylated multi walled carbon nanotubes with Pb + Cd resulted in an increase in the content of Pb and Cd and decrease in oxidative damages as compared to treatments with carboxylated multi walled carbon nanotubes or Pb + Cd alone. The results showed that exposure to carboxylated multi walled carbon nanotubes caused oxidative stress and biochemical and subcellular damages as a result of treatment with Pb + Cd in the leaves. The results obtained from the investigation suggested that the continuous release of carboxylated multi walled carbon nanotubes into the environment may result in phytotoxicity and aggravate the ecological risks due to their combination with heavy metals.

The systemic toxicity and potential influence of multi walled carbon nanotubes was evaluated in red spinach by Begum and Fugetsu (2012). After 15 days of exposure under hydroponic conditions, the multi walled carbon nanotubes exposed plants showed inhibition of growth, cell death, and changes in root and leaf morphology. Reactive oxygen species generation and cytotoxicity were greatly increased in red spinach 15 days post-exposure to multi walled carbon nanotubes. However, supplementation of ascorbic acid reversed the adverse effects of multi walled carbon nanotubes exposure. It was concluded that excess reactive oxygen species generation and oxidative stress are the primary mechanism of toxicity of multi walled carbon nanotubes in red spinach.

The effect of multi walled carbon nanotubes in the presence and absence of NaCl was studied in broccoli (Martínez-Ballesta et al. 2016). Uptake and higher accumulation of multi walled carbon nanotubes was observed in cells demonstrating that multi walled carbon nanotubes can enter into the cells of adult plants under salt stress. Increased water uptake, growth and assimilation of CO_2 were observed in plants treated with multi walled carbon nanotubes. The authors hypothesized that assimilation of multi walled carbon nanotubes promoted more-favorable energetic forces and enhanced net assimilation of CO_2. Multi walled carbon nanotubes also enhanced aquaporin transduction which improved water uptake and transport, alleviating the negative effects of salt stress probably due to changes on plasma membrane properties of the cell wall.

5.3.6 Phytotoxic Effects of Graphene Nanomaterials

Few studies have been conducted to understand the effect of graphene nanomaterials on plants. Begum et al. (2011) investigated the interactions of graphene and its derivatives on root and shoot growth, biomass, shape, cell death and reactive oxygen species generation in cabbage, tomato, red spinach and lettuce. Morphological and physiological analyses indicated that exposure to graphene significantly inhibited plant growth and reduced the biomass. A dose dependent reduction in the number and size of leaves were observed in graphene-treated plants. A concentration-dependent increase in reactive oxygen species generation and cell death was observed in cabbage, tomato and red spinach indicating that the toxic effects were mediated through oxidative stress. Morphological analysis indicated that the epidermis of graphene treated roots of tomato and red spinach was loosely or completely detached.

Anjum et al. (2013) evaluated the effect of graphene oxide on the fava bean (*Vicia faba*) glutathione redox system, a major determinant of cellular redox homeostasis. A concentration dependent stress-response as well as decreased oxidative enzyme activity was observed. In a later study, Anjum et al. (2014) investigated the effect of single-bilayer grapheme oxide sheet in germinating faba bean seedlings to study their impact and potential mechanisms of toxicity. The results showed a concentration dependent decrease in the growth parameters and the activity of ascorbate peroxidase (*APX*) and catalase (*CAT*) enzymes and the levels of electrolyte leakage, H_2O_2 generation, lipid peroxidation and protein oxidation increased. Liu et al. (2015a, b) studied the effects of different concentrations of graphene on the germination and growth of rice seeds. Delay on the germination rate of rice seeds was observed with increasing graphene concentration. Inhibition of radicle and plumule growth and changes in the root length, stem length, adventitious number, root fresh weight, fresh weight of over ground part and root cap ratio of rice seedlings was also observed.

Short term and long term exposure studies were conducted in wheat (*Triticum aestivum* L.) to evaluate the phytotoxic effects of graphene on its growth and nutritional levels (Zhang et al. 2016). The plants were exposed different concentrations i.e. 250, 500, 1000 and 1500 mg/L of graphene under hydroponic culture conditions for 48 h and 30 days. Significant improvement in root elongation was observed but the root hair production was impaired. The phytotoxic effects were correlated with graphene induced-oxidative stress indicated by enhanced lipid peroxidation and antioxidant enzyme activity. Long term exposure to graphene resulted in reduction in shoot biomass, chlorophyll content, PSII activity and levels of nutrient elements viz. N, K, Ca, Mg, Fe, Zn and Cu. The results indicated that exposure to graphene inhibited plant growth and photosynthesis and caused an imbalance of nutrient homeostasis. The authors concluded that graphene has growth-limiting effects on plants, including root hair reduction, oxidative burst, inhibition of photosynthesis and nutritional disorder.

To understand the effects of graphene oxide on plant growth and development, *B. napus* cv. Zhongshuang 11 was treated with various concentrations of graphene oxide (Cheng et al. 2016). Treatment with 25–100 mg/L of graphene oxide resulted in shorter seminal root length compared to controls. Decrease in fresh root weight was observed when treated with 50–100 mg/L of graphene oxide. No significant change in lipid peroxidation as indicated by malondialdehyde content was observed. An increase in transcript levels of genes involved in absicic acid (ABA) biosynthesis (NCED, AAO and ZEP) and indole 3 acetic acid (IAA) biosynthesis (*ARF2*, *ARF8*, *IAA2* and *IAA3*) was observed after treatment with 50 mg/L of graphene oxide. However, an inhibition of transcript levels of IAA4 and IAA7 was observed. As compared with the control, exposure to graphene oxide treatment resulted in a higher ABA and lower IAA content. Overall, the results indicated that exposure to graphene oxide modulated the root growth of *B. napus* and affected ABA and IAA biosynthesis and concentration.

In a co-exposure study with graphene oxide and arsenate [As(V)], Hu et al. (2014) found that 0.1–10 mg/L of graphene oxide exposure enhanced the adverse effects of As (V) in wheat seedlings. They revealed that exposure to grapheme oxide greatly amplified the phytotoxicity of arsenic (As) in wheat causing a decrease in biomass and root numbers. Graphene oxide also triggered damage to cellular structures and enhanced the uptake of graphene oxide and arsenate. Co-transport of graphene oxide-loading arsenic (As) and transformation of As (V) to high-toxicity As (III) by graphene oxide were observed. The generation of dimethyl arsinate, produced from the detoxification of inorganic arsenic, was inhibited by graphene oxide in plants. Significant reduction in the fresh mass, shoot length and chlorophyll content was also observed in treated plants. In addition, the activity of peroxidase and superoxide dismutase, likely biomarkers for stress response, were increased in a concentration-dependent manner. Graphene oxide also regulated phosphate transporter gene expression and arsenate reductase activity to influence the uptake and transformation of As, respectively. The authors concluded that the indirect nanotoxicity of graphene oxide should be carefully considered in food safety. The results obtained from studies conducted on the effect of different types of carbon nanomaterials on various types of plants species has been concluded in Table 5.1.

Table 5.1 The phytotoxic effects of different carbon nanomaterials on different plant species is summarized in the table

Nanoparticle(s)	Plant	Toxic growth effects	Reference(s)
C_{60} fullerenes	Corn, soybean	Reduced biomass	Torre-Roche et al. (2013)
Functionalized carbon nanotube	Lettuce	Reduced root length	Cañas et al. (2008)
ws-C_{70}	Tobacco BY-2 cells	Cell boundary disruption and growth inhibition	Liu et al. (2013)
SWCNTs	Rice	Delayed flowering, decreased yield	Lin et al. (2009a, b)
Non-functionalized and functionalized SWCNTs	Carrot, cabbage, cucumber, lettuce, onion, tomato	Functionalized SWCNTs inhibited root growth in tomato. Non-functionalized SWCNTs inhibited lettuce root growth	Ca~nas et al. (2008)
MWCNTs	Rice cells	Increased ROS generation and decreased cell viability	Tan et al. (2009)
MWCNTs	Rice	Delayed flowering and seed setting. Reduced seed weight	Lin et al. (2009a, b)
MWCNTs	Zucchini	Negatively affected biomass production and transpiration	Stampoulis et al. (2009)
MWCNTs	Wheat	Enhanced the uptake of phenantrene to the living cells	Wild and Jones (2009)
MWCNTs	Garden cress, sorghum, tomato, radish, cucumber	Influenced seed germination and root growth depending of the type of sewage sludge	Oleszczuk et al. (2011)
MWCNTs	Onion	Chromosomal aberrations, DNA fragmentation and apoptosis in root cells	Ghosh et al. (2011)
MWCNTs	Onobrychis	Enhanced the POD activity	Smirnova et al. (2012)
MWCNTs	Lettuce	Reduced root length	Lin and Xing (2007)
MWCNTs	Rice	Chromatin condensed inside the cytoplasm and caused cell death, plasma membrane detachment from cell wall and cell shrinkage	Tan et al. (2009)
Oxidized MWCNTs	Mustard	Shorter germination time, enhanced root growth, and seedling stem development	Mondal et al. (2011)
CNTs	Red spinach	Growth inhibition, changes to tissue structure	Begum and Fugetsu (2012)
GO	Faba bean	Concentration dependent decrease in oxidative enzyme activity.	Anjum et al. (2014)
Single bilayer GO Sheet	Faba bean	Stress in plant development and growth. Reduction on peroxidase enzyme activity	Anjum et al. (2014)

(continued)

Table 5.1 (continued)

Nanoparticle(s)	Plant	Toxic growth effects	Reference(s)
Graphene	Arabidopsis	Fragmented nuclei, membrane damage and mitochondrial dysfunction	Begum and Fugetsu (2013)
Water-soluble graphene oxide (ws-GO)	Lettuce, cabbage, red spinach, tomato	Reduced plant growth, biomass, the number and size of leaves, increased ROS along with necrotic symptoms	Begum et al. (2011)

sWCNTs single walled carbon nanotubes, *MWCNTs* multi walled carbon nanotubes, *CNTs* carbon nanotubes, *GO* graphene oxide

Species names: – Corn – *Zea mays*; Soybean – *Glycine max*; Lettuce – *Lactuca sativa*; Rice – *Oryza sativa*; Carrot – *Daucus carota*; Cabbage – *Brassica oleracea*; Cucumber – *Cucumis sativus*; Onion – *Allium cepa*; Tomato – *Solanum lycopersicum*; Zucchini – *Cucurbita pepo*; Wheat – *Triticum aestivum*; Garden cress – *Lepidium sativum*; Sorghum – *Sorghum bicolor*; Tomato – *Solanum lycopersicum*; Radish – *Raphanus sativus*; Onobrychis – *Onobrychis arenaria*; Mustard – *Brassica juncea*; Red spinach – *Amaranthus dubius*; Faba bean – *Vicia faba*; Arabidopsis – *Arabidopsis thaliana*

Overall, analysis of the available reports indicate that the ability of different types of carbon nanomaterials to enter into plant cells and translocate to various parts poise different types of risks both in terms of plant health and as well as raises environmental issues. Following exposure, the different types of carbon nanomaterials were able to penetrate and enter through the seed coats, decreased seed germination, root growth and elongation. For example, from the studies it was observed that multiwalled carbon nanotubes were able to penetrate into the root system and translocated to the leaves and the fruits. The carbon nanomaterials were also able to modulate the expression of genes related to root development and also affect the synthesis of hormones related to root growth and development. The entry of carbon nanotubes to the plant tracheary elements may affect plant defense responses as well as wood development. Apart from this, mixture toxicity studies have found that carbon nanomaterials will enhance the toxicity and uptake of other environmental pollutants in plants.

5.4 Conclusions and Future Work

The application of carbon nanomaterials are growing. However, the production and release of different types of nanomaterials possessing various properties have complicated the evaluation of different types of carbon nanomaterials on plants. Due to safety considerations, several studies are being conducted to evaluate the effects of different types of carbon nanomaterials either alone or in combination with other environmental pollutants and heavy metals in plants. Results from the available studies indicate that exposure to different types of carbon nanomaterials caused toxic responses in plants and are depended on the concentration, exposure media and plant species. However, one limitation is that most of the studies were conducted under *in vitro* or controlled growth conditions thereby making it difficult to predict the effect of carbon nanomaterials under natural field conditions. Therefore, further studies need to done to understand the toxic effects of various carbon nanomaterials under natural conditions in plants.

References

Adhikari T, Biswas AK, Kundu S (2012) Nano-fertilizer-a new dimension in agriculture. Indian J Fert 6:22–24

Anjum NA, Singh N, Singh MK, Shah ZA, Duarte AC, Pereira E, Ahmad I (2013) Single-bilayer graphene oxide sheet tolerance and glutathione redox system significance assessment in faba bean (Vicia faba L.) J Nanopart Res 15:1770

Anjum NA, Singh N, Singh MK, Sayeed I, Duarte AC, Pereira E, Ahmad I (2014) Single-bilayer graphene oxide sheet impacts and underlying potential mechanism assessment in germinating faba bean (*Vicia faba* L.) Sci Total Environ 15:834–841

Avanasi R, Jackson WA, Sherwin B, Mudge JF, Anderson TA (2014) C_{60} fullerene soil sorption, biodegradation, and plant uptake. Environ Sci Technol 48:2792–2797

Basiuk VA, Basiuk EV, Shishkova S, Dubrovsky JG (2013) Systemic phytotoxic impact of as-prepared carbon nanotubes in long-term assays: a case study of *Parodia ayopayana* (Cactaceae). Sci Adv Mater 5:1337–1345

Begum P, Fugetsu B (2012) Phytotoxicity of multi-walled carbon nanotubes on red spinach (*Amaranthus tricolor* L.) and the role of ascorbic acid as an antioxidant. J Hazard Mater 243:212–222

Begum P, Fugetsu B (2013) Induction of cell death by graphene in *Arabidopsis thaliana* (Columbia ecotype) T_{87} cell suspensions. J Hazard Mater 15(260):1032–1041

Begum P, Ikhtiari RI, Fugetsu B (2011) Graphene phytotoxicity in the seedling stage of cabbage, tomato, red spinach and lettuce. Carbon 49:3907–3919

Burlaka OM, Pirko YV, Yemets AI, Blume YB (2015) Plant genetic transformation using carbon nanotubes for DNA delivery. Cytol Genet 49:349–357

Canas JE, Long M, Nations S, Vadan R, Dai L, Luo M, Ambikapathi R, Lee H, Olszyk D (2008) Effects of functionalized and nonfunctionalized single-walled carbon nanotubes on root elongation of select crop species. Environ Toxicol Chem 27:1922–1931

Chai M, Shi F, Li R, Liu L, Liu Y, Liu F (2013) Interactive effects of cadmium and carbon nanotubes on the growth and metal accumulation in a halophyte *Spartina alterniflora* (Poaceae). Plant Growth Regul 71:171–179

Cheng F, Liu YF, GY L, Zhang XK, Xie LL, Yuan CF, BB X (2016) Graphene oxide modulates root growth of *Brassica napus* L. and regulates ABA and IAA concentration. Plant Physiol 193:57–63

Chichiriccò G, Poma A (2015) Penetration and toxicity of nanomaterials in higher plants. Nano 5:851–873

Choi W, Lahiri I, Seelaboyina R, Kang YS (2010) Synthesis of graphene and its applications: a review. Crit Rev Solid State Mater Sci 35:52–71

Dang S, Liu Q, Zhang X, He K, Wang C, Fang X (2012) Comparative cytotoxicity study of water-soluble carbon nanoparticles on plant cells. J Nanosci Nanotechnol 12:4478–4484

Dasgupta N, Ramalingam C (2016) Silver nanoparticle antimicrobial activity explained by membrane rupture and reactive oxygen generation. Environ Chem Lett 14(4):477–485

Dasgupta N, Ranjan S, Mundekkad D, Ramalingam C, Shanker R, Kumar A (2015) Nanotechnology in agro-food: from field to plate. Food Res Int 69:381–400

Dasgupta N, Ranjan S, Rajendran B, Manickam V, Ramalingam C, Avadhani GS, Kumar A (2016) Thermal co-reduction approach to vary size of silver nanoparticle: its microbial and cellular toxicology. Environ Sci Pollut Res 23(5):4149–4163

Dasgupta N, Ranjan S, Ramalingam C (2017) Applications of nanotechnology in agriculture and water quality management. Environ Chem Lett 15(4):591–605

De Heer WA, Chatelain A, Ugarte DA (1995) Carbon nanotube field-emission electron source. Science 270:1179

De Volder MF, Tawfick SH, Baughman RH, Hart AJ (2013) Carbon nanotubes: present and future commercial applications. Science 339:535–539

Franklin NR, Li YM, Chen RJ, Javey A, Dai HJ (2001) Biological properties of carbon nanotubes. Appl Phys Lett 79:571

Fuhrer MS, Nygård J, Shih L, Forero M, Yoon YG, Mazzoni MSC, Choi H, Ihm J, Louie SG, Zettl A, McEuen PL (2000) Crossed nanotube junctions. Science 288:494

Georgakilas V, Perman JA, Tucek J, Zboril R (2015) Broad family of carbon nanoallotropes: classification, chemistry, and applications of fullerenes, carbon dots, nanotubes, graphene, nanodiamonds and combined super structures. Chem Rev 115:4744–4822

Ghodake G, Seo YD, Park DH, Lee DS (2010) Phytotoxicity of carbon nanotubes assessed by *Brassica juncea* and *Phaseolus mungo*. J Nanoelectron Optoelectron 5:157–160

Ghosh M, Chakraborty A, Bandyopadhyay M, Mukherjee A (2011) Multi-walled carbon nanotubes (MWCNT): induction of DNA damage in plant and mammalian cells. J Hazard Mater 15:327–336

Ghosh M, Bhadra S, Adegoke A, Bandyopadhyay M, Mukherjee A (2015) MWCNT uptake in *Allium cepa* root cells induces cytotoxic and genotoxic responses and results in DNA hyper methylation. Mutat Res 774:49–58

Gogos A, Knauer K, Bucheli TD (2012) Nanomaterials in plant protection and fertilization: current state, foreseen applications, and research priorities. J Agric Food Chem 60:9781–9792

Gopal A, Kumar R, Goswami A (2012) Nano-pesticides-a recent approach for pest control. The JPPS 4:1–7

Gottschalk F, Sonderer T, Scholz RW, Nowack B (2009) Modeled environmental concentrations of engineered nanomaterials ($TiO_{(2)}$), ZnO, Ag, CNT, Fullerenes) for different regions. Environ Sci Technol 43:9216–9222

Hamdi H, De La Torre-Roche R, Hawthorne J, White JC (2015) Impact of non-functionalized and amino-functionalized multiwall carbon nanotubes on pesticide uptake by lettuce (*Lactuca sativa* L.) Nanotoxicol 9:172–180

Hu ZS, Oskam G, Searson PC (2003) Influence of solvent on the growth of ZnO nanoparticles. J Colloid Interface Sci 263:454–460

Hu X, Kang J, Lu K, Zhou R, Mu L, Zhou Q (2014) Graphene oxide amplifies the phytotoxicity of arsenic in wheat. Sci Rep 19:6122

Hurt RH, Monthioux M, Kane A (2006) Toxicology of carbon nanomaterials: status, trends, and perspectives on the special issue. Carbon 44:1028–1033

Husen A, Siddiqi KS (2014) Carbon and fullerene nanomaterials in plant system. J Nanobiotech 12:16

Islam AE, Rogers JA, Alam MA (2015) Recent progress in obtaining semiconducting single walled carbon nanotubes for transistor applications. Adv Mater 27:7908–7937

Jain A, Ranjan S, Dasgupta N, Ramalingam C (2016) Nanomaterials in food and agriculture: an overview on their safety concerns and regulatory issues. Crit Rev Food Sci Nutr 15:1–21

Janardan S, Suman P, Ragul G, Anjaneyulu U, Shivendu R, Dasgupta N, Ramalingam C, Swamiappan S, Vijayakrishna K, Sivaramakrishna A (2016) Assessment on the antibacterial activity of nanosized silica derived from hypercoordinated silicon(iv) precursors. RSC Adv 6 (71):66394–66406

Jarosz P, Schauerman C, Alvarenga J, Moses B, Mastrangelo T, Raffaelle R, Ridgley R, Landi B (2011) Carbon nanotube wires and cables: near-term applications and future perspectives. Nanoscale 3:4542–4553

Jiang Y, Hua Z, Zhao Y, Liu Q, Wang F, Zhang Q (2014) The effect of carbon nanotubes on rice seed germination and root growth (Conference Paper). Lect Notes in Elect Eng 250:1207–1212

Keller AA, Lazareva A (2014) Predicted releases of engineered nanomaterials: from global to regional to local. Environ Sci Technol Lett 1:65–70

Keller AA, McFerran S, Lazareva A, Suh S (2013) Global life cycle releases of engineered nanomaterials. J Nanopart Res 15:1–17

Kemp K, Kimball B, Rybczynski J, Huang ZP, PF W, Steeves D, Sennett M, Giersig M, Rao DVGLN, Carnahan DL, Wang DZ, Lao JY, Li WZ, Ren ZF (2003) Photonic crystals based on periodic arrays of aligned carbon nanotubes. Nano Lett 3:13

Khodakovskaya MV, de Silva K, Biris AS, Dervishi E, Villagarcia H (2012) Carbon nanotubes induce growth enhancement of tobacco cells. ACS Nano 6:2128–2135

Kikuchi K, Nakahara N, Wakabayashi T, Suzuki S, Shiromaru H, Miyake Y et al (1992) NMR characterization of isomers of C_{78}, C_{82} and C_{84} fullerenes. Nature 357:142–145

Klaine SJ, Alvarez PJJ, Batley GE, Fernandes TF, Handy RD, Lyon DY, Mahendra S, McLaughlin MJ, Lead JR (2008) Nanomaterials in the environment: behavior, fate, bioavailability and effects. Environ Toxicol Chem 27:1825–1851

Kroto HW, Heath JR, O'Brien SC, Curl RF, Smalley RE (1985) C_{60}: buckminsterfullerene. Nature 318:162–163

Lahiani MH, Dervishi E, Chen J, Nima Z, Gaume A, Biris AS, Khodakovskaya MV (2013) Impact of carbon nanotube exposure to seeds of valuable crops. ACS Appl Mater Interfaces 5:7965–7973

Larue C, Pinault M, Czarny B, Georgin D, Jaillard D, Bendiab N, Mayne-L'Hermite M, Taran F, Dive V, Carrière M (2012) Quantitative evaluation of multi-walled carbon nanotube uptake in wheat and rape seed. J Hazard Mater 227:155–163

Lijima S (1991) Helicoidal microtubules of graphitic carbon. Nature 354:56–58

Lin D, Xing B (2007) Phytotoxicity of nanoparticles, inhibition of seed germination and root growth. Environ Pollut 150:243–250

Lin C, Fugetsu B, Su Y, Watari F (2009a) Studies on toxicity of multi-walled carbon nanotubes on Arabidopsis T_{87} suspension cells. J Hazard Mater 170:578–583

Lin S, Reppert J, Hu Q, Hudson JS, Reid ML, Ratnikova TA, Rao AM, Luo H, Ke PC (2009b) Uptake, translocation, and transmission of carbon nanomaterials in rice plants. Small 5:1128–1132

Liu S, Wei L, Hao L, Fang N, Chang MW, Xu R, Yang Y, Chen Y (2009a) Sharper and faster 'nano darts' kill more bacteria: a study of antibacterial activity of individually dispersed pristine single-walled carbon nanotube. ACS Nano 3:3891–3902

Liu Q, Chen B, Wang Q, Shi X, Xiao Z, Lin J, Fang X (2009b) Carbon nanotubes as molecular transpoters for walled plant cells. Nano Lett 9:1007–1010

Liu Q, Chen B, Wang Q, Shi X, Xiao Z, Lin J, Fang X (2009c) Carbon nanotubes as molecular transporters for walled plant cells. Nano Lett 9:1007–1010

Liu Q, Zhao Y, Wan Y, Zheng J, Zhang X, Wang C, Fang X, Lin J (2010a) Study of the inhibitory effect of water-soluble fullerenes on plant growth at the cellular level. ACS Nano 4:5743–5748

Liu Q, Zhao Y, Wan Y, Zheng J, Zhang X, Wang C, Fang X, Lin J (2010b) Study of the inhibitory effect of water-soluble fullerenes on plant growth at the cellular level. ACS Nano 4:5743–5748

Liu Q, Zhang X, Zhao Y, Lin J, Shu C, Wang C, Fang X (2013) Fullerene-induced increase of glycosyl residue on living plant cell wall. Environ Sci Technol 47:7490–7498

Liu H, Wang J, Feng Z, Lin Y, Zhang L, Su D (2015a) Facile synthesis of au nanoparticles embedded in an ultrathin hollow graphene nanoshell with robust catalytic performance. Small 11:5059–5064

Liu S, Wei H, Li Z, Li S, Yan H, He Y, Tian Z (2015b) Effects of graphene on germination and seedling morphology in rice. J Nanosci Nanotechnol 15:2695–2701

Loutfy RO, Lowe TP, Moravsky AP, Katagiri S (2002) Commercial production of fullerenes and carbon nanoparticles. In: Perspectives of fullerene nanotechnology. Springer, Dordrecht, pp 35–46

Ma XM, Geiser-Lee J, Deng Y, Kolmakov A (2010) Interactions between engineered nanoparticles (ENPs) and plants: phytotoxicity, uptake and accumulation. Sci Total Environ 408:3053–3061

Maddinedi SB, Mandal BK, Ranjan S, Dasgupta N (2015) Diastase assisted green synthesis of size-controllable gold nanoparticles. RSC Adv 5(34):26727–26733

Maddinedi SB, Mandal BK, Patil SH, Andhalkar VV, Ranjan S, Dasgupta N (2017) Diastase induced green synthesis of bilayered reduced graphene oxide and its decoration with gold nanoparticles. J Photochem Photobiol B Biol 166:252–258

Martínez-Ballesta MC, Zapata L, Chalbi N, Carvajal M (2016) Multiwalled carbon nano tubes enter broccoli cells enhancing growth and water uptake of plants exposed to salinity. J Nanobiotech 14:42

Mauter M, Elimelech M (2008) Environmental applications of carbon-based nanomaterials. Environ Sci Technol 42:5843–5859

Mondal A, Basu R, Das S, Nandi P (2011) Beneficial role of carbon nanotubes on mustard plant growth: an agricultural prospect. J Nanopart Res 13:4519–4528

Mushtaq YK (2011) Effect of nanoscale $Fe_{(3)}O_{(4)}$, $TiO_{(2)}$ and carbon particles on cucumber seed germination. J Environ Sci Health A Tox Hazard Subst Environ Eng 46:1732–1735

Oleszczuk P, Josko I, Xing BS (2011) The toxicity to plants of the sewage sludges containing multiwalled carbon nanotubes. J Hazard Mater 186:436–442

Pourkhaloee A, Haghighi M, Saharkhiz MJ, Jouzi H, Doroodmand MM (2011) Carbon nanotubes can promote seed germination via seed coat penetration. Seed Technol 33:155–169

Ranjan S, Ramalingam C (2016) Titanium dioxide nanoparticles induce bacterial membrane rupture by reactive oxygen species generation. Environ Chem Lett 14(4):487–494

Ranjan S, Dasgupta N, Chakraborty AR, Melvin Samuel S, Ramalingam C, Shanker R, Kumar A (2014) Nanoscience and nanotechnologies in food industries: opportunities and research trends. J Nanopart Res 16(6):1–23

Ranjan S, Dasgupta N, Rajendran B, Avadhani GS, Ramalingam C, Kumar A (2016) Microwave-irradiation-assisted hybrid chemical approach for titanium dioxide nanoparticle synthesis: microbial and cytotoxicological evaluation. Environ Sci Pollut Res 23(12):12287–12302

Rasool K, Lee DS (2015) Influence of multi-walled carbon nanotubes on anaerobic biological sulfate reduction processes. J Nanoelectron Optoelectroni 10:485–489

Reddy PVL, Hernandez-Viezcasb JA, Peralta-Videaa JR, Gardea-Torresdeya JL (2016) Lessons learned: are engineered nanomaterials toxic to terrestrial plants? Sci Total Environ 568:470–479

Rico CM, Majumdar S, Duarte-Gardea M, Peralta-Videa JR, Gardea-Torresdey JL (2011) Interaction of nanoparticles with edible plants and their possible implications in the food chain. J Agric Food Chem 59:3485–3498

Roco MC (2003) Nanotechnology: convergence with modern biology and medicine. Curr Opin Biotechnol 14:337–346

Rueckes T, Kim K, Joselevich E, Tseng GY, Cheung CL, Lieber CM (2000) Carbon nanotube-based nonvolatile random access memory for molecular computing. Science 289:94

Samaj J, Baluska F, Voigt B, Schlicht M, Volkmann D, Menzel D (2004) Endocytosis, actin cytoskeleton, and signaling. Plant Physiol 135:1150–1161

Sarlak N, Taherifar A, Salehi F (2014) Synthesis of nanopesticides by encapsulating pesticide nanoparticles using functionalized carbon nanotubes and application of new nanocomposite for plant disease treatment. J Agri and Food Chem 62:4833–4838

Sekhon BS (2014) Nanotechnology in agri-food production: an overview. Nanotechnol Sci Appl 7:31–53

Serag MF, Kaji N, Gaillard C, Okamoto Y, Terasaka K, Jabasini M, Tokeshi M, Mizukami H, Bianco A, Baba Y (2011) Trafficking and subcellular localization of multiwalled carbon nanotubes in plant cells. ACS Nano 5:493–499

Serag MF, Braeckmans K, Habuchi S, Kaji N, Bianco A, Baba Y (2012a) Spatio temporal visualization of subcellular dynamics of carbon nanotubes. Nano Lett 12:6145–6151

Serag MF, Kaji N, Tokeshi M, Baba Y (2012b) Introducing carbon nanotubes into living walled plant cells through cellulase-induced nanoholes. RSC Adv 2:398–400

Serag MF, Kaji N, Tokeshi M, Bianco A, Baba Y (2012c) The plant cell uses carbon nanotubes to build tracheary elements. Integr Biol 4:127–131

Serag MF, Kaji N, Habuchi S, Bianco A, Baba Y (2013) Nanobiotechnology meets plant cell biology: carbon nanotubes as organelle targeting nanocarriers. RSC Adv 3:4856–4862

Sharon M, Choudhry A, Kumar R (2010) Nanotechnology in agricultural disease and food safety. J Phytology 2:83–92

Shen CX, Zhang QF, Li J, Bi FC, Yao N (2010) Induction of programmed cell death in *Arabidopsis* and rice by single-wall carbon nanotubes. Am J Bot 97:1602–1609

Sheng Z, Song L, Zheng J, Hu D, He M, Zheng M, Gao G, Gong P, Zhang P, Ma Y et al (2013) Protein-assisted fabrication of nano-reduced graphene oxide for combined in vivo photo acoustic imaging and photothermal therapy. Biomaterials 34:5236–5243

Shukla A, Dasgupta N, Ranjan S, Singh S, Chidambram R (2017) Nanotechnology towards prevention of anaemia and osteoporosis: from concept to market. Biotechnol Biotechnol Equip 31(5):863–879

Siripireddy B, Mandal BK, Ranjan S, Dasgupta N, Ramalingam C (2017) Nano-zirconia – evaluation of its antioxidant and anticancer activity. J Photochem Photobiol B Biol 170:125–133

Smimova EA, Gusev AA, Zaitseva ON, Lazareva EM, Onishchenko GE, Kuznetsova EV, Tkachev AG, Feofanov AV, Kirpichnikov MP (2011) Multiwalled Carbonnanotubes penetrate in to plant cells and affects the growth of *Onbrychis arenaria* seedlings. Acta Nat 3:99–106

Smirnova E, Gusev A, Zaytseva O, Sheina O, Tkachev A, Kuznetsova E, Lazareva E, Onishchenko G, Feofanov A, Kirpichnikov M (2012) Uptake and accumulation of multiwalled carbon nanotubes change the morphometric and biochemical characteristics of *Onobrychis arenaria* seedlings. Front Chem Sci Eng 6:132–138

Srivastava V, Gusain D, Sharma YC (2015) Critical review on the toxicity of some widely used engineered nanoparticles. Ind Eng Chem Res 54:6209–6233

Stampoulis D, Sinha SK, White JC (2009) Assay-dependent phytotoxicity of nanoparticles to plants. Environ Sci Technol 43:9473–9479

Sun X, Liu Z, Welsher K, Robinson JT, Goodwin A, Zaric S, Dai H (2008) Nano graphene oxide for cellular imaging and drug delivery. Nano Res 1:203–212

Tammina SK, Mandal BK, Ranjan S, Dasgupta N (2017) Cytotoxicity study of Piper nigrum seed mediated synthesized SnO2 nanoparticles towards colorectal (HCT116) and lung cancer (A549) cell lines. J Photochem Photobiol B Biol 166:158–168

Tan XM, Lin C, Fugetsu B (2009) Studies on toxicity of multi-walled carbon nanotubes on suspension rice cells. Carbon 47:3479–3487

Tolaymat T, El Badawy A, Sequeira R, Genaidy A (2015) A system-of-systems approach as abroad and integrated paradigm for sustainable engineered nanomaterials. Sci Total Environ 511:595–607

Torre-Roche RDL, Hawthorne J, Deng Y, Xing B, Cai W, Newman LA, Wang Q, Ma X, Hamdi H, White JC (2013) Multiwalled carbon nanotubes and C_{60} fullerenes differentially impact the accumulation of weathered pesticides in four agricultural plants. Environ Sci Technol 47:12539–12547

United States Environmental Protection Agency (2007) Nanotechnology (White Paper: Technical Report EPA 100/B-07/001). Science Policy Council, US EPA, Washington, DC

Walia N, Dasgupta N, Ranjan S, Chen L, Ramalingam C (2017) Fish oil based vitamin D nanoencapsulation by ultrasonication and bioaccessibility analysis in simulated gastro-intestinal tract. Ultrason Sonochem 39:623–635

Wang X, Han H, Liu X, Gu X, Chen K, Lu D (2012a) Multi-walled carbon nanotubes can enhance root elongation of wheat (*Triticum aestivum*) plants. J Nanopart Res 14:841

Wang X, Han H, Liu X, Gu X, Chen K, Lu D (2012b) Multi-walled carbon nanotubes can enhance root elongation of wheat (*Triticum aestivum*) plants. J of Nanopart Res 14:841

Wang C, Liu H, Chen J, Tian Y, Shi J, Li D, Guo C, Ma Q (2014) Carboxylated multi-walled carbon nanotubes aggravated biochemical and subcellular damages in leaves of broad bean (*Vicia faba* L.) seedlings under combined stress of lead and cadmium. J Hazard Mater 15:404–412

Wang C, Zhang H, Ruan L, Chen L, Li H, Chang XL, Zhang X, Yang ST (2016) Bioaccumulation of^{13}C-fullerenol nanomaterials in wheat. Environ Sci Nano 3:799–805

Westerhoff P, Song G, Hristovski K, Kiser MA (2011) Occurrence and removal of titanium at full scale wastewater treatment plants: implications for TiO_2 nanomaterials. J Environ Monit 13:1195–1203

Wild E, Jones KC (2009) Novel method for the direct visualization of in vivo nanomaterials and chemical interactions in plants. Environ Sci Technol 43:5290–5294

Yadav BC, Kumar R (2008) Structure, properties and applications of fullerenes. Int J Nanotechnol Appl 2:15–24

Yan S, Zhao L, Li H, Zhang Q, Tan J, Huang M, He S, Li L (2013) Single-walled carbon nanotubes selectively influence maize root tissue development accompanied by the change in the related gene expression. JHazard Mater 246–247:110–118

Yan S, Zhang H, Huang Y, Tan J, Wang P, Wang Y, Hou H, Huang J, Li L (2016) Single-wall and multi-wall carbon nanotubes promote rice root growth by eliciting the similar molecular pathways and epigenetic regulation. IET Nanobiotechnol 10:222–229

Yao Z, Postma HWC, Balents L, Dekker C (1999) Carbon nanotube intra molecular junctions. Nature 402:273

Zhai G, Gutowski SM, Walters KS, Yan B, Schnoor JL (2015) Charge, size, and cellular selectivity for multiwall carbon nanotubes by maize and soybean. Environ Sci Technol 49:7380–7390

Zhang W, Guo ZY, Huang DQ, Liu ZM, Guo X, Zhong HQ (2011a) Synergistic effect of chemo photothermal therapy using PEGylated graphene oxide. Biomaterials 32:8555–8561

Zhang Y, Yang Z, Yang Y, Wang S, Shi L, Xie W, Sun K, Zou K, Wang L, Xiong J et al (2011b) Production of transgenic mice by random recombination of targeted genes in female germ line stem cells. J Mol Cell Biol 3:132–141

Zhang M, Gao B, Chen J, Li Y, Creamer AE, Chen H (2014) Slow-release fertilizer encapsulated by graphene oxide films. Chem Eng J 255:107–113

Zhang P, Zhang R, Fang X, Song T, Cai X, Liu H, Du S (2016) Toxic effects of graphene on the growth and nutritional levels of wheat (*Triticum aestivum* L.): short- and long term exposure studies. J Hazard Mater 317:543–551

Zhao S, Wang Q, Zhao Y, Rui Q, Wang D (2015) Toxicity and translocation of graphene oxide in *Arabidopsis thaliana*. Environ Toxicol Pharmacol 39:145–156

Chapter 6
Sustainable Desalination Process and Nanotechnology

Saikat Sinha Ray, Shiao-Shing Chen, Dhanaraj Sangeetha, Nguyen Cong Nguyen, and Hau-Thi Nguyen

Abstract A wide variety of methods are used for water treatment and purification. The use of membranes allows efficient treatment by reverse osmosis, nanofiltration, ultrafiltration, microfiltration, membrane distillation and forward osmosis. Membrane technology has been researched extensively for water treatment and desalination. Desalination is the technology predominantly used to solve water scarcity. The sustainability of desalination processes aims at reducing energy costs and increasing water recovery. In recent years numerous large-scale seawater desalination plants have been built in water-stressed countries. Construction of new desalination plants with the latest emerging technology is expected to increase in the future. Despite major advances in desalination technologies, seawater desalination is still more energy intensive compared to conventional processes uszd for the treatment of fresh water. However, forward osmosis and membrane distillation are emerging for sustainable desalination.

In this chapter we review key points of membrane processes including advantages and disadvantages of forward osmosis and membrane distillation. The advances in membrane material and modules is also discussed elaborately. Drawbacks of forward osmosis are also highlighted within each part, including draw solution development, reverse solute diffusion, concentration polarisation and membrane fouling. This chapter discuss the capability of membrane distillation in treating highly concentrated aqueous solutions derived from other desalination processes. We review the fabrication and performance of membranes, and the optimization of membrane distillation. Finally, the sustainability and application of forward osmosis and membrane distillation in seawater desalination is elaborately analysed.

S.S. Ray • S.-S. Chen (✉)
Institute of Environmental Engineering and Management, National Taipei University of Technology, Taipei, Taiwan
e-mail: f10919@ntut.edu.tw

D. Sangeetha (✉)
Department of Chemistry, Vellore Institute of Technology, Vellore, Tamil Nadu, India
e-mail: dsangeetha@vit.ac.in

N.C. Nguyen • H.-T. Nguyen
Faculty of Environment and Natural Resources, Dalat University, Dalat, Vietnam

K M Gothandam et al. (eds.), *Nanotechnology, Food Security and Water Treatment*, Environmental Chemistry for a Sustainable World 11,
https://doi.org/10.1007/978-3-319-70166-0_6

Keywords Water treatment • Membrane • Desalination • Reverse osmosis • Forward osmosis • Draw solution • Membrane distillation • Recovery • Fouling

Contents

6.1 Introduction

Recently, water scarcity becomes one of the prominent environmental problems faced by many people because of limited quantity of fresh water and uneven distribution (Valipour 2015a, b, 2016; Yannopoulos et al. 2015). According to the current statistics, the total volume of water on Earth is around 1.4 billion km^3, in which 97.5% consists of seawater and only 2.5% is fresh water. Interestingly, it has been reported that out of 2.5% of fresh water, only 0.35% is in the form of liquid which is available for human consumption (Gleick 1996; Valipour 2015a, b). Due to enormous increase in global population, the shortage of fresh water has been one of the environmental issues that requires permanent solution in the near future. In addition to that, the shortage of fresh water issue has been intensified by water pollution because of global industrialization and overpopulation of developing countries (Yuan et al. 2015). This issue could be eradicated if an ideal treatment technology can be developed to convert salt water or polluted water into potable water. A lot of research was focused on developing water purification technology over the past decades, and one of the most advanced applicable technology is reverse osmosis. Typically, reverse osmosis is a membrane separation methodology that uses applied pressure across a semi-permeable membrane for water treatment. Further, the first ever practical salty water desalination by utilizing reverse osmosis has been demonstrated by Sidney Loeb in the 1960s (Loeb and Sourirajan 1964). Currently, forward osmosis emerges out as one of the best in desalination technology that has started to gain attention because of potentially lower energy and operating cost than reverse osmosis (Cath et al. 2006; McCutcheon et al. 2006; Cath 2010; Ray et al. 2016a; Dasgupta et al. 2015, 2017; Shukla et al. 2017; Jain et al. 2016; Ranjan et al. 2014).

Membrane distillation is a membrane separation technology that can overcome few limitations of other membrane processes. Typically, high solute concentrations can be easily achieved by overcoming the issue of concentration polarization and purified form of water as a permeate can be produced. Recently, membrane distillation is a relatively novel membrane separation technology that may overcome some drawbacks of some conventional membrane processes (Lawson and Lloyd 1997; Alkhudhiri et al. 2012). In particular, the possibility of an industrial advancement of this present methodology is related to the developing commercial availability of membranes of potential interest. Membrane Distillation can be defined as a thermally-driven separation technology, in which only water molecules in vapour form transfer across a microporous hydrophobic membrane (Schofield et al. 1987). The driving force in the membrane distillation technology is the vapour pressure difference produced due to the temperature difference across the hydrophobic membrane. Generally, this technology offers different applications, such as seawater and brackish water desalination, wastewater treatment and in the food industry (Souhaimi and Matsuura 2011). This research article reviews all the necessary details based on membrane characteristics, membrane-related heat and mass transfer theories, membrane fouling and its effects of various operating conditions.

Forward Osmosis over the past 10 years has gained more attention, both commercially and academically. Basically, it works by utilizing two solutions with different concentrations (in other words different osmolality) separated by a semi-permeable membrane. This potential application is widely utilized in nature, but unfortunately it has not been fully recognized in industrial areas. Additionally, this process can be utilized on its own or in combinations with other processes, for example concentrating desalination and renewable power generation (Zhao et al. 2012a, b). Forward osmosis or osmotically driven membrane processes may not recently be 'main stream', but it is quiet probable that these processes are increasingly becoming a topic of some interest. Over the past decades, this process has considerable efficiency across a wide variety of applications such as power generation, emergency drinks, enhanced oil recovery, water treatment, fluid concentration, desalination. Unfortunately, only few of them were currently commercialised (Kessler and Moody 1976; Chung et al. 2012). This research article demonstrates some of the important aspects of this process and its derivatives, with regard to key problems, theories and some potential applications.

Additionally, membrane distillation has the capability to use the alternative source of energy, such as solar energy (Banat et al. 2002; Koschikowski et al. 2003). Basically, membrane distillation technology is a competitive methodology for desalination of brackish water and sea water (Alklaibi and Lior 2005). In other words, membrane distillation is an effective methodology for removing organic and heavy metals from aqueous solution from contaminated water. Furthermore, membrane distillation has also been utilized to treat radioactive waste in low concentration (Zakrzewska-Trznadel et al. 1999). Nevertheless, membrane distillation is also having few limitations such as low permeate flux compared to reverse osmosis. Even, the trapped air within the membrane offers an additional mass transfer resistance, which became another drawback of membrane distillation process that indirectly influences the permeate flux (El-Bourawi et al. 2006; Alkhudhiri et al. 2012).

As far as the history of desalination technology is concerned, Fig. 6.1 indicates the evolution of desalination technology where forward osmosis and membrane distillation emerge out as one of the potential methodologies that can be considered as third generation desalination technology.

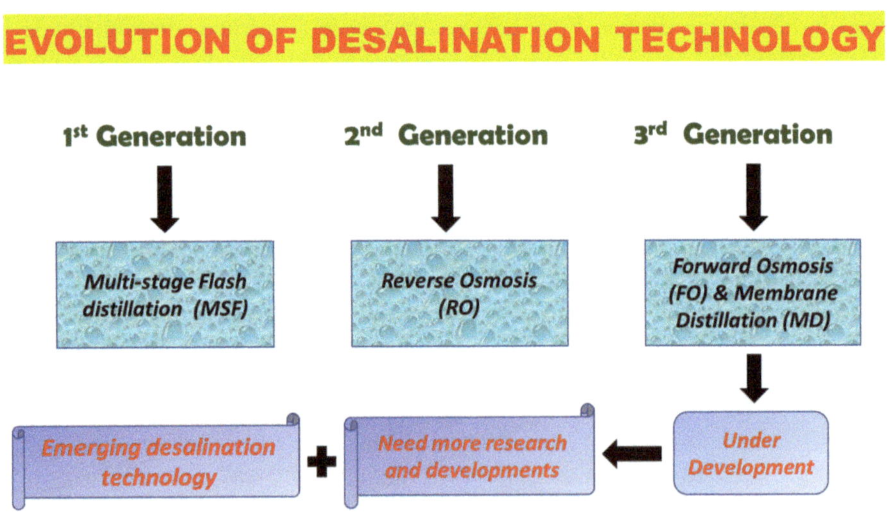

Fig. 6.1 Evolution of desalination technology. According to International Desalination Association (IDA), Multi-stage flash distillation plants produce about 60% of all desalinated water in the world (2001) whereas reverse osmosis was utilized in 66% of installed desalination capacity by the end of 2011. According to National Geographic magazine (March 2011), forward osmosis is one of three technologies that promised to reduce the energy requirements of desalination. Further, membrane distillation also emerges out as one of the efficient methodologies in the field of desalination technology

MSF Multi-stage flash distillation, *RO* Reverse osmosis, *FO* Forward osmosis, *MD* Membrane distillation

6.1.1 Background

Since 2000s and especially in current years, forward osmosis and membrane distillation have gained much attention because of high interest in membrane technology to produce ultra-pure water from seawater with these versatile technologies (Khayet 2011; Khayet and Matsuura 2011; Zhao et al. 2012a, b). A detailed survey has been presented by peer-reviewed publications (research articles and review articles) related to "forward osmosis" and "membrane distillation" of the last 10 years in Fig. 6.2. This presented database has been taken from the Scopus-based advanced scholar search system and the data clearly demonstrates that the forward osmosis and membrane distillation have regained much attention in recent research for the advancement of water treatment and desalination.

A schematic block diagram is indicated in Fig. 6.3 to demonstrate the process of forward osmosis and membrane distillation for water treatment and desalination.

Therefore, there is a need to address this research and development in order to explore the emerging technology for sustainable desalination in near future. In this research article, recent advances and developments in the application of forward osmosis and membrane distillation are discussed as a barrier layer for waste water treatment and desalination.

6.2 Potential of Forward Osmosis and Membrane Distillation

6.2.1 Efficiency of Forward Osmosis

As earlier mentioned, forward osmosis is a novel emerging membrane separation process with a number of major advantages compared to reverse osmosis. Recently, forward osmosis is considered as a complementary methodology but it has the efficiency to become the reference option in many industrial based applications. Forward osmosis is typically based on the natural phenomenon of a solvent passing from a portion of low osmotic pressure through a semipermeable membrane to one of high osmotic pressure. This phenomenon takes place continuously as far as nature is concerned such as in trees, plants, bacteria and animal cells.

Forward osmosis can be utilized to generate high quality permeate water from an aqueous effluent with varying level of contamination, a semipermeable forward osmosis membrane and a solution of high osmolality. Basically, this technology consumes very low energy (less operating cost), as it is maintained at very less pressure and at normal room temperature, which became one of the key advantages (Su et al. 2012; Lutchmiah et al. 2014). In addition to that, forward osmosis membranes are more resistant to membrane fouling and can tolerate chlorine,

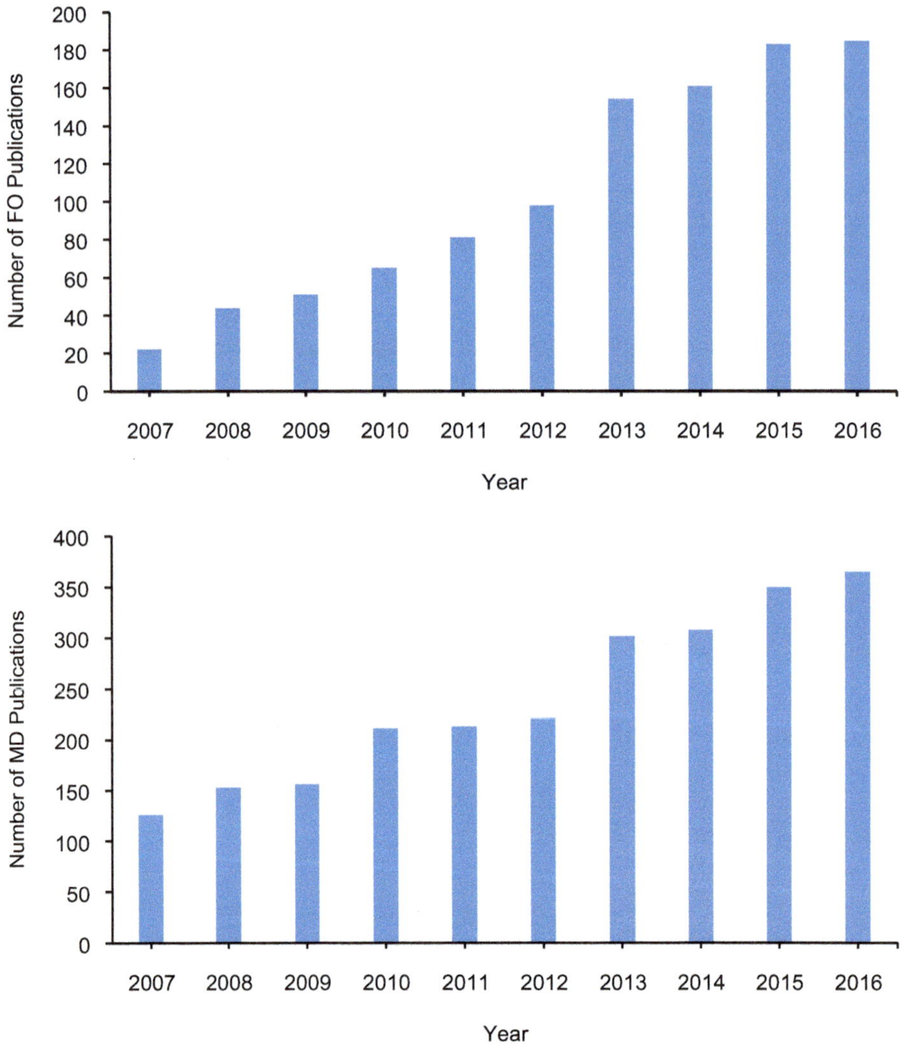

Fig. 6.2 Comparison of the number of peer reviewed publications since 2007. (Data analysis of publications has been done using the Scopus scholar search system with the term "Forward Osmosis" and "Membrane Distillation" with filter search of membrane and desalination, as on February 2017). The graphical data clearly explains that the forward osmosis and membrane distillation has gained attention in recent research for the advancement of water purification and desalination technology
FO Forward osmosis, *MD* Membrane distillation

thus less physical and chemical cleaning is required; thereby extending the life span of the forward osmosis membrane. But forward osmosis does not generate high quality permeate water suitable for utilization in only one single step, since the

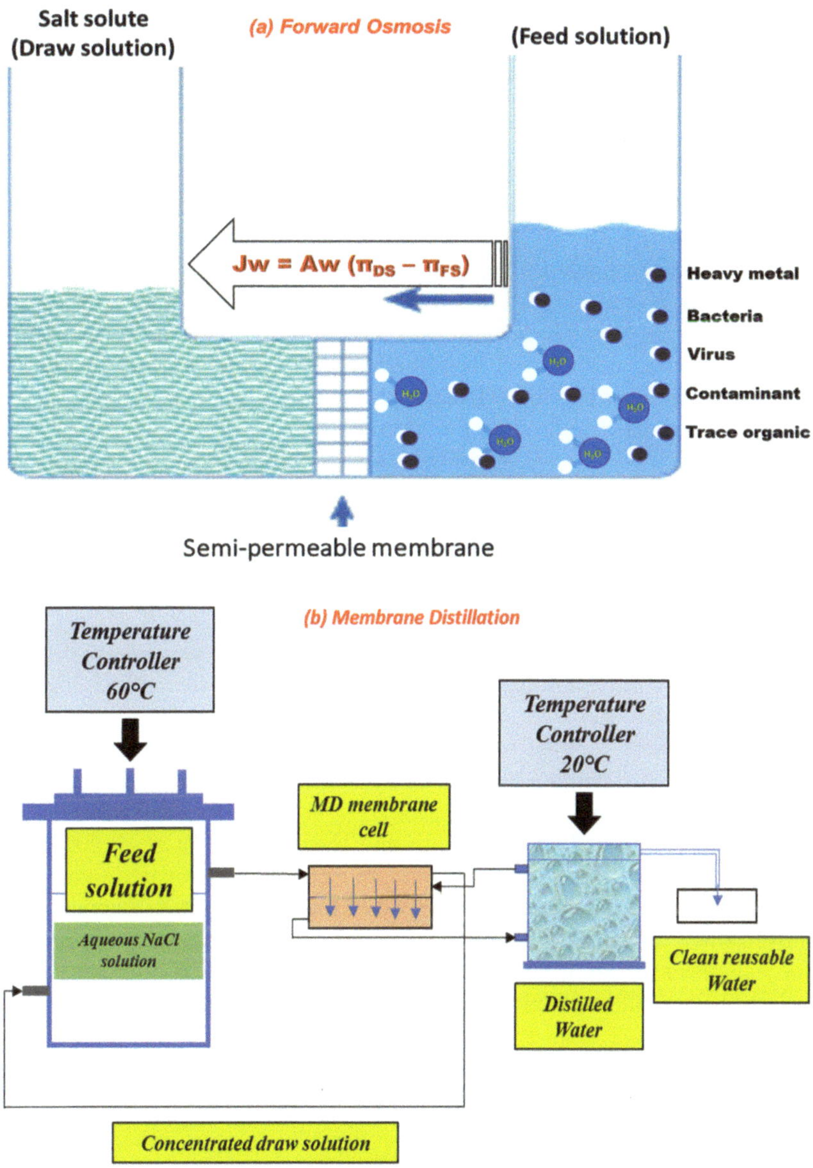

Fig. 6.3 (**a**) Typical pictorial form of Forward Osmosis (**b**) Lab set-up of Membrane Distillation. Forward osmosis is an osmotic process that uses a semi-permeable membrane to separate water from dissolved solutes. The driving force for this separation is an osmotic pressure gradient, whereas, for membrane distillation, the driving force of the process is given by a partial vapour pressure difference commonly triggered by a temperature difference

FO Forward osmosis, *MD* Membrane distillation

water produced adulterates with the draw solute. Hence, a second step is needed in order to separate the water generated from the draw solute (McGovern 2014; Shaffer et al. 2015). Further, in next stage, the draw solute is regenerated by leaving high quality of permeate water.

6.2.1.1 Basic Applications of Forward Osmosis

The most crucial advantageous applications of forward osmosis over conventional reverse osmosis process are as follows (Yip et al. 2010; Linares et al. 2014): (a) lower operating cost (Reduced energy consumption); especially when high osmolality solutions utilized; (b) less membrane fouling and high resistance to chlorine; (c) easier and more effective physical and chemical cleaning of the membrane; (d) longer life span of the membrane. Additionally, forward osmosis can be utilized in a wide range of applications, including the treatment of brackish water and sea water, water with various mineral salts and metal ions, effluents with silica, effluents with a high organic contamination etc. The most crucial examples are as follows (Kravath and Davis 1975; McGinnis and Elimelech 2008): (a) water generation in regions where there is a shortage of clean water; (b) effluent treatment for water reclamation and water reuse; (c) execution of a zero discharge system; (d) treatment of contaminated with complex effluents, that are usually difficult to treat by conventional technologies; (e) effective alternative methodology when lower energy needed.

Therefore, forward osmosis is an emerging methodology that is entirely reliable and feasible, and is a potential competitor for conventional technologies such as reverse osmosis and other separation processes. In brief, forward osmosis can be summarised as mentioned below (Miller and Evans 2006; Su et al. 2012; Altaee et al. 2014): forward osmosis is an alternative technology compared to the reverse osmosis process which reduces the energy consumption required; forward osmosis is a novel emerging process over conventional thermal evaporation technologies and has a wide variety of different advantageous applications in industries; forward osmosis is a potential methodology with the scope for further advancement and improved performance, has very low investment (operating costs) in zero discharge applications and reduces the need of pre-treatment process and further increases the potency as the process advances.

6.2.2 *Effectiveness of Membrane Distillation*

Although, the potentialities of membrane distillation are well known in water research area, its direct application in industrial area is still questionable due to high energy requirements. Hence, high water fluxes shall be obtained with reasonable energy consumption. However, membrane distillation was extensively recognized as one of the cost-effective technology for desalination operating at higher temperatures when waste form of heat can be employed to power the process (Alklaibi and Lior 2005; Susanto 2011). In general, the working temperature varies from 30–60 °C permit re-use of residual heat flows, and the utilization of alternative energy sources like solar, geothermics, and wind. Additionally, when compared with reverse osmosis, membrane distillation is less susceptible to flux limitations due to concentration polarisation, whereby a higher recovery of matter is attained on the feed stream. Typically, membrane distillation attempts 100% rejection for non-volatile dissolved substances. Comparatively, membrane distillation offers prominent advantages in membrane separation processes such lesser footprint, simple up-scaling, simple operational conditions (temperature and pressure), possibility to treat flows with heat-sensitive components or a high suspended particle-content at atmospheric pressure and a temperature below the boiling point of the feed (Hsu et al. 2002).

6.2.2.1 Basic Applications

There is no doubt that membrane distillation is versatile in nature. Therefore, membrane distillation is a well-known technology that can be utilized for the generation of clean and reusable water. Desalination of seawater and generation of pure water from the brackish salt water are the prominent membrane distillation applications as because non-volatile ionic matters are completely rejected (Souhaimi and Matsuura 2011). Moreover, membrane distillation also allows water purification in the pharmaceutical industries, textiles and chemical industries, for the concentration of fruit juices and milk processing, in bio-technology and bio-medical applications such as the removal of water from blood, in separating azeotropic aqueous mixes (alcohol-water), in the concentration of glycols, non-volatile acids and oil-in-water emulsions and applications where high temperature processing causes thermal degradation of the process flow. Further, membrane distillation permits greater efficiency where wastewater is contaminated with different organic dyes (Tomaszewska 2000; Camacho et al. 2013).

6.2.2.2 Efficiencies of Membrane Distillation

Membrane distillation can be utilized for the concentration/separation of non-volatile components such as ionic salts, acids, colloids and macro-molecules from aqueous solutions for the removal of traces of volatile organic compounds like choroform, benzene and tri-chloroethylene from water, and for the separation of other organic components like alcohols from diluted aqueous liquid (Lawson and Lloyd 1997; Martinez-Diez et al. 1999; Al-Obaidani et al. 2008). The advantages of membrane distillation process include: (a) lower operating temperature; (b) lower operating hydrostatic pressures than the pressure-driven processes such as reverse osmosis process; (c) high rejection factors achieved; (d) pure permeate water; (e) less demand of membrane mechanical properties/toughness; (f) less membrane fouling as compared to any pressure driven processes such as reverse osmosis; (g) possibility to utilize waste form of heat and renewable energy sources. (El-Bourawi et al. 2006; Souhaimi and Matsuura 2011). Therefore, based on its high compatibility, membrane distillation can be combined in various crucial industrial production cycles rising the potency of the entire technique.

6.3 Issues Associated with Forward Osmosis and Membrane Distillation

6.3.1 Concentration Polarization and Its Modelling

Concentration polarization is a typical and undeniable aspect in both processes such as pressure driven and osmotically driven membrane technologies. Generally, in osmotically driven membrane technologies, concentration polarization occurs due to difference in concentration between the feed solution stream and the draw solution stream through an asymmetric forward osmosis semipermeable membrane as indicated in Fig. 6.4. Thus, both type of concentration polarization i.e. (a) external concentration polarization and (b) internal concentration polarization can occur in forward osmosis as well as membrane distillation. In particular, external concentration polarization takes place at the surface of the dense active layer of the membrane and internal concentration polarization takes place within the porous support layer of the membrane (Martınez-Dıez and Vazquez-Gonzalez

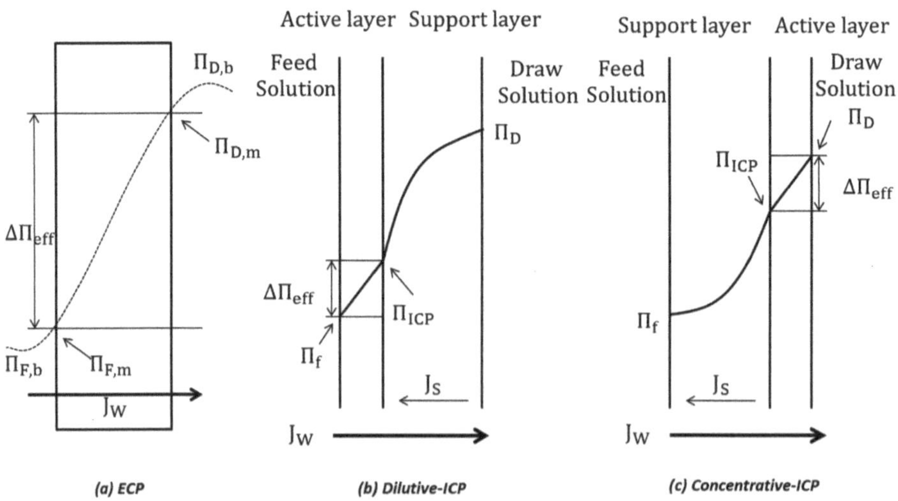

Fig. 6.4 (**a**) External concentration polarization. The line is the concentration profile, indicating concentrative (towards right) and dilutive (towards left) (**b**) Dilutive internal concentration polarization and (**c**) concentrative internal concentration polarization
ECP External concentration polarization, *ICP* Internal concentration polarization

1999; McCutcheon and Elimelech 2006). In addition to that, both external concentration polarization and internal concentration polarization have been illustrated.

Because of concentration polarization effect, the solute being retained by the membrane and the solvent passing the membrane. Consequently, the solute will accumulate and form a layer at the membrane interface with a relatively high concentration. The concentrated layer near the membrane is less permeable for the solvent (usually water) in comparison with an unaltered solution, which is an additional resistance. This phenomenon in parallel results in a (much) higher osmotic pressure $\Delta\prod$ at the membrane interface, even for cases where macromolecular solutions are used, and leads to a decrease in the driving force which then becomes $\Delta P - \Delta\prod$. Finally, the concentration at the membrane interface can reach such high values that the concentrated solution will change into a gel with a resistance. Gel layer formation occurs easily with protein containing liquids. The flux decline phenomena can be generally divided in fouling (irreversible and long term phenomena) and concentration polarization (reversible and directly occuring phenomena).

Concentration polarization is the growth of concentration gradients both internally and externally forward osmosis membranes during the running of system. As mentioned earlier, gradients lower the effective osmotic pressure difference across the membrane active layer and thus limit the achievable permeate water flux. According to Cath et al., concentration polarization takes place at lower water flux. It can be expressed mathematically by the performance ratio, $r_{\text{performance}}$ (Cath et al. 2006) as mentioned in Eq. 6.1;

$$r\ performance = \frac{Jw\ experimental}{Jw\ theoritical} \tag{6.1}$$

where $Jw_{\text{experimental}}$ is the measured flux experimentally whereas, $Jw_{,\text{theoretical}}$ is the estimated flux from the given equation Eq. 6.2;

$$Jw = A(\pi D - \pi F) \tag{6.2}$$

However, the performance ratio doesn't fully demonstrate the concentration polarization, as both external concentration polarization and internal concentration polarization take place (McCutcheon and Elimelech 2006). Typically, in membrane separation technologies there are four different types of concentration polarization that fall into two important categories, (1) external concentration polarization and (2) internal concentration polarization, and there are other two sub-categories in both cases; (a) dilutive and (b) concentrative. Figure 6.4 indicates the phenomena of external concentration polarization as well as internal concentration polarization:

6.3.1.1 Internal Concentration Polarization

According to Lee et al., the extent of the internal concentration polarization mainly depends up on the solute resistance to diffusion in the membrane's support layer, K (Lee et al. 1984) Eq. 6.3,

$$k = \frac{t\tau}{D}\in \tag{6.3}$$

where, t, τ and $\in$ represents the thickness, tortuosity and porosity of the support layer of membrane, and D represents the bulk diffusion coefficient of the solute. As per equation Eq. 6.3, the solute resistance also depends upon the solute diffusion coefficient. It indicates that the higher diffusion of solute leads to better compensation for the dilutive or concentrative internal concentration polarization respectively, as demonstrated in Fig. 6.4.

The solute resistance to diffusion, K, can be combined with equations describe by McCutcheon et al. (2006) to get the simplified form that describes the flux with dilutive internal concentration polarization, Eq. 6.4;

$$Jw = A\left(\pi Dexp\left(-\frac{JwK}{kD}\right) - \pi Fexp\left(\frac{Jw}{kF}\right)\right) \tag{6.4}$$

whereas, with concentrative internal concentration polarization, the expression can be written as, Eq. 6.5;

$$Jw = A\left(\pi Dexp\left(-\frac{Jw}{kD}\right) - \pi Fexp\left(\frac{JwK}{kF}\right)\right) \tag{6.5}$$

In general, for asymmetric membranes; that consist of both dense rejection layer and porous support layer – internal concentration polarization occur in the porous underlying support layer whereas external concentration polarization on the interphase in between the dense rejection layer and solutions (Gray et al. 2006):

1. Typically, when the rejection layer faces towards the feed solution side (which is also known as active layer towards feed solution (AL-FS) or forward osmosis (FO) mode), the water permeate through the porous underlying support layer dilutes the draw solution inside the support layer, that gives rise to dilutive internal concentration polarization. Whereas, concentrative external concentration polarization occurs on the dense rejection layer of the membrane (McCutcheon and Elimelech 2006).
2. On the other hand, when the rejection layer faces towards the draw solute (which is also known as active layer towards draw solution (AL-DS) or pressure retarded osmosis (PRO) mode), solutes inside the support layer are concentrated as water passes through the membrane, that gives rise to concentrative internal concentration polarization. Whereas, dilutive external concentration polarization occurs on the dense rejection layer of the membrane (Tang et al. 2010).

6.3.1.2 External Concentration Polarization

All membrane processes suffered from external concentration polarization effect at the membrane interfaces which are in contact with the bulk fluids because a thin layer of fluid at the interface may get polarized. For a fluid flowing via a rectangular channel, a thin layer of fluid close to the fluid-channel interface will be in laminar flow regardless of the nature of the free stream (Tan and Ng 2008). This thin layer is an interface between the membrane and the bulk solution. Within this thin layer of fluid, transport of water and other solutes are only based on convection (perpendicular to the membrane surface) and molecular diffusion. A concentration profile will develop and this phenomenon is known as external concentration polarization. The Fig. 6.4 demonstrates the concentration profile of osmotic agents along with a membrane. Since water diffuse from the feed stream solution into the membrane, a local concentration of the feed takes place. Typically, at the draw solution side of the membrane, a local dilution takes place when water from the membrane diffuse into the solution stream. This results in external concentration polarization in a boundary layer against the membrane, decreasing the effective osmotic gradient. Further, in this mass balance, the external concentration polarization at the feed side of the membrane is illustrated as Eq. 6.6;

$$JwC = JwCd - \frac{D\,dC}{dx} \tag{6.6}$$

where, C represents the concentration of solute in the layer at the length of x from the membrane, whereas, C_d represents the concentration of draw solute, D represents the diffusion coefficient of solute and dC/dx represents the concentration gradient within the boundary layer. By utilizing the boundary conditions, when $C = C_{membrane}$, $x = 0$ and $C = C_{bulk}$ when x =layer thickness, δ and presuming total rejection of salt, the following equation can be obtained Eq. 6.7;

$$\frac{C\ membrane}{C\ bulk} = \exp\left(\frac{Jw\ \delta}{D}\right) \tag{6.7}$$

This above-mentioned relationship is known as the polarization modulus. It may be concentrative or dilutive, mainly dependent on whether it is the feed or draw external concentration polarization studied. If considered dilutive external concentration polarization, the exponential term will be negative (−ve). δ/D can be substituted with the mass transfer coefficient 'kD' and 'kF' for external concentration polarization in draw stream and feed stream respectively.

As per McCutcheon and Elimelech et al., the mass transfer coefficient has been illustrated, k, as in given equation Eq. (McCutcheon and Elimelech 2006), Eq. 6.8;

$$k = \frac{ShD}{dh} \tag{6.8}$$

where, Sh represents the Sherwood number and dh is the hydraulic diameter. Since, osmotic pressure is considered as a colligative property, the polarization modulus can be expressed mathematically as given Eq. 6.9;

$$\frac{\pi\ membrane}{\pi\ bulk} = \exp\left(\frac{Jw}{k}\right) \tag{6.9}$$

where $\Pi_{membrane}$ represents the osmotic pressure at the membrane and Π_{bulk} represents the osmotic pressure in bulk. As per some researchers, the polarization modulus inclines with different conditions which are as follows (Blatt et al. 1970; McCutcheon et al. 2006): (1) increasing flux, (Jw); (2) thickness of boundary layer, (δ); (3) lower diffusion coefficient, (D).

External concentration polarization occurs on the inter phase in between the dense rejection layer and encompassing solutions. Basically, for symmetric, dense membranes which reject feed and draw solutions, external concentration polarization occurs on the membrane surfaces (Tan and Ng 2008, 2013): (1) Solutes are concentrated at the surface on the feed side, as water passes through the membrane, that gives rise to concentrative external concentration polarization; (2) On the other hand, solutes are diluted at the surface on the draw solution side, as water permeates from the feed side, that gives rise to dilutive external concentration polarization.

6.3.2 Membrane Fouling

The recent advancements of the membrane technologies fell far behind the initial anticipation, and one of the prominent hindrances that interfere more widespread of its application is that the filtration performance necessarily declines with respect to time. Typically, this aspect is commonly defined as membrane fouling, that refers to the blockage of membrane pores while filtration by the combination of filtering and adsorption of compounds and particulates onto the surface of membrane or within the pores of the membrane. Generally, blockage of pores lowers the permeate water flux and results in complexity of the membrane filtration performance. Therefore, membrane fouling became the most challenging and complicated issue for further membrane applications and development (Abdelrasoul et al. 2013). Figure 6.5 indicates membrane fouling and shows different types of membrane foulants responsible for fouling.

Membrane fouling can be broadly classified into two categories: (1) back-washable or non-back-washable, and (1) reversible or irreversible based on the type of attachment of particles to the membrane surface. Typically, back-washable membrane fouling can be easily eliminated by reversing the direction of flow at the end of each filtration cycle. However, non-back-washable membrane fouling cannot be eliminated by normal backwashing technique. Non-back-washable membrane fouling can be easily handled and removed by chemical cleaning technique. Furthermore, irreversible membrane fouling cannot be eliminated with flushing,

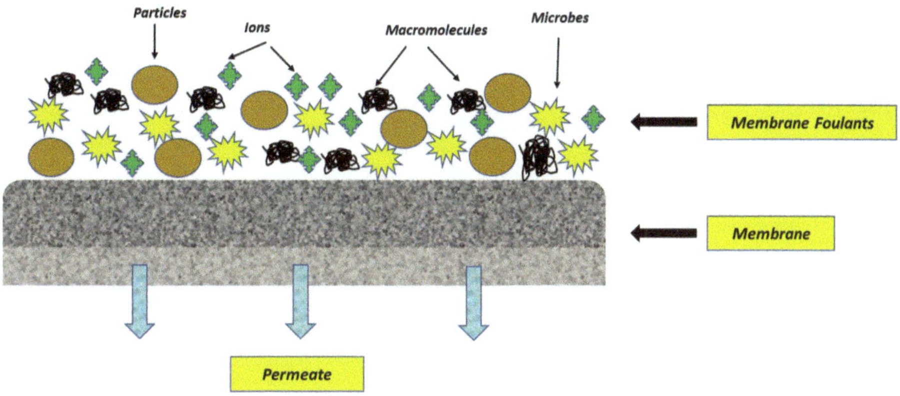

Fig. 6.5 A general membrane fouling pictorial presentation indicating different types of membrane foulants including particles, macromolecules, microbes and ions
There are other various types of foulants that may cause membrane fouling which are as follows: colloidal (clays, flocs), biological (bacteria, fungi), organic (oils, polyelectrolytes, humic) and scaling (mineral precipitates)

backwashing, physical cleaning, chemical cleaning, or by any other techniques, and hence the membrane cannot be recovered to its original permeate water flux. In addition to that, membrane fouling can also be subdivided, based on the type of fouling particulates and material, into four sub-categories: (1) inorganic fouling, (2) particle/colloidal fouling, (3) microbial/biological fouling, and (4) organic fouling. Membrane fouling is a continuous process where solutes or particulates/particles deposit onto the surface of membrane or into the membrane pores that results in an increase in the resistance to flow during filtration process. To understand the clear mechanism of membrane fouling, a pictorial form of fouling mechanism has been indicated in Fig. 6.6. Moreover, Table 6.1 indicates the basic description of different types of fouling.

Typically, natural organic matter is the organic substances present in ground or surface water and consists of various high molecular weight organic matters/compounds. Natural organic matter includes both humic and non-humic components. Humic fraction contains high molecular weight organic matters. On the other hand, non-humic natural organic matter foulants are proteins, amino sugars, polysaccharides, and polyoxyaromatics (Hong and Elimelech 1997; Lee et al. 2004). Recently, several studies have revealed that natural organic matter is one of the main membrane foulant in ultrafiltration. As per Makdissy et al., the organic colloidal fraction results in significant membrane fouling. In addition to that, polysaccharides are recognized as the dominant membrane foulant too. According to the recent research, it was reported that most of the membrane fouling were caused by hydrophobic natural organic matter components. However, as per some other researchers, neutral hydrophilic natural organic matter components were observed to be the major membrane foulants. Natural organic matter components, can be ranked in this order which is as follows: neutral hydrophilics natural organic

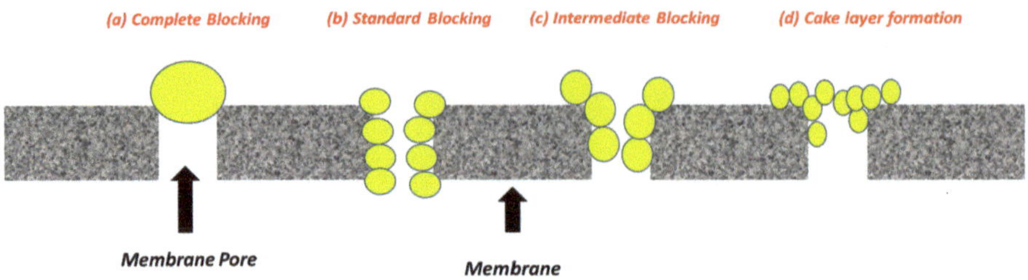

Fig. 6.6 Membrane fouling mechanism. This picture illustrates four kind of fouling namely: (**a**) Complete blocking; (**b**) Standard blocking; (**c**) Intermediate blocking; (**d**) Cake layer formation. Contamination of membranes causes a higher energy use, a higher cleaning frequency and a shorter life span of the membrane. Membrane contamination is usually called fouling
Adapted and modified from (Konieczny and Rafa 2000)

Table 6.1 A brief summarisation based on different types of fouling (Lee et al. 2005; Meng et al. 2009)

Nature of fouling	Brief description
Inorganic fouling	Accumulation of particulates matter when the concentration of the chemical species exceeds its saturation concentration. Several studies have shown that increased concentration of Ca^{2+} and Mg^{2+} caused more fouling
Particles or colloidal fouling	Algae, bacteria and certain natural organic matter fall into this category. To distinguish different fouling phenomena, particles and colloids here are referred to biologically inert particles and colloids that inorganic in nature and are originated from weathering of rocks
Microbial or biological fouling	Results due to formation of biofilms on the membrane surface. Once bacteria attach to the membrane surface, bacteria start multiplying and produce extracellular polymeric substances to form a viscous, slimy and hydrated form of gel
Organic fouling	Clogging of the membrane by organic substances, and organic carbons generally concentrate on the internal surface of the membrane

There are four categories of membrane fouling. They are (a) inorganic fouling or scaling, (b) particle or colloids fouling, (c) microbial fouling or biofouling, and (d) organic fouling

matter > hydrophobic acids natural organic matter> transphilic acids natural organic matter > charged hydrophilics natural organic matter (Seidel and Elimelech 2002; Lee et al. 2005). Therefore, it can be concluded that there are many conflicting results and hypothesis from different researchers and many facets of membrane fouling, so there would be no appropriate solution for membrane fouling control, but it has to be dealt with and designed specifically for a particular type of foulant and membrane in use, which has been discussed in next section. Table 6.2 indicates a brief discussion on nature of foulants along with the mode of fouling.

Table 6.2 Summarisation of nature of foulants and mode of fouling in membrane processes (Burggraaf 1996; Field 2010)

Nature of foulants	Mode of fouling
Large suspended particles	Particles present in the original feed (or developed due to concentration polarization) can block module channels as well as forming a cake layer on the surface
Small colloidal particles	Colloidal particles can develop a fouling layer (e.g., ferric hydroxide from brackish water can become a slimy brown fouling layer on membrane surface). In recovery of cells from fermentation broth, some colloids can be present
Macromolecules	Gel or cake formation on membrane. Macromolecular fouling within the structure of porous membranes
Small molecules	Some small organic molecules tend to have strong interactions with some polymeric membranes (e.g., anti-foaming agents, such as polypropylene glycols used during fermentation, adhere strongly to certain polymeric membranes)
Proteins	Interactions with surface or pores of membranes
Biological	Growth of bacteria on the membrane surface and excretion of extracellular polymers
Chemical reactions leading to scaling	Concentration increase and pH changes can lead to precipitation of salts and hydroxides

There are roughly four kinds of membrane foulants that have been distinguished: (i) organic precipitates (consists of macromolecules, biological substances, etc.); (ii) colloids; (iii) inorganic precipitates (consists of metal hydroxides, calcium salts, etc.) and (iv) particulates

6.3.2.1 Membrane Fouling Mechanism

Flux-time curve of membrane filtration process, as denoted in Fig. 6.7, initiates with (I) a rapid initial decrease of the water flux, (II) followed by a long duration of gradual flux decline, and (III) terminates with a steady-state flux. Typically, flux decrease in membrane filtration process is a consequence of the increase in the membrane resistance by the pore blockage and cake layer formation on the membrane surface. The pore blockage in membrane increases the membrane resistance, on the other hand, the cake layer formation creates an additional layer of resistance to the permeate flow. In addition to that, pore blockage on membrane and cake layer formation on membrane surface can be considered as two prominent mechanisms for membrane fouling (Choo and Lee 1996; Jiang et al. 2005).

According to the graph, the rapid initial decline of the permeate water flux can be attributed to pore blocking of membrane pores. However, the maximum permeate water flux occurs at the initial period of filtration because membrane pores are typically clean and opened at that time. Thus, flux decreases as pores of membrane are blocked by retained particulates/ particles. Basically, membrane pores are being blocked partially and the degree of pore blockage in membrane depends on the relative size of particles, shape and pores. As per previous report, the pore blockage

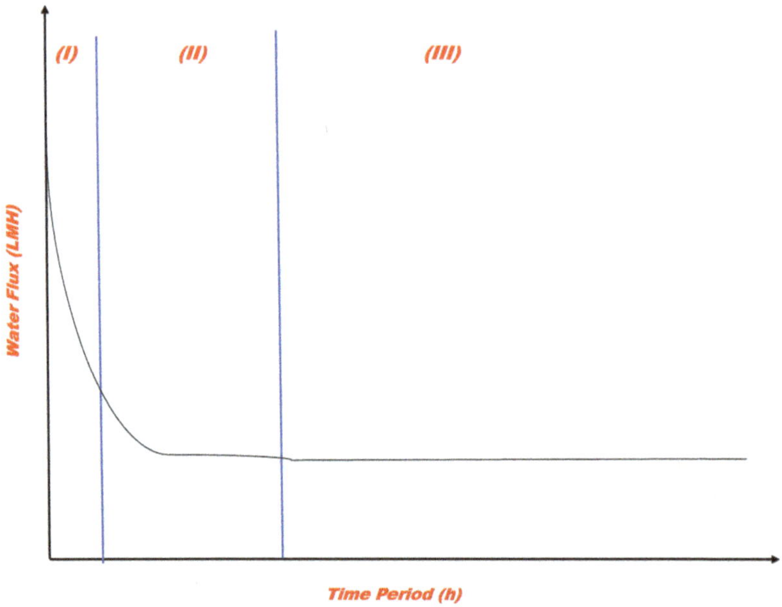

Fig. 6.7 Typical water flux-time curve during filtration processes. A general flux-time curve of filtration process, as indicated, starts with (I) a rapid initial decline of the permeate flux, (II) followed by a long period of gradual flux decrease, and (III) terminates with a steady-state flux

is generally more dominant when the size and shape of particles and pores are identical. Pore blockage in membrane is a rapid process compared to cake layer formation as less than one layer of particles is enough to attain the full pore blockage (Ho and Zydney 2000; Wang and Tarabara 2008; Guo et al. 2012).

Whereas, further permeate flux decreases after pores blocking because of cake layer formation growth on the membrane surface. The cake layer formation creates an additional membrane resistance to the permeate water flow and the membrane resistance of the cake layer inclines with the growth of cake layer thickness. As a result, the permeate water flux continues to decrease with time (Zhu and Elimelech 1997; Duclos-Orsello et al. 2006).

6.3.2.2 Membrane Fouling Control

After thorough analysis of modelling of permeate flux decline, the general ways to prevent and reduce membrane fouling have been mentioned in this section. Table 6.3 indicates an overview of various approaches to reduce the influence of membrane fouling. Typically, it has been categorized into direct and indirect techniques. Direct method includes the use of turbulence, which are clearly identical to the preventive measures applied to reduce the intensity of a concentration boundary layer (Field 2010). Whereas, the selection of proper operating models has

Table 6.3 Typical ways to reduce and prevent membrane fouling (Maartens et al. 2002; Field 2010)

Direct methodologies	Indirect methodologies
Turbulence promoters (eg. Modified membrane spacers)	Pretreatment by filtration
Pulsed or reverse flow	Treatment of the membrane surface
Rotating or vibrating membranes	Preparation of more hydrophilic Membranes
Stirred cells with rotating blades close to the Membrane	Selection of appropriate operating mode
Periodic maintenance cleaning	Selection of optimum operating conditions
(a) Chemical cleaning	
(b) Hydraulic cleaning	
(c) Mechanical cleaning	
Periodic backwash with permeate or gas	
Generation of a dynamic membrane layer	

An overview of several approaches to minimize the influence of membrane fouling. These have been divided into direct and indirect methods

been installed with indirect methodologies. In other words, it circumscribes the selection of whether one is selecting crossflow filtration or dead-end (direct) filtration with simultaneous periodic backflush (Vrouwenvelder and Van der Kooij 2001; Drews 2010; Field 2010).

Thus, from this brief introduction of minimizing and prevention of membrane fouling one can observe that a several approaches have been designed. While prevention is the objective, and this can be attained on the laboratory scale with well-defined feed solution, but the reality is that the primary aim is reduction. Currently, for new applications, pilot plant examines to establish the degree of membrane fouling rates and hence, the effectiveness of cleaning methods are necessary.

6.3.3 Reverse Solute Flux

Typically, in osmotically driven membrane technology, reverse flux of the solute from the draw solution stream through the membrane to the feed solution stream is also unavoidable due to differences in concentration. Earlier, Cath et al. reported that the reverse solute flux of the draw solution shall be considered thoroughly because it may increase the capital cost of the entire process (Cath et al. 2006; Hancock and Cath 2009). Current researches have correlated the reverse solute flux of the draw solute to membrane fouling. As per Lay et al. and Lee et al., suggested that reverse solute flux of the draw solution can increase the cake enhanced osmotic pressure effect and aggravate forward osmosis membrane fouling (Lay et al. 2010;

Lee et al. 2010; Qin et al. 2010). Thus, multivalent ion based solutions with lower diffusion coefficients might be favourable in certain applications in which high rejection is preferred (Cath et al. 2006). But unfortunately, few multivalent ions (for e.g. $Mg^{2+}$and Ca^{2+}) might interfere with the foulants in the feed stream solution after reverse solute diffusion, which is likely to aggravate membrane fouling (Zou et al. 2011). Additionally, multivalent ion solutions may also offer more severe internal concentration polarization effect due to lower solution diffusion coefficients and larger ion size (Zhao and Zou 2011a, b). The specific reverse solute diffusion, which is termed as the ratio of the reverse solute diffusion to the water flux, has been recommended as an evaluation of membrane selectivity(Hancock and Cath 2009). This factor offers another parameter for the measure of forward osmosis performance in addition to the water flux (Jw) and salt rejection %. In general, a higher value indicates a deterioration in membrane selectivity and a lower efficiency in forward osmosis process. Recently, a study has indicated that the above-mentioned factor is evaluated by the selectivity of the membrane active layer, but it is independent of the concentration of draw solute and the structure of the supportive layer of membrane (Phillip et al. 2010). This observation possesses significant implications as it offers another criterion for the development of a novel forward osmosis based membrane: high selectivity of the membrane active layer. This can reduce the reverse solute flux, and thus enhance forward osmosis performance. Moreover, recent research revealed that utilizing a multivalent ion solution as a novel draw solute may reduce the reverse solute flux (Hancock and Cath 2009) and therefore, minimize membrane fouling (Lay et al. 2010), but it may lead to higher internal concentration polarization (Zhao and Zou 2011a) which must be fully considered. Thus, it can be concluded that, reverse solute flux became one of the crucial issues in osmotically driven membrane processes and it must be thoroughly considered and can be reduced in the novel design/developments of both forward osmosis membranes and draw solutions.

6.4 Advancements in Forward Osmosis Process

6.4.1 Exploration of Draw Solutes

Forward osmosis emerges out as key technology for overcoming the water scarcity issue in the future. This continuous process possesses crucial potential in order to attain energy-efficient separations in many contexts, such as the desalination of seawater, and brackish water or the purification of contaminated water sources (Cath et al. 2006). Moreover, the key advantages of forward osmosis (compared with reverse osmosis) are the high amount of water recovery, much considerably low energy consumption, and less tendency for membrane fouling (Cath et al. 2006). The forward osmosis process can be operated inexpensively because of the absence of hydraulic pressure. Draw solutes is one of the most important component for forward osmosis but in many cases lower water flux, high reverse solute flux, toxic by-products and poor performance have been identified as the main drawbacks of draw agents (Zhao et al. 2012a, b; Ge et al. 2013; Ray et al. 2016a, b). Therefore, there is a need of new draw solute that can overcome these drawbacks. In addition to that, the draw solute must be nontoxic and compatible with forward osmosis membranes (Zhao et al. 2012a; Ge et al. 2013). However, the following criteria must be considered while selecting a draw agent which are as follows: (a) high osmolality, that may increase the water flux (Jw); (b) lower reverse solute flux (Js), which may reduce the replenishment cost of draw solute; and (c) high recovery and rejection, which may reduce operating costs and energy consumption during the process (Achilli et al. 2010; Ge et al. 2012). There are other major criteria that may include nontoxicity, favourable solubility, and limited membrane fouling (Phuntsho et al. 2011).

Many efforts have been executed to explore suitable draw solutions. Earlier, volatile solutions, such as sulfur dioxide, were utilized as draw solute in 1960s and could be recovered by a heating gas stripping operational condition. Then in 1970s, a draw solute made of aqueous aluminium sulfate has been prepared as an osmotically active agent and particular salts were added in solutions in order to facilitate the separation of salts by precipitation method, as well as to neutralize the solution (Ling and Chung 2011; Chung et al. 2012; Na et al. 2014).

6.4.2 *Draw Solute Characteristics Affect Forward Osmosis Performance*

The osmolality across the forward osmosis membrane is the driving force of the entire process, therefore draw solution should possess high osmolality and high solubility in water. Typically, the osmolality must be much higher than the osmo lality of the feed stream solution in order to produce high permeate water flux across the forward osmosis membrane. As proposed earlier by Van't Hoff (van't Hoff 1887), the osmotic pressure (π) of an ideal dilute solution can be indicated as mentioned below, Eq. 6.10:

$$\pi = \mathrm{n}\left(\frac{\mathrm{C}}{\mathrm{m}}\right)\mathrm{R.T} \tag{6.10}$$

where n represents the no. of moles of species formed by the dissociation of solutes in the solution; c represents the concentration of the solute in terms of g/L; M represents the molecular weight of the solute in terms of g/mol; R represents the gas constant (R = 0.0821); T represents the absolute temperature of the solution in terms of K.

Based on the above-mentioned equation, it can be stated that a draw solution of a low molecular weight and high water solubility will induce a high osmotic potential and finally, produce high permeate water flux across the forward osmosis membrane. Therefore, the draw solution concentration has the capability to influence the forward osmosis process performance.

The diffusion coefficient (Ds) of the draw solution may also affect the performance of the entire process and can be expressed as Eq. 6.11;

$$D_s = t\tau/\mathrm{K}\,\varepsilon \tag{6.11}$$

where K represents the solute resistance while diffusion within the membrane underlying support layer, t, τ and ε represent the thickness, tortuosity and porosity of the membrane underlying support layer respectively. The above-mentioned equation clearly indicates that a draw solution with a high diffusion coefficient (Ds) will easily diffuse across the membrane support layer and hence it decreases the internal concentration polarization effect. Typically, draw solution with less molecular weight possess higher diffusion coefficient (Ds) and can produce higher permeate water fluxes through the membrane. Nevertheless, it has been demonstrated that lower molecular weight draw solutes indicate high reverse solute flux. The reverse salt diffusion (Js) of a draw solute can be evaluated by using this equation, Eq. 6.12:

$$Js = \frac{VtCt - VoCo}{A.\Delta t} \tag{6.12}$$

where C_t (mol/L) and V_t (L) are the concentration of salt and volume of feed at a time t, C_0 (mol/L) and V_0 (L) are the concentration of salt and volume of feed at time 0, $\Delta t = t - t_0$ (h) represent the time duration of the experiment and Am (m^2) represents the effective membrane area utilized. Additionally, reverse solute flux of the draw solute decreases the performance of the process by decreasing the net osmotic potential across the forward osmosis membrane. The accumulation of draw solution in the feed stream solution may also have negative effect on entire treatment processes (Phillip et al. 2010).

In general, pressure-based membrane separation process, temperature is one of the crucial factors affecting the forward osmosis performance; and it has been observed that draw solution temperature may affect significantly on the efficiency of the forward osmosis process. It has been revealed that, increasing the temperature of draw solution can improve the permeate water flux because of reduction in solution viscosity and hence enhanced mass transfer coefficient through the membrane (Zhao and Zou 2011a, b). In addition to that, increasing the temperature of draw solution will incline the diffusion coefficient (Ds) in solution and thus decrease the K value (i.e. solute resistance to diffusion within the support layer of the membrane) based on the given equation that will finally enhance the permeate water flux (McCutcheon and Elimelech 2007; Wang et al. 2011). Nevertheless, it was also observed that increasing the temperature enhanced water flux only to a certain critical level; after that scaling of membrane can occur, that results in water flux decline and in overall decreases efficiency of the process(Zhao and Zou 2011a, b, Kim et al. 2014).

6.4.3 Classifications of Draw Solute

In the past few years, there has been a fast growth in the preparation of proper draw solutes for forward osmosis technology; but, very few reviews articles based on draw solutions have been published till date. Basically, draw solutes can be categorized into four classes which is as follows: (1) volatile compounds, (2) organic compounds, (3) inorganic compounds and (3) novel synthetic compounds including magnetic nanoparticles and polymer hydrogel (Ge et al. 2013; Nguyen et al. 2016). Therefore, an overview of various draw solutions that has been utilized till date, including some of limitations and applications is summarised in Table 6.4.

6.4.4 Development of High Performance Forward Osmosis Membranes

As far as forward osmosis membranes are concerned, there are very limited review articles on the advancements and development of forward osmosis membrane. Generally, forward osmosis membranes were casted by conventional technique i.e. phase inversion and thin-film composites membranes by interfacial polymerization technologies. Forward osmosis membranes fabricated from the layer-by-layer technique have been analysed but unfortunately, their reverse salt fluxes found to be extremely high (Qiu et al. 2011; Cui et al. 2013). In order to enhance water flux, using of hydrophilic materials as substrates for high performance forward osmosis membranes is necessary (Chung et al. 2015). Currently, thin film composite membranes fabricated on nano-fibrous and multi-bore substrates with high mechanical strength/properties have been performed. Membrane based research and development must consider membranes with minimum membrane fouling and lower internal concentration polarization. Till now, forward osmosis membranes with double skinned layers, have demonstrated promising efficiencies with reduced membrane fouling and lower internal concentration polarization (Wang et al. 2010; Duong et al. 2014).

The membrane morphology and structure for an 'ideal' forward osmosis membrane is different from a reverse osmosis membrane, one of the crucial factors that affects the performance is internal dilutive concentration polarisation within the porous supportive layer of the membrane. This indicates that the forward osmosis membrane must be thin, with a porous open structure and lower tortuosity. As far as the forward osmosis membrane performance is concerned, it is noteworthy to indicate that most forward osmosis membranes are of asymmetric composite type membrane which means that it consists of a thin rejection layer (thickness of 100–200 nm) combined with underlying supportive layer (thickness of 100–200 μm), that provides mechanical strength and overall support to the forward osmosis membrane (Yip et al. 2010). So, it is necessary to mention that forward

Table 6.4 Overview of various draw solution utilized in forward osmosis process

Type of draw solution	Draw solute	Recovery process	Limitations	Applications and commercial status
Volatile compounds	Ammonium Bicarbonate (NH_4HCO_3)	Heating – decomposition into NH_3 and CO_2	Low solubility; High reverse salt flux; High replenishment cost; Thermally not stable	Seawater desalination; Commercial (Oasys Water) for brine concentration
	Sulfur dioxide (SO_2)	Heating, air stripping or distillation	Volatile; Corrosive; Unstable	Seawater desalination
Organic compounds	Glucose ($C_6H_{12}O_6$)	Direct application, Lower RO pressure	High molecular size causes high ICP effect	Emergency drinking solutions
	Fructose ($C_6H_{12}O_6$)	Direct application, Low pressure RO	High molecular size causes high internal concentration polarization (ICP) effect	Nutritious drink production
	Sucrose ($C_{12}H_{22}O_{11}$)	Nanofiltration	Low water flux	Wastewater treatment
	Ethanol (C_2H_6O)	Pervaporation-based separation	High reverse salt diffusion and low water flux	Recovery of water from highly impaired sources
	Sodium formate (HCOONa)	RO process	High reverse salt flux; High replenishment cost	Wastewater treatment
	Sodium acetate ($C_2H_3NaO_2$)	RO process	Relatively high replenishment cost compared to inorganic salts	Wastewater treatment
	Switchable polarity solvents (SPS)	Polar to non-polar phase shift induced by CO_2	Degradation of FO membrane	Industrial purification of water Under development (Idaho National Laboratory)
Inorganic compounds	Sodium Chloride (NaCl)	RO process, Distillation/RO, Direct application	High reverse salt diffusion; Protein deterioration	Seawater desalination, Algal biodiesel production, Sucrose concentration, Reuse of domestic wastewater
				Pre-commercial (IDE Technologies, Porifera)

(continued)

Table 6.4 (continued)

Type of draw solution	Draw solute	Recovery process	Limitations	Applications and commercial status
	Diammonium Phosphate ($(NH_4)_2HPO_4$)	Direct application	Low water flux	Direct fertigation
	Potassium Chloride(KCl)	Direct application	High reverse salt diffusion	Direct fertigation
	Ammonium Chloride (NH_4Cl)	Direct application	High reverse salt diffusion	Direct fertigation
	Ammonium Nitrate (NH_4NO_3)	Direct application	High reverse salt diffusion	Direct fertigation
	Potassium Bromide (KBr)	RO process	Very high reverse salt diffusion; High replenishment cost	Desalination
	Sodium Bicarbonate ($NaHCO_3$)	RO process	Low water solubility; Contain scale precursor ions	Desalination
	Potassium Bicarbonate ($KHCO_3$)	RO process	Contain scale precursor ions; Not easily recovered by RO	Desalination
	Magnesium Chloride ($MgCl_2$)	Nanofiltration or direct application	High viscosity; Low diffusion coefficient; Mg^{2+} may affect membrane fouling by complexing with some functional group	Seawater desalination; Emergency drinking solutions
	Calcium Chloride ($CaCl_2$)	RO process	Contain scale precursor ions	Tomato juice concentration; Emergency drinking solutions
	Ammonium Sulfate ($(NH_4)_2SO_4$)	Direct application	High replenishment cost	Direct fertigation
	Magnesium Sulfate ($MgSO_4$)	Nanofiltration	High viscosity; Low water solubility; Contain scale precursor ions	Desalination
	Calcium Nitrate ($Ca(NO_3)_2$)	Direct application	High replenishment cost; Poor water extraction capacity	Direct fertigation
	Ammonium Phospate ($NH_4H_2PO_4$)	Direct application	Low water flux	Direct fertigation

(continued)

Table 6.4 (continued)

Type of draw solution	Draw solute	Recovery process	Limitations	Applications and commercial status
Novel synthetic material	Polyacrylic acid MNPs (PAA MNPs)	Magnetic field separator, FO process using RO brine as draw solution, Ultrafiltration	Slightly drop of water flux due to agglomeration of the magnetic nanoparticles	Protein enrichment, desalination
	Nano size dextran coated Ferric oxide MNPs (Fe_3O_4)	External magnet	Slightly drop of water flux due to agglomeration of the magnetic nanoparticles	Brackish water desalination
	Polyelectrolytes of polyacrylic acid sodium (PAA-Na)	FO-MD integrated process, Ultrafiltration	High viscosity	Wastewater treatment
	Thermo-sensitive polyelectrolytes	Hot ultrafiltration	Poor water flux	Desalination
	Polymer hydrogels	Direct application, Heating, Pressure Stimuli	Energy intensive; Poor water flux	Seawater desalination
	Polymer-graphene composite hydrogels	Heating	Poor water flux	Seawater desalination
	Na_3PO_4 + Triton X100	Two stage: UF-NF system	Lower water flux	Desalination
	EDTA–2Na + Trion X100	NF-TS80	Rejection = 95% (Recovery stage)	Desalination

The driving force for this separation is an osmotic pressure gradient between a solution of high concentration, often referred to as a "draw" and a solution of lower concentration, referred to as the "feed"

FO Forward osmosis, *RO* Reverse osmosis, *MD* Membrane distillation, *ICP* Internal concentration polarization, *UF* Ultrafiltration, *NF* Nanofiltration

osmosis membrane performance is based on the physical properties of both the thin rejection skin and the support skin of the membrane.

6.4.4.1 Pure Water Permeability Coefficient (A-value)

A-value which is also known as the pure water permeability coefficient. Typically, it is a feature of the active layer of membrane which helps to evaluate the permeate water flux performance at particular osmotic pressure difference across the membrane's active layer. Future researchers must focus to increase the membrane

A-value in order to enhance the permeate water flux across the membrane during forward osmosis process. Hence, it can be concluded that higher the A-value of forward osmosis membrane, better the performance. The A-value can be evaluated by utilizing the given equation (Ong et al. 2015), Eq. 6.13;

$$A = Jw/(\Delta\pi - \Delta P) \tag{6.13}$$

where, Jw is the permeate water flux across the membrane, $\Delta\pi$ is the osmotic pressure difference across the membrane whereas ΔP is the hydrostatic pressure difference.

6.4.4.2 Salt Permeability Coefficient (B-value)

B-value which is also known as the salt permeability coefficient. Generally, it is another feature of the active layer of membrane that helps to evaluate the reverse diffusion of a particular draw solution at a given concentration difference of the solute across the membrane's active layer. The researchers must focus to decrease the membrane B-value in order to minimize the quantity of draw solution being lost into the feed solution stream during forward osmosis process. Hence, it can be concluded that lower the B-value of forward osmosis membrane, better the performance. The B-value can be evaluated by utilizing the given equation (Shaffer et al. 2015), Eq. 6.14;

$$B = \frac{Jw(1 - R)}{R} \exp\left(-\frac{Jw}{k}\right) \tag{6.14}$$

Where, R is the salt rejection, was evaluated from the difference between the bulk feed (C_f) and permeate (C_p) salt concentrations, and $R = 1 - C_p/C_f$., whereas k is the mass transfer coefficient for the crossflow channel of the forward osmosis membrane module.

6.4.4.3 Structural Parameter (S-value)

S-value which is also known as the structural parameter. Basically, it is another crucial analysis of the resistance of the support layer of membrane towards the solute diffusion. Future researchers must focus to minimize the S-value of forward osmosis membrane because lower the S-value, the easier for solutes to diffuse inside the membrane's support layer, and hence higher the permeate water flux during forward osmosis operation. According to some researchers, for high performance forward osmosis membrane, future research and development must aim for S-values lower than 300 μm. The S-value can be evaluated by utilizing the given equation (Wei et al. 2011), Eq. 6.15;

$$S = \frac{D}{Jw} . \ln \left(\frac{B + A.\pi D}{B + Jw + A.\pi F} \right) \tag{6.15}$$

where, D is the diffusion coefficient of the draw solute used for Jw, is the osmotic pressure of bulk draw solute and is the osmotic pressure of feed solution. The lower S-value is the indication of less internal concentration polarization effect.

6.5 Recent Developments in Membrane Distillation Application

Membrane distillation is an emerging non-isothermal membrane technology that utilizes thermal energy in order to maintain a vapor phase of volatile molecules existing in the feed solution (water molecules) and condensing of the permeated vapor in the permeate stream. The difference in temperature leads to a vapor pressure difference across the membrane. Typically, only vapor can pass across the membrane because of hydrophobic nature of the membrane (El-Bourawi et al. 2006). Furthermore, there are four prominent configurations for the membrane distillation system, the difference being in the process to impose a vapor pressure difference across the membrane's pores to drive the permeate water flux. These four configurations have been discussed in Table 6.5. In direct contact membrane distillation, an aqueous solution colder than the feed solution is maintained in direct contact with the distillate stream of the hydrophobic membrane. Both the hot feed solutions and cold distillate solutions are circulated to the membrane surfaces by utilizing pumps. In general, direct contact membrane distillation is the simplest and widely studied membrane distillation configuration (Wang and Chung 2015). Whereas, a stationary air-gap is inserted in between the hydrophobic membrane and a condensing surface in air gap membrane distillation configuration. In this configuration, the distillate volatile molecules (i.e. water molecules) permeates both the porous hydrophobic membrane and the stagnant air-gap to finally condense over a cold surface inside the membrane module (Meindersma et al. 2006). On the other hand, in the third membrane distillation configuration, i.e. sweeping gas membrane distillation, a cold inert gas (mostly dried air) sweeps the distillate stream of the membrane carrying the vapor molecules and condensation occurs outside of the membrane module. In this configuration, because of heat transfer from the hot feed stream via the membrane, the sweeping gas temperature in the distillate side inclines constantly along the membrane module length (Khayet 2011). In order to establish the driving force across the membrane distillation membrane, vacuum is also utilized in the distillate stream by a vacuum based pump. The applied vacuum pressure must be lower than the saturation pressure of the volatile molecules (mostly water molecules) to be separated from the hot feed solution stream. Basically, in this module, condensation also occurs outside of the membrane configuration (Ding et al. 2006).

Table 6.5 Comparative study of four major categories of membrane distillation (Summers et al. 2012; Shirazi et al. 2015)

Membrane distillation configurations	Conditions	Summarization
Direct contact membrane distillation	Membrane is in direct contact with process liquids, i.e. hot feed stream and cold distillate streams	(a) High permeation water flux
		(b) Low energy efficiency
		(c) Simplest membrane distillation mode
Air-gap membrane distillation	A stationary air-gap in the permeate side is inserted in between the membrane and a condensing plate	(a) Highest energy efficiency
		(b) Low permeation flux
		(c) Air-gap is around 2–10 mm
Sweeping gas membrane distillation	Stripping cold inert gas or air is utilized as a carrier for the produced vapor molecules in the permeate stream	(a) Useful for concentrating of non-volatile compounds
		(b) Condensation happens outside the module
Vacuum membrane distillation	Permeate stream is vapor or air under vacuum conditions	(a) Useful for removal of volatile compounds
		(b) Permeate is condensed outside the module

The basic four techniques mainly differ by the arrangement of their distillate channel or the manner in which this channel is operated. The following technologies are as follows: Direct Contact Membrane Distillation; Air Gap Membrane Distillation; Vacuum Membrane Distillation; Sweeping Gas Membrane Distillation

6.5.1 Operating Parameters of Membrane Distillation Process

In this part of this article, the influence of feed temperature (T_f), effect of concentration and type of membrane has been thoroughly reviewed and key observations will be discussed.

6.5.1.1 Feed Temperature Dependence

The feed temperature has a major impact on the permeate water flux which has been discussed in Table 6.6. As per Antoine's equation, the vapour pressure inclines exponentially with increasing temperature. Hence, it can be concluded that the operating temperature in membrane distillation has an influence on the permeate water flux (Alklaibi and Lior 2005; Alkhudhiri et al. 2012). It has been observed that, the permeate water flux increases exponentially when the temperature of the hot feed stream solution rises (Gunko et al. 2006). According to some authors, it

Table 6.6 Effect of temperature on water flux

Membrane distillation type	Membrane type	Solution type	Feed temperature (Tf) (°C)	Permeate Flux (Kg/m^2. h)	References
AGMD	PVDF	Artificial seawater	40–70	1–7	Banat and Simandl (1998)
DCMD	PVDF	Pure water	40–70	3.6–16	Phattaranawik et al. (2003)
DCMD	PTFE	NaCl (2 M)	17–35	3–25	Martinez-Diez et al. (1999)
SGMD	PTFE	Pure water	40–70	5–16.5	Khayet et al. (2000)
DCMD	PVDF	Pure water	40–70	7–33	Srisurichan et al. (2006)
AGMD	PTFE	NaCl (1%)	5–45	1–6	Hsu et al. (2002)

Typically, the vapour pressure increases with increasing temperature difference that results in higher permeate flux
AGMD Air Gap Membrane Distillation, *DCMD* Direct Contact Membrane Distillation, *SGMD* Sweeping Gas Membrane Distillation, *PVDF* Polyvinylidene fluoride, *PTFE* Polytetrafluoroethylene

was revealed that increasing the temperature difference, that results in increased vapour flux. Furthermore, the temperature polarization declines with increasing temperature of feed stream solution (Phattaranawik and Jiraratananon 2001).

6.5.1.2 Effect of Concentration

There has been a significant decline in the permeate water flux when feed concentration increases due to decrease in vapour pressure and increase in temperature polarization (Martínez 2004). Similarly, for long term operation, there is reduction in permeate water flux with time. Furthermore, some authors pointed that the permeate water flux decreases slightly with increase in feed concentration. There is reduction of about 12% in permeate water flux when the feed (NaCl) concentration increased from 0 to 2 molar concentration (Qtaishat et al. 2008). This decline in the permeate water flux is because of the reduction in vapour pressure. Moreover, key findings for this decrease were reported which are as follows: (a) water activity, that is a function of temperature, declines when the concentration inclines (b) mass transfer coefficient of the boundary layer of the membrane at the feed stream side declines due to increased impact of concentration polarization, and (c) heat transfer coefficient declines at the boundary layer, due to reduction in the membrane surface temperature (Banat and Simandl 1994). Hence, vapour pressure of the feed stream solution decreases, that causes reduction in performance of membrane distillation.

6.5.1.3 Type of Membranes

Typically, the membrane permeate water flux is proportional to the porosity, whereas it is inversely proportional to the thickness and tortuosity of the membrane (Garcıa-Payo et al. 2000; Ray et al. 2016a, b). Some authors revealed that for a higher pore size membrane, higher permeate water flux can be achieved. Additionally, higher water flux is attained utilizing a membrane without support layer, compared to the same kind of membrane pore size with support layer (Garcıa-Payo et al. 2000). Similarly, for a more efficient membrane distillation process, lower thermal conductivity material must be utilized (without support layer membrane) (Izquierdo-Gil et al. 1999; Alklaibi and Lior 2005).

6.5.2 Advancements in Fabrication of High Performance Membrane

One of the key issues in membrane distillation process is the lack of efficient high performance hydrophobic membrane distillation membranes. Few researchers have pointed certain requirements for a potential membrane distillation membranes which are as follows: (a) single layered or multi-layered with at least one must be hydrophobic in nature; (b) pore size varies from nanometers to few micrometers with a narrow distribution; (c) tortuosity factor must be small; (d) hydrophobic layer must have high porosity; (e) thickness must be optimum so as to increase the mass transport and reduce the heat loss across the membrane; (f) thermal conductivity must be low; (g) surface of the membrane must be smooth in order to be avoid membrane fouling; (h) thermal stability must be good for long term application; (i) must be resistant to chemicals like acid (to avoid scaling issue); (j) high performance of membrane distillation membranes for long term operation.

Recently, nanotechnology has more impact on membrane based desalination techniques including membrane distillation. Electrospun nanofiber based membranes have been demonstrated in many experiments and have shown interesting outcomes. Table 6.7 indicates the contribution of electrospinning technology for fabricating high performance membranes. Due to the control in operational conditions based on fiber size, shape, and morphology, electrospun membranes have been utilized for wastewater filtration and membrane distillation applications (Ray et al. 2016a, b; 2017).

Typically, the modification assured the operational stability of the application and a permeate water of high quality has been attained during long time operation. According to some reported results, the membrane surface of polyetherimide hollow fiber membrane was modified with fluorinated silica layer. Interestingly, there is an increase in hydrophobicity of the membrane due to increase in surface roughness and decrease in surface energy of the fabricated membrane (Zhang and Wang 2013). On the other hand, Fang et al. utilized fluoroalkylsilane to make

Table 6.7 Electrospun nanofibrous membranes fabricated for membrane distillation

Technique used	Membrane material	Porosity (%)	Contact angle (°)	Flux (LMH)	References
Electrospinning with post treatment	PVDF	80	150	31.6	Liao et al. (2013a, b)
Electrospinning followed by hot pressing	PVDF-HFP	63	125	21	Lalia et al. (2013)
Electro–spinning followed by hot pressing	PVDF	71	136	20.6	Liao et al. (2013a, b)
Electro-spinning followed by sintering	PTFE	82	136	16	Zhou et al. (2014)
Electrospinning	PVDF	72	143	39	Essalhi and Khayet (2014)
Electrospinning	PVDF-HFP	80	127	13	Lalia et al. (2014)

Contribution of electrospinning technology for casting high performance membrane distillation membranes
PVDF Polyvinylidene fluoride, *PVDF-HFP* Poly(vinylidene fluoride-co-hexafluoropropylene), *PTFE* Polytetrafluoroethylene

hydrophobic property to the surface of porous alumina ceramic hollow fiber membrane for membrane distillation process. Finally, the modified membrane has been utilized for vacuum membrane distillation and showed that the performance was comparable with the commercial polymeric membrane (Fang et al. 2012). A brief discussion has been summarised of various modifications applied on membrane surface in Table 6.8.

6.6 Sustainability in Desalination Application

Reverse osmosis is the most recent energy efficient sea-water cum brackish-water desalination process that has been widely utilized. Unfortunately, energy costs are responsible for up to 75% of the total operating expenditure in reverse osmosis plants. Typically, energy utilization in reverse osmosis desalination plants varies from 0.5–3 kWh/m^3 for brackish water desalination whereas 3–7 kWh/m^3 for seawater desalination (Akther et al. 2015). In general, brackish water plants often need more energy due to higher water recovery (Elimelech and Phillip 2011). In order to dilute seawater before it is fed to reverse osmosis stream, sources such as impaired water or wastewater effluent can be utilized. In addition to that, the direct mixing of such feed streams with the seawater would be counterproductive due to the fact that more toxic pollutants can be introduced into the feed solution stream, thus further limiting the effective processing of the impaired water and seawater. Therefore, the utilization of forward osmosis can assure that the seawater is effectively diluted by the impaired water without mixing directly and this will have a significant energy minimization in the whole plant as less energy is required

Table 6.8 Attempts for modification of membrane surface

Base of the polymer	Modification	Key objectives	References
PVDF	Immobilization of detonation nano-diamonds	To avoid wetting	Bhadra et al. (2014)
PVDF	Grafting of polyethylene glycol followed by deposition of TiO_2 particles	Incorporation of anti-oil fouling properties	Zuo and Wang (2013)
PVDF	Hydrophobic modified $CaCO_3$ nanoparticles	Improvement in pore size distribution, surface roughness and porosity	Hou et al. (2012)
PVDF	Deposition of TiO_2 nanoparticles on microporous membrane followed by flourosilanization	Improvement in hydrophobic character	Razmjou et al. (2012)
Polyetherimide	Blending followed by surface segregation of surface modifying macromolecule	Fabrication of hydrophobic/hydrophilic membrane	Essalhi and Khayet (2012)
CNT-bucky paper membranes	Thin layer coating of PTFE	Improvement of hydrophobicity and mechanical strength	Dumée et al. (2011)

Modifications based on pore size, surface area to volume ratio and morphology enhances the performance of membranes
PVDF Polyvinylidene fluoride, *CNT* Carbon Nano Tube, *PTFE* Polytetrafluoroethylene

to desalinate diluted saline water when compared with sufficient strength seawater. The technology that can be utilized, depend on the nature of feed stream solution and draw solution, the kind of forward osmosis membrane used, and the scale of production (lab scale/pilot scale/full scale) (Hoover et al. 2011; Akther et al. 2015). Recovery of draw solution is a prominent step and a sustainable utilization of forward osmosis for desalination of sea water would be easily attained by the integration of forward osmosis with other desalination methodologies. However, there are many opportunities of reduced energy requirements while the diluted saline draw solute of forward osmosis is utilized as feed stream in membrane distillation, electro-dialysis, reverse osmosis, microbial desalination cell, and thermal based distillation. Capital cost can be saved in the near future if diluted saline water from forward osmosis with lower osmolality is fed into full-scale reverse osmosis plants Additionally, water recovery from oily waste-water sources through forward osmosis–membrane distillation hybrid process offers promising advantages for forward osmosis process capital cost in the near future. Therefore, development of various draw solutions to enhance higher permeate water flux and higher recovery rate via novel developed draw solutes and newly developed membranes would also expand a perspective of new scopes for future forward osmosis applications. Compared to traditional draw solute i.e. NaCl, the utilization of hydroacids complexes that can minimize reverse solute flux, concentration polarization-inhibiting micellar solutions, simple-regenerative switchable polarity solvents, polymer hydrogels, polyelectrolytes based draw solutes, and hydrophilic based nanoparticles has assured the achievement of good permeate water fluxes (Elimelech and Phillip 2011; Akther et al. 2015). Despite these currently prepared draw solutes could reduce concentration polarization and reverse salt transport associated with NaCl, efforts to boost the lower water flux and high regeneration energy needed by them would be a significant advancement for future forward osmosis applications. Development in current efforts on the utilization of forward osmosis membranes in order to remove toxic heavy metals and hazardous materials/substances from waste-water feed stream such as boron would lead to the generation of high-quality permeate water and ensure sustainable health. Forward osmosis membrane alternative costs could be decreased by the minimization of membrane structural parameter by preparing membrane supportive layer that would safeguard the active layer and minimize internal concentration polarization. The current efforts on the doping of nanomaterials/nanoparticles and hydrophilic sulfonated based polymers into thin film composite, and preparation of double skinned membranes hold promising substitute for the applications in forward osmosis (Yip et al. 2010; Blandin et al. 2015).

Forward osmosis process has been less utilized for applications in industry, however for many years the potential scientists have performed experiments that lead to potential industrial applications. In this present article, most of the potential applications such as: hydration bags, seawater and brackish water desalinisation and power generation. Additionally, it offers much more applications in various areas such as: food processing industry, pharmaceutical industry or wastewater treatment plant and water purification plant (Jacob). Furthermore, the most famous examples in these research areas are as follows: concentrating fruit juices in food processing industries; osmotic pumps in pharmaceutical industries; and in wastewater treatment and water purification: (a) concentrating dilute industrial based waste-water; (b) concentrating landfill leachate; (c) direct potable reuse and reclamation for advanced life support systems; (d) concentrating digested sludge liquids; (e) forward osmosis for source water treatment/purification – hydration bags.

There are various applications which have been put into practice (utilized in industrial field), though on a relatively smaller scale or with few real references, however it has been seen that currently a significant rise in the number of research papers published in both form peer and non-peer reviewed articles. Interestingly, few of these have mentioned with the clear objective of indicating the diversified range of applications that forward osmosis processes may be implemented. However, some of them can be in the form of single step processes, where only forward osmosis has been implemented with no recovery of the draw solute and others have incorporated a recovery methodology of the draw solute. Table 6.9 indicates the commercial status of forward osmosis based on primary current applications.

As far as the applications are concerned, pharmaceutical based industries, energy and waste-water treatment plants could be interested by recent advancements and applications by forward osmosis.

Table 6.9 Recent commercial status along with primary applications

Company name	Application description	Current status
Forward Water Technologies	Desalination technology	Under Development
Hydration Technology Innovations (HTI)	Emergency drinks production unit	Commercial
	Flow back water concentration technique	Commercial
IDE Technologies	Osmotic Power processing	Pre-commercial
Oasys Water	Brine concentration technique	Commercial
Statkraf	Osmotic Power processing	Pre-commercial
Modern Water	Forward Osmosis or Reverse Osmosis processes	Commercial
	Forward osmosis – cooling tower make-up technique	Commercial
	Thermal desalination feed water softening	Under Development
Trevi Systems	Desalination technology	Commercial

A wide range of implementations that either have been put into practice is categorized based on their current status and applications

6.7 Research Challenges

The ideology of forward osmosis was well known for 100 years, but it was during the twentieth century that forward osmosis process emerges out as the water recycle and reclamation method. Nevertheless, due to lack of proper membranes and membrane modules, the research and development on forward osmosis was virtu ally dormant. However, during the meantime, reverse osmosis became the remarkable technology for desalination since the invention of commercially available membranes. Since 2000s, forward osmosis was drawing much attention as a future desalination technology. Recently, the crucial challenges of forward osmosis process are as follows (1) lack of effective membranes (2) lack of cost effective draw agents which can be easily recycled and regenerated (3) limited analysis of membrane fouling (Cath et al. 2006; Achilli et al. 2010; Chung et al. 2012; Alsvik and Hägg 2013). Despite of all the efforts put in during the last 20 years, forward osmosis is far from commercial progress because of the following reasons:

6.7.1 Loss of Driving Force Due to Concentration Polarization

To reduce the filtration resistance, the skin layer should be thin as much as possible. Forward osmosis as well as reverse osmosis are not exceptional cases. The skin layers of the membrane are cast on porous support layers that includes an intermediate layer and supportive fabric (Song and Elimelech 1995).

In reverse osmosis, feed stream solution comes in contact with the skin layer and the permeate fills in porous support layer of the membrane, hence concentra tion polarization takes place only in feed stream side of the membrane. In forward osmosis, concentration polarization takes place in both side of the membrane. If skin layer is facing towards the feed solution, concentration polarization takes place in feed side just like in reverse osmosis. In addition to that, concentration polarization also takes place in the porous support layer of the membrane, where fresh water contacts with saline water. Since the turbulence in the bulk stream does not effectively relieve the concentration polarization in the porous support layer of the membrane, concentration polarization leads to a loss of driving force. On the other hand, if the support layer faces towards feed solution, foulants will fill up the pores of support layer of the membrane leading to a serious filtration resistance. In meantime, the permeate emerging on the skin layer dilutes the draw solute and minimize the driving force (Sablani et al. 2001; Mulder 2012).

The loss of driving force resulted by concentration polarization is detrimental with respect to the energy potency in forward osmosis process. For example, even if 0.5 M NaCl solution is utilized as draw solute, that is equal to 25 bar of transmembrane pressure, permeate water flux is generally not higher than 10 LMH. This starkly contrasts with reverse osmosis, where water flux is basically 20–40 LMH at 10 bar. If permeability is compared, <0.4 LMH/bar and 2–4 LMH/bar for forward osmosis and reverse osmosis, respectively (Elimelech 2007; Akther et al. 2015).

6.7.2 *Reverse Salt Flux/Diffusion*

In the continuous process, salts permeate through the membrane by means of diffusion. A large amount of salt diffuses back to the feed stream because of extremely high salt concentration in the draw solution stream. As feed solution, must remain in contact with the membrane for a longer period of time in forward osmosis than in reverse osmosis because of lower flux, the quantity of reverse salt diffusion towards feed stream becomes even more. As a consequence, a large amount of salts will be lost from the draw solution stream. Recently, forward osmosis membranes have been utilized and applied to Membrane Bioreactor, where typically the draw solution is circulated during the continuous process, however, a large quantity of salts diffuse to biological system and hence damage the biological stability in membrane bioreactor system.

6.7.3 Expensive Production/Capital Cost

There are few impressions that forward osmosis is less vulnerable with membrane fouling, but the low membrane fouling is just a natural result of the lower permeate flux. As reported earlier, particulates in feed solution stream tend not to be retained on membrane surface at low flux due to the critical permeate flux caused by the reverse transport mechanism. However, even if some particles are retained, they tend not to be compacted too much at low permeate flux and would not lead to filtration resistances. Thus, as a result, membranes appear not to foul easily.

A drawback of the less operating permeate flux is a larger membrane surface areas essential to treat the same feed stream. As for example, if the permeate flux is 5 LMH in forward osmosis process, 100% more membrane surface areas are essential comparing to the reverse osmosis operating at 10 LMH permeate flux. It is noteworthy that, if the operating capital expenditure for draw solutes recovery will be considered, then the overall capital costs will incline even further.

6.8 Conclusion

In this paper, an elaborate review was conducted to analyse the recent trend and advancements in forward osmosis and membrane distillation desalination processes. Typically, publications from last 10 years were thoroughly reviewed in order to understand the current research and developments to render the forward osmosis and membrane distillation process more economical and practical. In this review paper, forward osmosis membrane modules, draw solutions, membrane distillation configuration, membrane characteristics, membrane fouling and operating parameters were examined. The draw solutes for application of forward osmosis process would be those that can be easily regenerated and re-concentrated utilizing minimal energy. Advancements in draw solutes for forward osmosis process that can possess high osmolality and lower toxicity. Thus, an extensive research can be performed in order to minimize the reverse solute flux and maximize the water flux in forward osmosis process. Recently, the development of new membranes became the most important measure to develop the field of forward osmosis and membrane distillation technology. In general, the membranes must possess high water permeability, high rejection, minimal internal concentration polarization, high chemical stability, and high mechanical and tensile strength are desired for the optimization of forward osmosis desalination efficiency. Compared to other desalination techniques, the crucial advantages of membrane distillation process include 100% theoretical rejection of ions, macromolecules and colloids; lower operating temperature and pressure; requirement of lower mechanical strength of the membrane; lesser footprint compared to conventional processes. However, besides these advantages, membrane distillation still faces difficulties for commercialization. Therefore, in case of membrane distillation process, membrane technology must be emphasised for higher rejection and water flux. Finally, the recent developments suggest that forward osmosis and membrane distillation were found to be sustainable process for practical application in large scale in near future.

Acknowledgements The authors would like to acknowledge the financial assistance from the Ministry of Science and Technology (MOST), Republic of China (Taiwan) and Institute of Environmental Engineering and Management (IEEM), National Taipei University of Technology under the grant number: 104-2221-E-027-004-MY3.

References

Abdelrasoul A, Doan H Lohi A (2013) Fouling in membrane filtration and remediation methods. Mass transfer-advances in sustainable energy and environment oriented numerical modeling 195–218

Achilli A, Cath TY, Childress AE (2010) Selection of inorganic-based draw solutions for forward osmosis applications. J Membr Sci 364(1):233–241

Akther N, Sodiq A, Giwa A, Daer S, Arafat H, Hasan S (2015) Recent advancements in forward osmosis desalination: a review. Chem Eng J 281:502–522

Alkhudhiri A, Darwish N, Hilal N (2012) Membrane distillation: a comprehensive review. Desalination 287:2–18

Alklaibi A, Lior N (2005) Membrane-distillation desalination: status and potential. Desalination 171(2):111–131

Al-Obaidani S, Curcio E, Macedonio F, Di Profio G, Al-Hinai H, Drioli E (2008) Potential of membrane distillation in seawater desalination: thermal efficiency, sensitivity study and cost estimation. J Membr Sci 323(1):85–98

Alsvik IL, Hägg M-B (2013) Pressure retarded osmosis and forward osmosis membranes: materials and methods. Polymer 5(1):303–327

Altaee A, Zaragoza G, van Tonningen HR (2014) Comparison between forward osmosis-reverse osmosis and reverse osmosis processes for seawater desalination. Desalination 336:50–57

Banat FA, Simandl J (1994) Theoretical and experimental study in membrane distillation. Desalination 95(1):39–52

Banat FA, Simandl J (1998) Desalination by membrane distillation: a parametric study. Sep Sci Technol 33:201–226

Banat F, Jumah R, Garaibeh M (2002) Exploitation of solar energy collected by solar stills for desalination by membrane distillation. Renew Energy 25(2):293–305

Bhadra M, Roy S, Mitra S (2014) Nanodiamond immobilized membranes for enhanced desalination via membrane distillation. Desalination 341:115–119

Blandin G, Verliefde AR, Tang CY, Le-Clech P (2015) Opportunities to reach economic sustainability in forward osmosis–reverse osmosis hybrids for seawater desalination. Desalination 363:26–36

Blatt WF, Dravid A, Michaels AS, Nelsen L (1970) Solute polarization and cake formation in membrane ultrafiltration: causes, consequences, and control techniques. In: Flynn JE (ed) Membrane science and technology. Plenum Press, New York, pp 47–97

Burggraaf AJ (1996) Important characteristics of inorganic membranes. Membr Sci Technol 4:21–34

Camacho LM, Dumée L, Zhang J, Li J-d, Duke M, Gomez J, Gray S (2013) Advances in membrane distillation for water desalination and purification applications. Water 5(1):94–196

Cath TY (2010) Osmotically and thermally driven membrane processes for enhancement of water recovery in desalination processes. Desalin Water Treat 15(1–3):279–286

Cath TY, Childress AE, Elimelech M (2006) Forward osmosis: principles, applications, and recent developments. J Membr Sci 281(1):70–87

Choo K-H, Lee C-H (1996) Membrane fouling mechanisms in the membrane-coupled anaerobic bioreactor. Water Res 30(8):1771–1780

Chung T-S, Zhang S, Wang KY, Su J, Ling MM (2012) Forward osmosis processes: yesterday, today and tomorrow. Desalination 287:78–81

Chung T-S, Luo L, Wan CF, Cui Y, Amy G (2015) What is next for forward osmosis (FO) and pressure retarded osmosis (PRO). Sep Purif Technol 156:856–860

Cui Y, Wang H, Wang H, Chung T-S (2013) Micro-morphology and formation of layer-by-layer membranes and their performance in osmotically driven processes. Chem Eng Sci 101:13–26

Dasgupta N, Ranjan S, Mundekkad D, Ramalingam C, Shanker R, Kumar A (2015) Nanotechnology in agro-food: from field to plate. Food Res Int 69:381–400

Dasgupta N, Ranjan S, Ramalingam C (2017) Applications of nanotechnology in agriculture and water quality management. Environ Chem Lett 15(4):591–605

Ding Z, Liu L, Li Z, Ma R, Yang Z (2006) Experimental study of ammonia removal from water by membrane distillation (MD): the comparison of three configurations. J Membr Sci 286 (1):93–103

Drews A (2010) Membrane fouling in membrane bioreactors—characterisation, contradictions, cause and cures. J Membr Sci 363(1):1–28

Duclos-Orsello C, Li W, Ho C-C (2006) A three mechanism model to describe fouling of microfiltration membranes. J Membr Sci 280(1):856–866

Dumée L, Campbell JL, Sears K, Schütz J, Finn N, Duke M, Gray S (2011) The impact of hydrophobic coating on the performance of carbon nanotube bucky-paper membranes in membrane distillation. Desalination 283:64–67

Duong PH, Chung T-S, Wei S, Irish L (2014) Highly permeable double-skinned forward osmosis membranes for anti-fouling in the emulsified oil–water separation process. Environ Sci Technol 48(8):4537–4545

El-Bourawi M, Ding Z, Ma R, Khayet M (2006) A framework for better understanding membrane distillation separation process. J Membr Sci 285(1):4–29

Elimelech M (2007) Yale constructs forward osmosis desalination pilot plant. Membr Technol 2007(1):7–8

Elimelech M, Phillip WA (2011) The future of seawater desalination: energy, technology, and the environment. Science 333(6043):712–717

Essalhi M, Khayet M (2012) Surface segregation of fluorinated modifying macromolecule for hydrophobic/hydrophilic membrane preparation and application in air gap and direct contact membrane distillation. J Membr Sci 417:163–173

Essalhi M, Khayet M (2014) Self-sustained webs of polyvinylidene fluoride electrospun nanofibers: Effects of polymer concentration and desalination by direct contact membrane distillation. J Membr Sci 454:133–143

Fang H, Gao J, Wang H, Chen C (2012) Hydrophobic porous alumina hollow fiber for water desalination via membrane distillation process. J Membr Sci 403:41–46

Field R (2010) Fundamentals of fouling. Membr Water Treat 4:1–23

Garcıa-Payo M, Izquierdo-Gil MA, Fernández-Pineda C (2000) Air gap membrane distillation of aqueous alcohol solutions. J Membr Sci 169(1):61–80

Ge Q, Su J, Amy GL, Chung T-S (2012) Exploration of polyelectrolytes as draw solutes in forward osmosis processes. Water Res 46(4):1318–1326

Ge Q, Ling M, Chung T-S (2013) Draw solutions for forward osmosis processes: developments, challenges, and prospects for the future. J Membr Sci 442:225–237

Gleick PH (1996) Water resources. Encycl Clim Weather 2:817–823

Gray GT, McCutcheon JR, Elimelech M (2006) Internal concentration polarization in forward osmosis: role of membrane orientation. Desalination 197(1–3):1–8

Gunko S, Verbych S, Bryk M, Hilal N (2006) Concentration of apple juice using direct contact membrane distillation. Desalination 190(1–3):117–124

Guo W, Ngo H-H, Li J (2012) A mini-review on membrane fouling. Bioresour Technol 122:27–34
Hancock NT, Cath TY (2009) Solute coupled diffusion in osmotically driven membrane processes. Environ Sci Technol 43(17):6769–6775
Ho C-C, Zydney AL (2000) A combined pore blockage and cake filtration model for protein fouling during microfiltration. J Colloid Interface Sci 232(2):389–399
van't Hoff J (1887) The function of osmotic pressure in the analogy between solutions and gases. Proc Phys Soc Lond 9(1):307
Hong S, Elimelech M (1997) Chemical and physical aspects of natural organic matter (NOM) fouling of nanofiltration membranes. J Membr Sci 132(2):159–181
Hoover LA, Phillip WA, Tiraferri A, Yip NY, Elimelech M (2011) Forward with osmosis: emerging applications for greater sustainability. ACS Publications, Washington DC
Hou D, Wang J, Sun X, Ji Z, Luan Z (2012) Preparation and properties of PVDF composite hollow fiber membranes for desalination through direct contact membrane distillation. J Membr Sci 405:185–200
Hsu S, Cheng K, Chiou J-S (2002) Seawater desalination by direct contact membrane distillation. Desalination 143(3):279–287
Izquierdo-Gil M, Garcıa-Payo M, Fernández-Pineda C (1999) Air gap membrane distillation of sucrose aqueous solutions. J Membr Sci 155(2):291–307
Jacob A (2006) A critical review of the history, development and future prospects for Forward Osmosis, Elsevier 28th November, p 1–3
Jain A, Ranjan S, Dasgupta N, Ramalingam C (2016) Nanomaterials in food and agriculture: an overview on their safety concerns and regulatory issues. Crit Rev Food Sci Nutr 15:1–21
Jiang T, Kennedy M, Guinzbourg B, Vanrolleghem PA, Schippers J (2005) Optimising the operation of a MBR pilot plant by quantitative analysis of the membrane fouling mechanism. Water Sci Technol 51(6–7):19–25
Kessler J, Moody C (1976) Drinking water from sea water by forward osmosis. Desalination 18 (3):297–306
Khayet M (2011) Membranes and theoretical modeling of membrane distillation: a review. Adv Colloid Interf Sci 164(1):56–88
Khayet M, Matsuura T (2011) Chapter 4-Formation of hollow fibre MD membranes. Membrane distillation. Elsevier, Amsterdam
Khayet M, Godino P, Mengual JI (2000) Nature of flow on sweeping gas membrane distillation. J Membr Sci 170(2):243–255
Kim Y, Elimelech M, Shon HK, Hong S (2014) Combined organic and colloidal fouling in forward osmosis: fouling reversibility and the role of applied pressure. J Membr Sci 460:206–212
Konieczny K, Rafa J (2000) Modeling of the membrane filtration process of natural waters. Pol J Environ Stud 9(1):57–64
Koschikowski J, Wieghaus M, Rommel M (2003) Solar thermal driven desalination plants based on membrane distillation. Water Sci Technol Water Supply 3(5–6):49–55
Kravath RE, Davis JA (1975) Desalination of sea water by direct osmosis. Desalination 16 (2):151–155
Lalia BS, Guillen-Burrieza E, Arafat HA, Hashaikeh R (2013) Fabrication and characterization of polyvinylidenefluoride-co-hexafluoropropylene (PVDF-HFP) electrospun membranes for direct contact membrane distillation. J Membr Sci 428:104–115
Lalia BS, Guillen E, Arafat HA, Hashaikeh R (2014) Nanocrystalline cellulose reinforced PVDF-HFP membranes for membrane distillation application. Desalination 332(1):134–141
Lawson KW, Lloyd DR (1997) Membrane distillation. J Membr Sci 124(1):1–25
Lay WC, Chong TH, Tang CY, Fane AG, Zhang J, Liu Y (2010) Fouling propensity of forward osmosis: investigation of the slower flux decline phenomenon. Water Sci Technol 61 (4):927–936
Lee S, Aurelle Y, Roques H (1984) Concentration polarization, membrane fouling and cleaning in ultrafiltration of soluble oil. J Membr Sci 19(1):23–38

Lee N, Amy G, Croué J-P, Buisson H (2004) Identification and understanding of fouling in low-pressure membrane (MF/UF) filtration by natural organic matter (NOM). Water Res 38 (20):4511–4523

Lee S, Cho J, Elimelech M (2005) Combined influence of natural organic matter (NOM) and colloidal particles on nanofiltration membrane fouling. J Membr Sci 262(1):27–41

Lee S, Boo C, Elimelech M, Hong S (2010) Comparison of fouling behavior in forward osmosis (FO) and reverse osmosis (RO). J Membr Sci 365(1):34–39

Liao Y, Wang R, Tian M, Qiu C, Fane AG (2013a) Fabrication of polyvinylidene fluoride (PVDF) nanofiber membranes by electro-spinning for direct contact membrane distillation. J Membr Sci 425:30–39

Liao Y, Wang R, Fane AG (2013b) Engineering superhydrophobic surface on poly (vinylidene fluoride) nanofiber membranes for direct contact membrane distillation. J Membr Sci 440:77–87

Linares RV, Li Z, Sarp S, Bucs SS, Amy G, Vrouwenvelder JS (2014) Forward osmosis niches in seawater desalination and wastewater reuse. Water Res 66:122–139

Ling MM, Chung T-S (2011) Desalination process using super hydrophilic nanoparticles via forward osmosis integrated with ultrafiltration regeneration. Desalination 278(1):194–202

Loeb S, Sourirajan S (1964) UCLA Engineering Report 60-60 (1960). US Patents 3 (3,133,137):196

Lutchmiah K, Verliefde A, Roest K, Rietveld LC, Cornelissen ER (2014) Forward osmosis for application in wastewater treatment: a review. Water Res 58:179–197

Maartens A, Jacobs E, Swart P (2002) UF of pulp and paper effluent: membrane fouling-prevention and cleaning. J Membr Sci 209(1):81–92

Martínez L (2004) Comparison of membrane distillation performance using different feeds. Desalination 168:359–365

Martınez-Dıez L, Vazquez-Gonzalez MI (1999) Temperature and concentration polarization in membrane distillation of aqueous salt solutions. J Membr Sci 156(2):265–273

Martinez-Diez L, Florido-Diaz F, Vazquez-Gonzalez M (1999) Study of evaporation efficiency in membrane distillation. Desalination 126(1–3):193–198

McCutcheon JR, Elimelech M (2006) Influence of concentrative and dilutive internal concentration polarization on flux behavior in forward osmosis. J Membr Sci 284(1):237–247

McCutcheon JR, Elimelech M (2007) Modeling water flux in forward osmosis: implications for improved membrane design. AIChE J 53(7):1736–1744

McCutcheon JR, McGinnis RL, Elimelech M (2006) Desalination by ammonia–carbon dioxide forward osmosis: influence of draw and feed solution concentrations on process performance. J Membr Sci 278(1):114–123

McGinnis RL, Elimelech M (2008) Global challenges in energy and water supply: the promise of engineered osmosis. Environ Sci Technol 42(23):8625–8629. doi:https://doi.org/10.1021/es800812m

McGovern RK (2014) On the potential of forward osmosis to energetically outperform reverse osmosis desalination. J Membr Sci 469:245–250

Meindersma G, Guijt C, De Haan A (2006) Desalination and water recycling by air gap membrane distillation. Desalination 187(1–3):291–301

Meng F, Chae S-R, Drews A, Kraume M, Shin H-S, Yang F (2009) Recent advances in membrane bioreactors (MBRs): membrane fouling and membrane material. Water Res 43(6):1489–1512

Miller JE, Evans LR (2006) Forward osmosis: a new approach to water purification and desalination, Sandia Report, SAND2006-4634, Sandia National Laboratories. United States Department of Energy, Albequerque/Livermore

Mulder J (2012) Basic principles of membrane technology. Springer Science & Business Media, Berlin

Na Y, Yang S, Lee S (2014) Evaluation of citrate-coated magnetic nanoparticles as draw solute for forward osmosis. Desalination 347:34–42

Nguyen NC, Chen S-S, Nguyen HT, Ray SS, Ngo HH, Guo W, Lin P-H (2016) Innovative sponge-based moving bed–osmotic membrane bioreactor hybrid system using a new class of draw solution for municipal wastewater treatment. Water Res 91:305–313

Ong RC, Chung T-S, de Wit JS, Helmer BJ (2015) Novel cellulose ester substrates for high performance flat-sheet thin-film composite (TFC) forward osmosis (FO) membranes. J Membr Sci 473:63–71

Phattaranawik J, Jiraratananon R (2001) Direct contact membrane distillation: effect of mass transfer on heat transfer. J Membr Sci 188(1):137–143

Phattaranawik J, Jiraratananon R, Fane A (2003) Heat transport and membrane distillation coefficients in direct contact membrane distillation. J Membr Sci 212(1):177–193

Phillip WA, Yong JS, Elimelech M (2010) Reverse draw solute permeation in forward osmosis: modeling and experiments. Environ Sci Technol 44(13):5170–5176

Phuntsho S, Shon HK, Hong S, Lee S, Vigneswaran S (2011) A novel low energy fertilizer driven forward osmosis desalination for direct fertigation: evaluating the performance of fertilizer draw solutions. J Membr Sci 375(1):172–181

Qin J-J, Kekre KA, Oo MH, Tao G, Lay CL, Lew CH, Cornelissen ER, Ruiken CJ (2010) Preliminary study of osmotic membrane bioreactor: effects of draw solution on water flux and air scouring on fouling. Water Sci Technol 62(6):1353–1360

Qiu C, Qi S, Tang CY (2011) Synthesis of high flux forward osmosis membranes by chemically crosslinked layer-by-layer polyelectrolytes. J Membr Sci 381(1):74–80

Qtaishat M, Matsuura T, Kruczek B, Khayet M (2008) Heat and mass transfer analysis in direct contact membrane distillation. Desalination 219(1–3):272–292

Ranjan S, Dasgupta N, Chakraborty AR, Melvin Samuel S, Ramalingam C, Shanker R, Kumar A (2014) Nanoscience and nanotechnologies in food industries: opportunities and research trends. J Nanopart Res 16(6):1–23

Ray SS, Chen S-S, Nguyen NC, Nguyen HT, Li C-W, Wang J, Yan B (2016a) Forward osmosis desalination by utilizing chlorhexidine gluconate based mouthwash as a reusable draw solute. Chem Eng J 304:962–969

Ray SS, Chen S-S, Li C-W, Nguyen NC, Nguyen HT (2016b) A comprehensive review: electrospinning technique for fabrication and surface modification of membranes for water treatment application. RSC Adv 6(88):85495–85514

Ray SS, Chen S-S, Nguyen NC, Hsu H-T, Nguyen HT, Chang C-T (2017) Poly (vinyl alcohol) incorporated with surfactant based electrospun nanofibrous layer onto polypropylene mat for improved desalination by using membrane distillation. Desalination 414:18–27

Razmjou A, Arifin E, Dong G, Mansouri J, Chen V (2012) Superhydrophobic modification of TiO 2 nanocomposite PVDF membranes for applications in membrane distillation. J Membr Sci 415:850–863

Sablani S, Goosen M, Al-Belushi R, Wilf M (2001) Concentration polarization in ultrafiltration and reverse osmosis: a critical review. Desalination 141(3):269–289

Schofield R, Fane A, Fell C (1987) Heat and mass transfer in membrane distillation. J Membr Sci 33(3):299–313

Seidel A, Elimelech M (2002) Coupling between chemical and physical interactions in natural organic matter (NOM) fouling of nanofiltration membranes: implications for fouling control. J Membr Sci 203(1):245–255

Shaffer DL, Werber JR, Jaramillo H, Lin S, Elimelech M (2015) Forward osmosis: where are we now? Desalination 356:271–284

Shirazi A, Mahdi M, Kargari A (2015) A review on applications of Membrane Distillation (MD) process for wastewater treatment. J Membr Sci Res 1(3):101–112

Shukla A, Dasgupta N, Ranjan S, Singh S, Chidambram R (2017) Nanotechnology towards prevention of anaemia and osteoporosis: from concept to market. Biotechnol Biotechnol Equip 31(5):863–879

Song L, Elimelech M (1995) Theory of concentration polarization in crossflow filtration. J Chem Soc Faraday Trans 91(19):3389–3398

Souhaimi MK, Matsuura T (2011) Membrane distillation: principles and applications. Elsevier, Amsterdam
Srisurichan S, Jiraratananon R, Fane A (2006) Mass transfer mechanisms and transport resistances in direct contact membrane distillation process. J Membr Sci 277(1):186–194
Su J, Zhang S, Ling MM, Chung T-S (2012) Forward osmosis: an emerging technology for sustainable supply of clean water. Clean Techn Environ Policy 14(4):507–511
Summers EK, Arafat HA, Lienhard JH (2012) Energy efficiency comparison of single-stage membrane distillation (MD) desalination cycles in different configurations. Desalination 290:54–66
Susanto H (2011) Towards practical implementations of membrane distillation. Chem Eng Process Process Intensif 50(2):139–150
Tan CH, Ng HY (2008) Modified models to predict flux behavior in forward osmosis in consideration of external and internal concentration polarizations. J Membr Sci 324(1):209–219
Tan CH, Ng HY (2013) Revised external and internal concentration polarization models to improve flux prediction in forward osmosis process. Desalination 309:125–140
Tang CY, She Q, Lay WC, Wang R, Fane AG (2010) Coupled effects of internal concentration polarization and fouling on flux behavior of forward osmosis membranes during humic acid filtration. J Membr Sci 354(1):123–133
Tomaszewska M (2000) Membrane distillation-examples of applications in technology and environmental protection. Pol J Environ Stud 9(1):27–36
Valipour M (2015a) Future of agricultural water management in Africa. Arch Agron Soil Sci 61 (7):907–927
Valipour M (2015b) Land use policy and agricultural water management of the previous half of century in Africa. Appl Water Sci 5(4):367–395
Valipour M (2016) Variations of land use and irrigation for next decades under different scenarios. IRRIGA 1(01):262–288
Vrouwenvelder J, Van der Kooij D (2001) Diagnosis, prediction and prevention of biofouling of NF and RO membranes. Desalination 139(1):65–71
Wang P, Chung T-S (2015) Recent advances in membrane distillation processes: membrane development, configuration design and application exploring. J Membr Sci 474:39–56
Wang F, Tarabara VV (2008) Pore blocking mechanisms during early stages of membrane fouling by colloids. J Colloid Interface Sci 328(2):464–469
Wang KY, Ong RC, Chung T-S (2010) Double-skinned forward osmosis membranes for reducing internal concentration polarization within the porous sublayer. Ind Eng Chem Res 49 (10):4824–4831
Wang KY, Teoh MM, Nugroho A, Chung T-S (2011) Integrated forward osmosis–membrane distillation (FO–MD) hybrid system for the concentration of protein solutions. Chem Eng Sci 66(11):2421–2430
Wei J, Qiu C, Tang CY, Wang R, Fane AG (2011) Synthesis and characterization of flat-sheet thin film composite forward osmosis membranes. J Membr Sci 372(1):292–302
Yannopoulos SI, Lyberatos G, Theodossiou N, Li W, Valipour M, Tamburrino A, Angelakis AN (2015) Evolution of water lifting devices (pumps) over the centuries worldwide. Water 7 (9):5031–5060
Yip NY, Tiraferri A, Phillip WA, Schiffman JD, Elimelech M (2010) High performance thin-film composite forward osmosis membrane. Environ Sci Technol 44(10):3812–3818
Yuan H, Abu-Reesh IM, He Z (2015) Enhancing desalination and wastewater treatment by coupling microbial desalination cells with forward osmosis. Chem Eng J 270:437–443
Zakrzewska-Trznadel G, Harasimowicz M, Chmielewski AG (1999) Concentration of radioactive components in liquid low-level radioactive waste by membrane distillation. J Membr Sci 163 (2):257–264
Zhang Y, Wang R (2013) Fabrication of novel polyetherimide-fluorinated silica organic–inorganic composite hollow fiber membranes intended for membrane contactor application. J Membr Sci 443:170–180

Zhao S, Zou L (2011a) Effects of working temperature on separation performance, membrane scaling and cleaning in forward osmosis desalination. Desalination 278(1):157–164
Zhao S, Zou L (2011b) Relating solution physicochemical properties to internal concentration polarization in forward osmosis. J Membr Sci 379(1):459–467
Zhao Z, Zheng J, Wang M, Zhang H, Han CC (2012a) High performance ultrafiltration membrane based on modified chitosan coating and electrospun nanofibrous PVDF scaffolds. J Membr Sci 394:209–217
Zhao S, Zou L, Tang CY, Mulcahy D (2012b) Recent developments in forward osmosis: opportunities and challenges. J Membr Sci 396:1–21
Zhou T, Yao Y, Xiang R, Wu Y (2014) Formation and characterization of polytetrafluoroethylene nanofiber membranes for vacuum membrane distillation. J Membr Sci 453:402–408
Zhu X, Elimelech M (1997) Colloidal fouling of reverse osmosis membranes: measurements and fouling mechanisms. Environ Sci Technol 31(12):3654–3662
Zou S, Gu Y, Xiao D, Tang CY (2011) The role of physical and chemical parameters on forward osmosis membrane fouling during algae separation. J Membr Sci 366(1):356–362
Zuo G, Wang R (2013) Novel membrane surface modification to enhance anti-oil fouling property for membrane distillation application. J Membr Sci 447:26–35

Chapter 7
Fungal-Based Nanotechnology for Heavy Metal Removal

Manisha Shakya, Eldon R. Rene, Yarlagadda V. Nancharaiah, and Piet N.L. Lens

Abstract Heavy metal pollution, cleaning and recycling are a major environmental issue. In particular, there is a need for efficient techniques to treat wastewaters. Conventional technologies to treat industrial waters are limited by stringent health policies and emerging contaminants. Fungi-based nanotechnology is rapidly emerging as an effective technology to treat industrial wastewaters. This chapter reviews the recent developments in fungal biosorption, biological synthesis of nanoparticles using fungi, and the application of fungi-based nanosorbents for heavy metals removal.

Keywords Heavy metals • Fungus • Remediation • Nanotechnology • Bioaugmentation

Contents

M. Shakya (✉) • E.R. Rene • Y.V. Nancharaiah • P.N.L. Lens
UNESCO-IHE Institute for Water Education, Delft, The Netherlands
e-mail: manisha_shakya@yahoo.com

K M Gothandam et al. (eds.), *Nanotechnology, Food Security and Water Treatment*, Environmental Chemistry for a Sustainable World 11,
https://doi.org/10.1007/978-3-319-70166-0_7

7.1 Introduction

Heavy metal pollution is considered to be one of the persisting global environmental problems. Despite the rapid advancements made in the last few decades in the field of engineering sciences and medical health, heavy metals present in water sources and ambient air continually engender adverse effects on public health and the environment. Apart from its natural occurrence, heavy metals are widely used in different applications in industry, agriculture and defence operations. Industrial effluents are contaminated with heavy metals such as zinc (Zn), lead (Pb), copper (Cu), cadmium (Cd) and nickel (Ni), which are toxic in nature (Liu et al. 2008; Fu and Wang 2011). Metals are non-biodegradable and can be bio-accumulated in the food chain, leading to carcinogenic effects on plants, animals and humans. In order to limit the exposure of toxic heavy metals to the environment and public health, standards and guidelines have been established by different organizations such as the World Health Organization (WHO), European Union (EU), and the United States Environmental Protection Agency (USEPA).

Apart from the grievous environmental issues, heavy metals are non-renewable natural resources. Natural reserves of heavy metals are being depleted due to their excessive use in industrial applications. Therefore, the removal of heavy metals should be focused along with their recovery. Different physicochemical technologies for the removal of heavy metals from wastewater, both at the lab and industrial scale, have been proposed: membrane filtration, chemical precipitation, ion-exchange, reverse osmosis, coagulation–flocculation, flotation and electrochemical methods (Fu and Wang 2011). Even though extensive research has been done on the use of these technologies, there are still some drawbacks related to their applications. Especially in the case of low concentrations of heavy metals, most of the conventional technologies are expensive and inefficient (Say et al. 2001).

Because of global concern towards environmental protection and the development of greener remediation techniques for pollution remediation, biosorption has become one of the promising techniques for removing metal and metalloid ions from wastewater (Gautam et al. 2014). The use of nanoparticles for the removal of heavy metals from wastewater has also received great interest. Different studies have demonstrated successful application of nanotechnology for the remediation of metal pollution. Research has been conducted on different nanomaterials such as carbon nanotubes, dendrimers, nanostructured catalytic membranes, nanosorbents, nanocatalysts, bioactive nanoparticles, biomimetic membranes and molecularly imprinted polymers (MIP) for removing pollutants. Use of biologically synthesised nanoparticles for the remediation of heavy metals from wastewater is an emerging research topic in the field of nanotechnology for wastewater management (Dasgupta et al. 2015, 2017; Jain et al. 2016; Ranjan et al. 2014). Considering this line of progressive research, this chapter deals with the application of nanosorbents with a special focus on fungi based nanotechnology for the removal of heavy metals from wastewater.

7.2 Heavy Metals

Heavy metals are elements that have a specific density over 5 g cm^{-3} and relative atomic mass above 40 (Järup 2003; Srivastava and Majumder 2008). A large number of elements from the periodic table fall into this category. The most important heavy metals relevant in the environmental context are: arsenic (As), cadmium (Cd), chromium (Cr), cobalt (Co), copper (Cu), lead (Pb), mercury (Hg), nickel (Ni), tin (Sn), vanadium (V) and zinc (Zn). Existing in different mineral forms such as sulphides, oxides, carbonates and silicates naturally, most of these metals are insoluble in water (Dean et al. 1972; Nriagu and Pacyna 1988), whereas some salt forms are soluble in water.

7.2.1 Sources of Heavy Metals Pollution

Metals are widely distributed in rocks and soil naturally and as trace elements, they also form an essential part of various processes (Srivastava and Majumder 2008). Natural weathering of the rocks and soil break them down and releases the metals into water bodies. Rapid industrialization and technical advancement have influenced the geochemical metal cycles by exploiting the heavy metals for various industrial uses and dumping them in various toxic forms as a waste matrix into the natural environment causing heavy metal pollution. Major sources of heavy metal pollution are industries, including the mining activities, tanneries, fertilizer production, batteries, electroplating, oil refineries and paper industries (Järup 2003). Some of the anthropogenic sources of heavy metals are summarized in Table 7.1.

7.2.2 Health and Environmental Risk of Heavy Metal Pollution

Heavy metals, in trace quantities, have nutritional value and are essential for the functioning of living cells. However, recent development in mining technologies to utilize and disperse mineral resources has altered their natural geochemical cycles.

Table 7.1 Significant anthropogenic sources and causes of major heavy metal pollution in the environment

Source	Heavy metals released	Causes of pollution
Batteries	Cd, Hg, Ni, Pb, Sb, Zn	Waste battery fluid
Electronics	As, Au, Cd, Cr, Hg, Mn, Ni, Pb	Wastewater from manufacturing and recycling process
Electroplating	Cu, Cr, Ni, Zn	Effluents from plating processes
Fertilizers	Cd, Cr, Mo, Pb, U, V, Zn	Run-off
Landfill	Cd, Cr, Cu, Hg, Ni, Pb, Zn	Leachate
Manures sewage sludge	Zn, Cu, Ni, Pb, Cd, Cr, As, Hg	Land spreading
Metalliferous mining	Cd, Cu, Ni, Cr, Co, Zn, As	Acid mine drainage, slag heaps
Paints and pigments	Pb, Cr, As, Ti, Ba, Zn	Effluent from manufacture, old paint deterioration
Paper and pulp	Zn, Cu, Cd, Pb, Ni, Fe, Mn	Effluents

Modified from Gautam et al. (2014)

The environment and human health are in risk because of exposure to heavy metals and their toxicity. Heavy metals discharged into the environment can be transported through a long distance by air or water movement. They have adverse impacts on human health, as well as on terrestrial and aquatic ecosystems, due to their toxic and bioaccumulation characteristics (Wang and Chen 2006).

Heavy metals can alter the physical and chemical properties of air and water transporting them. Once they enter the food chain, they can be biomagnified in the higher trophic levels causing significant changes in the biochemical cycle of living cells (Srivastava and Majumder 2008). Very well known diseases such as "Itai Itai" due to Cd pollution in Japan (Gautam et al. 2014), "Arsenecosis" due to As pollution in Bangladesh, and "Minimata" due to Hg pollution in Japan are caused due to poisoning by heavy metals (Volesky 1990). Thus, heavy metal pollution has become a serious problem to the present society. Regarding the potential hazards to human health, Cr, Ni, Zn, Cu and Cd are considered as "priority metals". Cr in the hexavalent oxidation state is highly toxic to humans and animals. Similarly, Cu, Ni and Zn are moderately toxic and all these heavy metals have shown toxic effects on plants (Álvarez-Ayuso et al. 2003).

Table 7.2 summarizes the toxic effects of some heavy metals to human health. Because of their acute toxicity, bioaccumulation and non-biodegradable nature, wastewater loaded with heavy metals should be treated well before their environmental discharge. Different regulatory bodies have set the maximum limits for the discharge of toxic heavy metals and their permissible concentration in drinking water in order to avoid risks due to the presence of such elements in the environment (Table 7.3).

Table 7.2 Toxic effects of heavy metals to humans

Heavy metals	Toxicity	References
Cu	Irritation of mucus membrane, capillary damage, renal degradation, problem in nervous system	Acheampong et al. (2010)
Cd	Diarrhoea, nausea, muscular cramps, renal degradation, chronic pulmonary problems, skeletal deformity	Acheampong et al. (2010) and Boparai et al. (2011)
Cr	Respiratory cancer, asthma, skin ulcerations, irritation and ulceration of the nasal septum	Kirman et al. (2012) and Das et al. (2015)
Pb	Behavioural disturbances, kidney damage, anaemia and toxicity to the reproductive system	Xu et al. (2012a, b) and Moghadasali et al. (2013)
Ni	Skin allergies, lung fibrosis, cardiovascular And kidney diseases and cancer of the respiratory track	Gupta et al. (2010) and Kanold et al. (2016)
Zn	Nausea, vomiting, stomach cramps, anaemia, skin irritation, muscle stiffness, restlessness	González and Pliego-Cuervo (2014)
Hg	Mental and motor dysfunction, kidney damage, fatigue, memory loss, headache, muscle and joint pain, tingling around the mouth, muscle and joint pain	Falih (1997) and Silbernagel et al. (2011)

Table 7.3 Permissible limits for inland discharge of different heavy metals according to the United States Environmental Protection Agency (USEPA), and their permissible concentration in drinking water according to the European Union (EU) and World Health Organization (WHO)

Heavy metal	USEPA, 2002 (mg L^{-1})	EU, 1998 (mg L^{-1})	WHO, 2011 (mg L^{-1})
Cr	2.0	0.05	0.05
Cd	1.0	0.005	0.003
Cu	5.0	2.0	2.0
Ni	5.0	0.02	0.07
Zn	5.0	–	–

7.2.3 Technologies for the Removal of Heavy Metals from Wastewater

In order to follow the regulatory limits of discharge, wastewater generated by industries need to be treated prior to their discharge. The growing necessity for efficient and environmentally sustainable technologies for the removal of heavy metals from wastewater has resulted in marked improvements in conventional treatment processes. Various technologies such as ion exchange, adsorption, oxidation, membrane filtration, ozonation, coagulation/flocculation, photochemical methods and adsorption combined with magnetic separation have been used for the treatment of wastewater contaminated with heavy metals (Mahdavian and Mirrahimi 2010; Gautam et al. 2014). The advantages and disadvantages of different technologies to treat polluted effluents with heavy metals are summarized in Table 7.4.

Table 7.4 Advantages and disadvantages of different treatment technologies for the removal of heavy metals from wastewater

Technology	Advantages	Disadvantages	References
Adsorption	Flexibility in design and operation, high capacity, fast kinetics	Performance depends on the type of adsorbent, physical or chemical activation might be required	Loukidou et al. (2003), Crini (2005)
Biological treatment	Feasibility of removing certain metals	Not yet commercialized	Ahmaruzzaman (2011)
Chemical precipitation	Cheap, simple and non-metal selective	Production of sludge with high cost for sludge disposal	Aderhold et al. (1996), Rashed et al. (2013)
Coagulation-flocculation	Efficient, simple to operate, characteristics of good sludge settling and dewatering	Large amount of chemical is used and sludge generation	Aderhold et al. (1996)
Flotation	Small particles can be removed, and low retention time	Expensive	Rubio et al. (2002)
Ion exchange	Metal selective and high regeneration capacity	Initial capital cost and maintenance costs are high	Rengaraj et al. (2003)
Membrane filtration	Can be operated in small space, less chemical consumption and waste generation	Expensive due to high investment, maintenance and operational costs, membrane fouling	Qin et al. (2002), Madaeni and Mansourpanah (2003)
Photo chemical	Does not produce sludge	Production of by-products	Ahmaruzzaman (2011)

7.3 Bioremediation of Heavy Metals with Fungal Biosorption

Bioremediation is one of the emerging technologies for the clean-up of pollutants like heavy metals using microorganisms. In terms of economy and efficiency, bioremediation processes for treating toxic pollutants are better compared to chemical and physical methods and the potential of fungal communities for bioremediation processes has recently been realized. Biosorption is one of these bioremediation methods which is discussed in the following sections.

7.3.1 Biosorption

Biosorption refers to "many modes of non-active metal uptake by (microbial) biomass which may be dead" (Volesky 1990). It is the property of certain bio-molecules to bind certain ions or molecules from aqueous medium (Gautam et al. 2014). The adsorption of metals occurs on the biological material due to the biological activities in living cells or metal binding functional groups of cells in both living and dead cells (Bakircioglu et al. 2010).

Previous studies have shown the dependency of metal cation removal on the interactions of these metals with specific functional groups present in microbial cells (Loukidou et al. 2003). Generally, the binding mechanisms that cause biosorption include ion exchange, micro-precipitation and electrostatic interactions (Acheampong et al. 2010). In the mechanism of ion exchange, the metal ion within the wastewater is replaced by another similarly charged ion (Han et al. 2006). According to Tan and Cheng (2003), ion exchange is the dominant mechanism for the removal of Cu(II), Ni(II), Zn(II), Pb(II), and Cr(III) by *Penicillium chrysogenum*. The same mechanism is also involved in removing Cd(II) by *Saccharomyces cerevisiae* (Romero-González et al. 2001). The biosorption of Pb (II) and Cd(II) onto *Amanita rubescens* biomass was also mainly due to the ion-exchange between the hydrogen atoms of amine (−NH), hydroxyl (−OH) and carboxyl (−COOH) groups of the biomass and the metal ions (Sarı and Tuzen 2009). Chelation is another metal binding mechanism in biosorption where the metal ions bind with an organic molecule to form a ring structure (Acheampong et al. 2010). Different functional groups present in cells that take part in chelation are carboxyl, sulphydryl, amino, sulphate, phosphate, thioether and carbonyl groups. According to Xu et al. (2012a, b), a chelation process is responsible for the biosorption of Pb(II) onto iron oxide nanoparticles immobilized *Phanerochaete chrysosporium* cells. The same mechanism took place in the removal of Cr (VI) using surface modified *P. chrysosporium* (Chen et al. 2011).

The use of microorganisms as adsorbents is efficient, cost-effective and safe for the removal of heavy metals from soils, sediments and water. Biosorption using microorganisms exploits bacteria, yeast, algae and fungi to remove heavy metals from water. Microbes can be produced in large quantities and the process can be operated under different conditions of inoculum size, pH, temperature, ionic strength and metal concentrations. In addition to this, low cost, free availability, good regeneration capacity and less use of chemicals also make microbial biomass very attractive for biosorption processes (Bakircioglu et al. 2010).

7.3.2 Use of Fungi as Biosorbents for Heavy Metal Removal

Although different microorganisms like bacteria, algae and fungi have been used in biosorption processes, fungi are conceived as a prospective candidate because they are easy to handle, produce large quantities of biomass and are easy to wash and separate from aqueous solutions due to their hyphae network (Gopal et al. 2002, Çeribasi and Yetis 2004, Xu et al. 2012a, b, Akar et al. 2013). Particularly, fungi are attractive because of the following reasons: (i) they can be easily manipulated genetically and morphologically, (ii) ability to tolerate heavy metals, and (iii) withstand low pH conditions (Sarı and Tuzen 2009). The surface of the fungal biomass is coated with polymeric substances (carbohydrates, proteins, lipid and nucleic acids) which consist of negatively charged functional groups such as carboxylate and phosphate. Due to these negatively charged functional groups, they are able to adsorb cations (Say et al. 2001). Different fungal strains have shown the potential to be used as biosorbents (Dias et al. 2002; Yan and Viraraghavan 2003; Bayramoglu and Arıca 2008; Sarı and Tuzen 2009). Table 7.5 overviews the adsorption capacity of different fungal strains with their optimum pH conditions to adsorb heavy metals from aqueous solutions.

7.3.3 Role of Fungal Morphology on the Biosorption Process

Fungal cultures normally grow by hyphae extension. Fungi can grow in the form of mycelia to form pellets (Grimm et al. 2005), which depends on factors such as level and type of inoculum, shear forces, medium composition, pH, cell physiology, agitation rate, temperature and dissolved oxygen concentrations (Grimm et al. 2005; Fu and Wang 2011; Espinosa-Ortiz et al. 2016). Fungal morphology plays a very important role in the biosorption of heavy metals. Both filamentous and pelleted forms of fungi have been applied in the biosorption process. The main advantages of using fungal pellets in biosorption processes can be summarized as follows (Xu et al. 2012a, b): (i) high mechanical strength compared to dispersed mycelium, (ii) porous characteristics and increased biosorption capacity, and (iii) stronger resistance towards environmental perturbations.

7.4 Nanotechnology in Wastewater Treatment

Nanotechnology has been recognized as one of the frontier technologies that offer a good potential for treating wastewater in a more effective and efficient manner than commercial methods that have shown to pose technical and operational challenges. Some of the conventional wastewater treatment techniques require additional costs when dealing with heavy metals. Recently, in the field of bio-nanotechnology, the application of various novel nanomaterials produced *in-situ* for the treatment of industrial wastewater contaminated with different pollutants, including toxic heavy metal ions, has been tested. Due to their inimitable action toward contaminants, the field of nanotechnology and bio-nanotechnology is under active research and development for application in the treatment of wastewater. Nanoparticles based technologies applied in water treatment consists of different methods such as reverse osmosis, nanofiltration and ultrafiltration membranes. Indeed, various nanoparticles, filters with nanofibers, and carbon nanotubes are among the emerging products used in nanotechnology.

Table 7.5 Application of different fungal strains for the biosorption of heavy metals

Fungal strain	Treatment	Heavy metal	Adsorption capacity (mg g^{-1})	pH	References
Amanita rubescents	Dried biomass	Pb (II)	38.4	5.0	Sarı and Tuzen (2009)
		Cd (II)	27.3		
Aspergillus terreus	Immobilized in polyurethane foam	Fe (II)	164.5	4.5	Dias et al. (2002)
		Cr (VI)	96.5		
		Ni (II)	19.6		
Aspergillus flavus	Dried biomass	Pb (II)	12.44	–	Dwivedi et al. (2013)
		Ni (II)	0.53		
		Cr (II)	0.05		
	Dried biomass	Zn (II)	287.8	5.0	Aftab et al. (2013)
Claviceps paspali	Dead biomass	Zn (II)	1.0	–	Luef et al. (1991)
Lentinus edodes	Washed with saline solution and heat inactivated live pellets	Hg (II)	336.3	6.0	Falih (1997)
		Cd (II)	78.6		
		Zn (II)	33.7		
Mucor rouxii	Biomass treated with NaOH	Pb (II)	35.69	5.0	Yan and Viraraghavan (2003)
		Ni (II)	11.09		
		Cd (II)	8.46		
		Zn (II)	7.75		
	Dead biomass	Pb (II)	25.22	5.0	Yan and Viraraghavan (2003)
		Ni (II)	6.34		
		Cd (II)	8.36		
		Zn (II)	16.62		
Phanerochaete chrysosporium	Inactivated and boiled pellets in alkaline for 45 min	Cd (II)	15.2	4.5	Li et al. (2004)
	Inactivated and boiled pellets in alkaline for 45 min	Pb (II)	12.34	4.5	Li et al. (2004)
	Surface modified pellets	Cr (VI)	279.9	3.0	Chen et al. (2011)
	Resting cells	Ni (II)	77.96	4.0	Çeribasi and Yetis (2004)
		Pb (II)	73.56		
	Dried biomass	Pb (II)	45.25	6.0	Say et al. (2001)
		Cd (II)	13.24		
		Cu (II)	10.72		
	Washed cells	Pb (II)	90.0	5.0	Gopal et al. (2002)
		Cd (II)	17.0		
		Cu (II)	43.0		
	Iron oxide nanoparticles immobilized pellets	Pb (II)	176.33	5.0	Xu et al. (2012a, b)
	Live biomass	Pb (II)	1.33	–	Dey et al. (1995)

7.4.1 Nanoparticles

Nanoparticles are particles which have a size of 0.1 to 100 nm which can be formed by two techniques: a top-down and a bottom-up approach (Fig. 7.1). In the first approach, nanoparticles are formed by breaking bulk particles into nanosize range particles. In the second approach, atoms are build into molecules to synthesize nanosized particles. They have specific properties which are different from the bulk materials. As a result, nanoparticles have a wide array of applications in the field of electronics, medical and health sciences, information technology, biotechnology and the environmental sector (Buzea et al. 2007). Nanoparticles are also used in the wastewater treatment sector.

7.4.2 Biological Synthesis of Nanoparticles

Different physical and chemical technologies are available for the synthesis of nanoparticles (Murray et al. 1993; Qu and Peng 2002) as shown in the Fig. 7.2. However, the use of flammable and harmful materials, the requirement of high temperature, pressure and oxygen free environment make these technologies unmanageable and non-feasible. To rise above these issues, the biosynthesis of nanomaterial has been recognized as the best alternative to synthesize nanoparticles due to its dependency on active and efficient microorganisms for the transformation of metal(loid) ions to nanoparticles (Suresh 2014). Several organisms have been tested for this approach, such as *Fusarium oxysporum*, *Rhodobacter sphaeroides*, *Schizosaccharomyces pombe*, *Rhodopseudomonas palustris* and *Escherichia coli*.

As a self defence mechanism under metal stress situations, microorganisms can produce nanoparticles when they are exposed to the metal ions. When they seize the target ions from their environment, the enzymes generated by the cell activities act as reducing agents and convert the metal ions into elemental metal nanoparticles. This biogenic method is carried out at room temperature. Other advantages of these methods are avoiding the excess use of chemicals, production of the required size of quantum dots and eventually the particles can be solubilised in water. Nanoparticles produced biologically are reported to be biocompatible and environmentally benign. However, they have some disadvantages like more downstream processing is required for the purification and shortage of control on the surface and morphology traits (Narayanan and Sakthivel 2010; Syed and Ahmad 2013). Despite these qualities, the rate of synthesis of biologically synthesised nanoparticles is slow and is not mono-disperse.

To overcome these problems, special effort has to be considered in the sector of microbial cultivation methods and the extraction techniques. Another solution may be the photo-biological methods of nanoparticles production which require further optimization studies (Narayanan and Sakthivel 2010). A summary of the

Fig. 7.1 Different approaches for the formation of nanoparticles

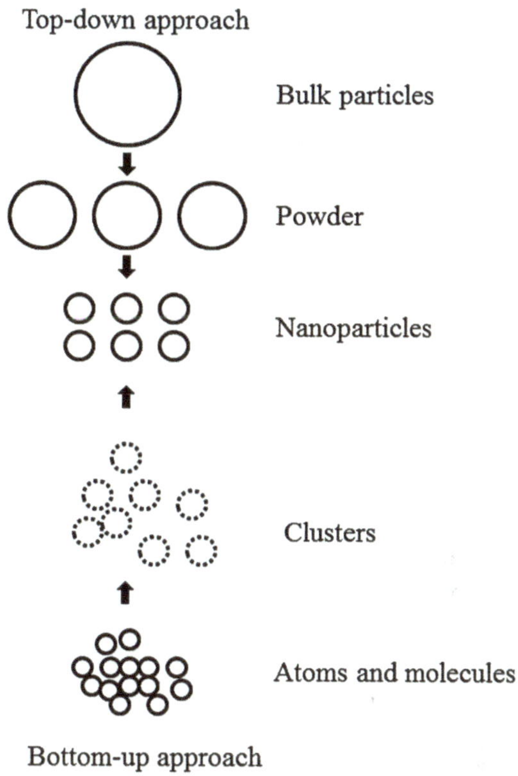

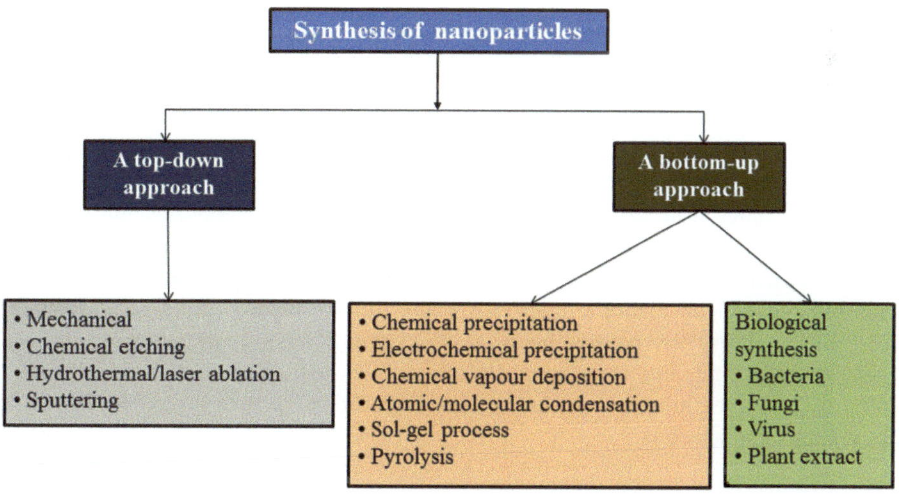

Fig. 7.2 Technologies for the synthesis of nanoparticles

Table 7.6 Advantages and disadvantages of biological and physico-chemical methods for nanoparticle synthesis

Method	Advantages	Disadvantages
Physical methods	Highly controlled particle shape, size and narrow size distribution can be achieved.	Uses extreme conditions and high end facilities. High cost is associated with the production.
Chemical methods	Highly controlled particle shape, size and distribution is possible. Cost effective compared to top down synthesis of physical methods.	Uses extremely hazardous reducing agents, stabilizing agents, and high temperatures. These chemicals are highly toxic to living organisms.
Bacteria	Inexpensive source of reducing and stabilizing agents (enzymes and biomolecules) in nanoparticle synthesis. Production is performed at ambient conditions. Scale up is possible.	Particles with a wide range of size distribution are produced. Difficulties in controlling the desired shape, size and distribution of particles. Problems with repeatability. Organic material is always associated with the nanoparticles.
Fungi	Inexpensive source of reducing and stabilizing agents in nanoparticle biosynthesis. Synthesis occurs at ambient conditions. Smaller sized particles can be obtained using fungi compared to bacteria. Scale up is possible.	Particles are formed with a wide range of particle size distribution. There are problems with reproducibility and less control on particle size, shape and distribution. Particles are associated with biogenic organic material.

advantages and disadvantages of biological and physico-chemical methods for nanoparticles synthesis is shown in Table 7.6.

7.4.2.1 Intracellular Biosynthesis of Nanoparticles

Biosynthesis of nanoparticles can be classified into intracellular and extracellular synthesis according to the location in which the nanoparticles are formed. In the intracellular method, the metal ions are transported into the microbial cell with the help of specialized transporters and converted into nanoparticles in the presence of enzymes (Zhang et al. 2011). There have been several studies on the intracellular synthesis of nanoparticles. For instance, 1–1.5 nm CdS nanoparticles were produced using the yeast *Schizosaccharomyces pombe* (Kowshik et al. 2002). Intracellularly produced nanoparticles are more homogeneous in size and shape distribution; however, harvesting of product and recovery of nanoparticles are more troublesome and expensive (Basavaraja et al. 2008). To overcome this difficulty, extracellular biosynthesis of nanoparticles is recommended for high throughput applications.

7.4.2.2 Extracellular Biosynthesis of Nanoparticles

This method is used to entrap metal ions on the surface of cells, and these metal ions are reduced with the help of certain enzymes (Zhang et al. 2011). The production of CdTe quantum dots with 2–3.6 nm size extracellularly was demonstrated by Syed and Ahmad (2013). Suresh (2014) also demonstrated the extracellular synthesis of CdSe nanoparticles using *Fusarium oxysporum*. The thermophilic fungus *Humicola* sp. was also reported to produce extracellular protein capped silver nanoparticles (Syed et al. 2013). In addition, different types of nanoparticles have been produced extracellularly using different kinds of microorganisms such as *Aspergillus flavus* (Vigneshwaran et al. 2007), *Rhodopseudomonas* (Bai et al. 2009), *Lactobacillus* (Prasad and Jha 2010), *Helminthosporum solani* (Suresh 2014), *Aspergillus clavatus*, *Aspergillus sydowii*, *Fusarium semitectum*, *Penicillium brevicompactum* and *Shewanella oneidensis* (Kitching et al. 2015).

7.4.3 Fungi as Biological Factory for Nanoparticle Production

Even though different microorganisms have been explored for the synthesis of nanoparticles including bacteria, yeast and algae, fungi are very promising for nanoparticles production because they are easy to culture and handle at the laboratory and industrial scale (Table 7.6). While biosynthesis of nanoparticles by bacteria is well researched, fungal biosynthesis of nanoparticles need much more investigation. For instance, only around 30 species of fungi have been reported to produce gold nanoparticles (Kitching et al. 2015). Generally, fungi can tolerate higher metal concentrations compared to bacteria and secrete plentiful extracellular redox proteins to synthesize nanoparticles after reducing the soluble metal ions to their insoluble form. Especially at low pH, the biosorption capacity of fungi is higher than that of bacteria. A big biological diversity of fungi may provide novel metal reductases for metal detoxification and bioreduction. The biomass of fungi is also available as a by-product of different industries which reduces the costs for procuring the biomass. A thorough understanding of the biosynthetic mechanism of nanoparticles in fungi is needed to reduce the time of biosynthesis and to scale up their production process. Fungi show promises for industrial biosynthesis of nanoparticles because of large biomass production, easy handling, large protein secretion compared to bacteria. Additionally, their morphology can be manipulated according to the need. They also produce stable protein coated nanoparticles (Vigneshwaran et al. 2007; Syed et al. 2013).

7.5 Role of Nanosorbents for Heavy Metals Removal

Deployment of new water treatment technologies using nanomaterials has received much attention due to their unique physico-chemical and biological properties, such as exceptionally small size, high surface area to volume ratio, surface modifiability, magnetic properties, short intraparticle diffusion distance, changeable surface chemistry and biocompatibility (Amin et al. 2014). Generally, adsorption is considered as a polishing step to remove the organic and inorganic contaminants present in wastewater. The application of conventional adsorbents is limited due to limited active sites, surface area, lack of selectivity and adsorption kinetics. Moreover, conventional adsorbents require more raw materials for processing and preparation and they are considered to be energy intensive if the adsorbents are to be modified for their surface properties. Waste reduction can be achieved by employing green nanotechnology in order to synthesize nanomaterials and use them as adsorbents. However, more quantitative research should be done in order to validate these claims. Comparative studies on energy and material consumption are still missing in the literature to replace traditional materials with nanoparticles. Moreover, research gaps still exist on the topics such as the safety of using nanosorbents and their potential impact on the environment. If these issues are properly addressed, nanosorbents could be a cheap, green and sustainable alternative for the developing countries to treat industrial wastewater instead of using the expensive conventional methods.

7.6 Mechanism of Heavy Metals Removal Using Nanosorbents

Nanosorbents are nanoscale particles made from organic or inorganic materials that have a high affinity to absorb substances. They have different physical, chemical and biological properties than their normal size equivalents. For example, nanomaterials of metals and metal oxides deliver a higher ratio of surface area to particle size that leads to the exhibition of different optical, magnetic and electrical properties (El Saliby et al. 2008). They contain unsaturated surface atoms which help them to form strong chemical bonds with the metal ions (Lemos et al. 2008). Recent studies have shown that nanoparticles have the capacity to penetrate in contaminated zones where the micro particles cannot penetrate (Gao et al. 2011; Sheet et al. 2014). They can also be functionalized with various chemical groups to increase their affinity towards a particular compound, i.e. enhanced selective properties (Bhattacharya et al. 2013). These physio-chemical characteristics prove the effectiveness of nanomaterials in water purification systems. Application of these particles as adsorbent lead to the availability of a high number of atoms or molecules on the surface of contaminants, thereby enhancing the adsorption capacities of sorbent materials. Different studies have demonstrated the successful application of nanosorbents such as carbon nanotubes, nanoscale metal oxides and nanofibers for the removal of heavy metals from wastewater (Sheet et al. 2014). Typical examples include the removal of Cu (Bystrzejewski et al. 2009; Yalcinkaya et al. 2011; Al-Rashdi et al. 2012), Pb (Yalcinkaya et al. 2011), Cd (Afkhami et al. 2011), Cr (Hu et al. 2005a, b) and Ni (Karimi and Kafi 2015) from wastewater.

Among the different nanosorbents reported in the literature, the carbon based nanosorbents have revieved great attention in the field of heavy metal remediation. Lee et al. (2012) synthesised carbon-based nanosorbents by ethylene decomposition on stainless steel mesh without the use of external catalyst for the treatment of water containing Ni(II) ions. A number of batch sorption tests were performed by the authors to determine the effects of initial pH, initial metal concentration and contact time on Ni(II) removal by the nanosorbents. The kinetic data fitted well to a pseudo second-order model indicating that the process was of the chemisorption type whereas the equilibrium data fitted to the Freundlich isotherm. Further analysis by the Boyd kinetic model revealed that the main mechanism of adsorption was controlled by boundary layer diffusion. This study suggests that the prepared carbon based nanosorbent is a promising sorbent for the sequestration of Ni (II) from aqueous solutions.

A similar study was done by Di et al. (2006), wherein Cr adsorption on Ceria nanoparticles supported on aligned carbon nanotubes (CeO_2/ACNTs) was studied. This novel adsorbent was prepared by the chemical reaction of $CeCl_3$ with NaOH in a carbon nanotube solution followed by heat treatment. The maximum adsorption capacity of this adsorbent at pH 7.0 was found to be 30.2 mg g^{-1} at an equilibrium Cr(VI) concentration of 35.3 mg l^{-1}. Lu and Chiu (2006) and Li et al. (2005) also studied the adsorption capacity of Zn and Pb onto purified carbon nanotubes. The maximum adsorption capacity was found to be 43.7 mg g^{-1} for Zn and 30.3 mg g^{-1} for Pb, respectively. According to the authors, the adsorption data for Zn best fitted the Langmuir isotherm, whereas Pb adsorption was found to be pH dependent and it followed the Freundlich isotherm model.

Magnetic nanoparticles are also very promising in removing heavy metal ions from water. Liu et al. (2008) developed humic acid (HA) coated Fe_3O_4 nanoparticles for the removal of toxic Hg(II), Pb(II), Cd(II), and Cu(II) from water. The sorption of the heavy metals to Fe_3O_4/HA reached equilibrium in less than 15 min, following the Langmuir adsorption model with maximum adsorption capacities ranging between 46.3 and 97.7 mg g^{-1}. This nanosorbent was able to remove 99% of Hg(II) and Pb(II) and greater than 95% of Cu(II) and Cd(II) in natural and tap water, respectively, at the optimized pH conditions. According to Xu et al. (2012a, b), the contaminants can be adsorbed in magnetic nanoparticles by various mechanisms such as surface sites binding, magnetic selective adsorption, electrostatic interaction and modified ligands combination. The performance of some of the chemically prepared nanoadsorbents and their maximum adsorption capacities for heavy metal removal are listed in Table 7.7.

Table 7.7 Adsorption capacities of different types of chemically synthesized nano adsorbents for heavy metal removal

Nanoadsorbents	Metal adsorbed	pH	Best fitted model	Adsorption capacity (mg g^{-1})	References
Maghemite nanoparticles	Cr (VI)	8.0	Freundlich	1.9	Hu et al. (2005a, b)
Fe_2O_3	Pb (II)			36	Nassar (2010)
Magnetic nanoparticles	Cu (II)				Bystrzejewski et al. (2009)
Jacobsite ($MnFe_2O_4$)	Cr (VI)	2.0	Langmuir	31.5	Hu et al. (2005a, b)
B_2O_3/TiO_2 composite	Cd(II)			49	Kalfa et al. (2009)
Titanium dioxide	Cr (III)	7.0		7.0	Liang et al. (2006)
Carboxymethyl-β-cyclodextrin modified Fe_3O_4 nanoparticles	Cu (II)		Langmuir	47.2	Badruddoza et al. (2011)
Amino-functionalized Fe_3O_4 nanoparticles	Cu (II)		Langmuir	12.4	Huang and Chen (2009)
	Cr (VI)		Langmuir	11.2	
Anatase nanoadsorbent	Pb (II)		Langmuir	31.3	Kocabaş-Ataklı and Yürüm (2013)
	Cu (II)		Freundlich	23.7	
Alumina-coated magnetite nanoparticles	Ni(II)		Langmuir	9.7	Karimi and Kafi (2015)
3-Mercaptopropionic acid functionalized bio magnetite nanoparticles	Ni(II)		Langmuir	42.0	Venkateswarlu et al. (2015)
Ceria nanoparticles supported on aligned carbon nanotubes	Cr (IV)	7.0	Langmuir	30.2	Di et al. (2006)
Single-walled carbon nanotubes	Zn (II)		Langmuir	43.7	Lu and Chiu (2006)

7.7 Fungal Based Nanosorbents for Heavy Metals Removal

For industrial applications, the immobilization of the fungal biomass is essential to maintain microbial cell activity in a toxic environment, the re-use of biomass, the use of conventional reaction systems, and for the lack of biomass-liquid separation requirements. In order to achieve the right size, mechanical strength, rigidity and porosity of the biosorbent, immobilization of fungal biomass on different materials is performed. Different techniques have been practised for the immobilization of fungal biomass. For example reticulated foams, activated carbon and glass raschig rings have been used as inert supports for the immobilization of biomass. For instance polyurethane foams were used as supporting materials to immobilise the biomass of *Aspergillus terreus* in order to treat heavy metals such as iron, chromium and nickel (Dias et al. 2002). Polymeric matrices such as calcium alginate, polyacrylamide, polysulfone, polyethylenimine and polyhydroxoethylmethacrylate (polyHEMA) are also used for the immobilization of biomass. Immobilization of fungal biomass with nanoparticles is a new research line in the field of biosorption of heavy metals. Iron oxide nanomaterials have also shown good potential for the immobilization of biomass due to their chemical inertness and favourable biocompatibility (Xu et al. 2012a, b).

In general, nanomaterials should be stable to avoid aggregation and achieve low deposition rates. Even though nanoparticles are effective for the adsorption of heavy metals, they have some drawbacks such as low mechanical strength and clogging of filters in continuous flow through reactors. It is reported that nanomaterials tend to aggregate in solution. More commonly, the electrostatic and van der Waals interactions affect the stability of colloidal nanoparticles. Different solutions such as the use of stabilizer, electrostatic surfactant, and steric polymers have been widely proposed. For instance, TiO_2 and SiO_2 nanoparticles immobilised on different supporting materials have been used to treat Cd(II) where the supporting material helped to increase the permeability of the solution and the aggregates did not pass through the filter paper (Kalfa et al. 2009). Fungal biomass has attracted attention in frontier research areas in order to obtain effective nano bio-composites for the removal of metal pollutants because they offer advantages such as good physical, chemical and biological stability (Perullini et al. 2010).

Bakircioglu et al. (2010) tested filamentous fungal biomass loaded TiO_2 nanoparticles and used this mixture as a sorbent for Pb removal. The adsorbent had a longer life span compared to the pure nanoparticles of TiO_2 and the modified biosorbent was found to be cheap because the fungal biomass used was obtained as a by-product from an oil plant. Along with this, the preconcentration level for this nanobiocomposite was found to be much higher (868 times) compared to other biocomposites of fungi such as *Aspergillus fumigatus* immobilized Diaion HP-2MG (Soylak et al. 2006), *Penicillium italicum* loaded on Sepabeads SP (Tuzen and Soylak 2008), *Pseudomonas aeruginosa* immobilized onto multi-walled carbon (Tuzen et al. 2008) and *Aspergillus niger* loaded on silica gel (Baytak et al. 2007).

In another study, iron oxide nanoparticles immobilized onto *Phanerochaete chrysosporium* were successfully tested for the adsorption of Pb and a maximum adsorption capacity of 176.3 mg g^{-1} was reported (Xu et al. 2012a, b). This value is much higher compared to the adsorption capacity of a magnetic adsorbent formed by encapsulating magnetic functionalized nanoparticles by calcium-alginate beads (100 mg g^{-1}) (Bée et al. 2011) or dried biomass of *P. chrysosporium* (69.8 mg g^{-1}) (Say et al. 2001). The fungal nanocomposite was reused 5 times with greater than 90% recovery of the metals in each cycle. According to the authors, immobilization of nanoparticles with fungal biomass provided stability to the biosorbent and also increased the adsorption capacity.

Espinosa-Ortiz et al. (2016) studied the Zn(II) removal efficiency of novel core shell based selenium nanoparticles immobilized fungal pellets of *Phanerochaete chrysosporium* from aqueous solution. The fungus produced elemental selenium nanoparticles intracellularly by reducing selenite from wastewater streams. The influence of different operational parameters such as pH, initial metal ion concentration, ionic strength and biosorbent dosage on the removal of Zn(II) was investigated and a Zn(II) removal efficiency of ~80% was reported. In another study by Peng et al. (2010), *Saccharomyces cerevisiae* immobilized chitosan coated magnetic nanoparticles (Fig. 7.3) was tested for the adsorption of Cu(II) from aqueous solution. The authors observed that this nano-biocomposite was quite efficient for the adsorption of Cu(II) and a removal efficiency of >90% was noticed within 20 min, with a maximum adsorption capacity of 134 mg g^{-1}.

As shown in several recent literature reports, these types of novel nanobiosorbents will have broad applications in the field of heavy metal removal from wastewater. Such immobilization techniques help to pre-condition the fungal biosorbent for application in large-scale systems. Although several immobilization media and techniques have been tested, little information is available on combining nanotechnology with existing biotechniques for long term practical applications. In addition, screening studies should be performed to select the best microorganism that can be used for immobilization and possibility to scale up these nano-biocomposites has to be assessed.

7.8 Desorption of Metal Ions and the Removal of Nanoparticles After Water Treatment

Studies conducted by Lu and Chiu (2006) and Li et al. (2005) has revealed that metal ions adsorbed onto nanotubes can be easily removed by altering the pH values of the solution using both HCl and HNO_3. However, the applications of nanosorbents in wastewater treatment will invariably lead to the release of nanoparticles into the environment. Tracking their mobility, bioavailability and persistence in the environment is essential to assess their potential risk in the environment. Traditional methods for the removal of particulate matter during

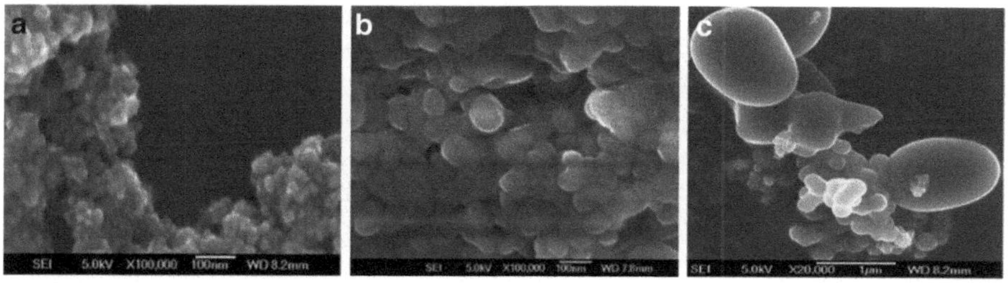

Fig. 7.3 Transmission electron micrograph of (**a**) Pure Fe_3O_4 magnetic nanoparticles, (**b**) Chitosan-coated magnetic nanoparticle and (**c**) *S. cerevisiae* immobilized on the surface of Chitosan-coated magnetic nanoparticles (Source: Peng et al. 2010)

wastewater treatment like flocculation, sedimentation and filtration will not be effective to remove these nanoparticles due to their small size (Bhattacharya et al. 2013). Hence, detailed research is essential to quantify how much nanomaterials are released to the environment and sustainable eco-technologies should be developed to treat those released nanoparticles in treated water.

7.9 Technical Challenges

Presently, novel nanosorbents are being synthesized and tested for the remediation of industrial wastewater contaminated with heavy metals. In order to scale up this technology and meet the requirements of the real world, there are several challenges that need to be addressed. For example, the novel nanosorbent must be economically beneficial and socially acceptable and at the same time it should be able to comply with the different water quality regulations to guarantee safety of human health and environment. Scaling up the process is challenging because the commercial production of nanoparticles still requires more intense research. Specifically, in the case of bionanoparticles, more studies on the stability, longevity of operation, regeneration and reuse of the bionanoparticles as well as a cost-benefit analysis should be performed in order to scale up its production. The treatment procedure must ensure that no nanosorbent is released to the environment.

7.10 Conclusions

The use of nanotechnology based biosorbents have shown promising results to treat wastewater polluted with heavy metals. Although most of the nanobiosorbents have so far been investigated only at the laboratory scale, for full scale applications, the selection of a proper fungal biomass, economic and energy consumption aspects of immobilization, pilot-scale testing to confirm the reuse and regeneration capacity of the nanosorbent should be investigated. From a green technology view point, the toxicity and environmental fate of nanosorbents are areas of concern during material selection and design for wastewater treatment. In continuous systems, parameters such as inoculum size, pH, temperature, ionic strength and heavy metal concentrations play a major role in determining the kinetics of the heavy metal removal.

References

Acheampong MA, Meulepas RJ, Lens PN (2010) Removal of heavy metals and cyanide from gold mine wastewater. J Chem Technol Biotechnol 85:590–613. https://doi.org/10.1002/jctb.2358

Aderhold D, Williams C, Edyvean R (1996) The removal of heavy-metal ions by seaweeds and their derivatives. Bioresour Technol 58:1–6. https://doi.org/10.1016/S0960-8524(96)00072-7

Afkhami A, Madrakian T, Ahmadi R, Bagheri H, Tabatabaee M (2011) Chemically modified alumina nanoparticles for selective solid phase extraction and preconcentration of trace amounts of Cd (II). Microchim Acta 175:69–77. https://doi.org/10.1007/s00604-011-0660-x

Aftab K, Akhtar K, Jabbar A, Bukhari IH, Noreen R (2013) Physico-chemical study for zinc removal and recovery onto native/chemically modified *Aspergillus flavus* NA9 from industrial effluent. Water Res 47:4238–4246. https://doi.org/10.1016/j.watres.2013.04.051

Ahmaruzzaman M (2011) Industrial wastes as low-cost potential adsorbents for the treatment of wastewater laden with heavy metals. Adv Colloid Interf Sci 166:36–59. https://doi.org/10.1016/j.cis.2011.04.005

Akar T, Celik S, Gorgulu Ari A, Tunali Akar S (2013) Removal of Pb^{2+} ions from contaminated solutions by microbial composite: combined action of a soilborne fungus *Mucor plumbeus* and *alunite matrix*. Chem Eng J 215–216:626–634. https://doi.org/10.1016/j.cej.2012.11.001

Al-Rashdi B, Tizaoui C, Hilal N (2012) Copper removal from aqueous solutions using nano-scale diboron trioxide/titanium dioxide (B_2O_3/TiO_2) adsorbent. Chem Eng J 183:294–302. https://doi.org/10.1016/j.cej.2011.12.082

Álvarez-Ayuso E, Garcıa-Sánchez A, Querol X (2003) Purification of metal electroplating waste waters using zeolites. Water Res 37:4855–4862. https://doi.org/10.1016/j.watres.2003.08.009

Amin M, Alazba A, Manzoor U (2014) A review of removal of pollutants from water/wastewater using different types of nanomaterials. Adv Mater Sci Eng 2014:1–24. https://doi.org/10.1155/2014/825910

Badruddoza AZM, Tay ASH, Tan PY, Hidajat K, Uddin MS (2011) Carboxymethyl-β-cyclodextrin conjugated magnetic nanoparticles as nano-adsorbents for removal of copper ions: synthesis and adsorption studies. J Hazard Mater 185:1177–1186. https://doi.org/10.1016/j.jhazmat.2010.10.029

Bai H, Zhang Z, Guo Y, Yang G (2009) Biosynthesis of cadmium sulfide nanoparticles by photosynthetic bacteria *Rhodopseudomonas palustris*. Colloids Surf B: Biointerfaces 70:142–146. https://doi.org/10.1016/j.colsurfb.2008.12.025

Bakircioglu Y, Bakircioglu D, Akman S (2010) Biosorption of lead by filamentous fungal biomass-loaded TiO_2 nanoparticles. J Hazard Mater 178:1015–1020. https://doi.org/10.1016/j.jhazmat.2010.02.040

Basavaraja S, Balaji S, Lagashetty A, Rajasab A, Venkataraman A (2008) Extracellular biosynthesis of silver nanoparticles using the fungus *Fusarium semitectum*. Mater Res Bull 43:1164–1170. https://doi.org/10.1016/j.materresbull.2007.06.020

Bayramoglu G, Arıca MY (2008) Removal of heavy mercury (II), cadmium (II) and zinc (II) metal ions by live and heat inactivated *Lentinus edodes* pellets. Chem Eng J 143:133–140. https://doi.org/10.1016/j.cej.2008.01.002

Baytak S, Koçyiğit A, Türker AR (2007) Determination of lead, iron and nickel in water and vegetable samples after preconcentration with *Aspergillus niger* loaded on silica gel. Clean-Soil Air Water 35:607–611. https://doi.org/10.1002/clen.200700124

Bée A, Talbot D, Abramson S, Dupuis V (2011) Magnetic alginate beads for Pb(II) ions removal from wastewater. J Colloid Interface Sci 362:486–492. https://doi.org/10.1016/j.jcis.2011.06.036

Bhattacharya S, Saha I, Mukhopadhyay A, Chattopadhyay D, Ghosh UC, Chatterjee D (2013) Role of nanotechnology in water treatment and purification: potential applications and implications. Int J Chem Sci Technol 3:59–64

Boparai HK, Joseph M, O'Carroll DM (2011) Kinetics and thermodynamics of cadmium ion removal by adsorption onto nano zerovalent iron particles. J Hazard Mater 186:458–465. https://doi.org/10.1016/j.jhazmat.2010.11.029

Buzea C, Pacheco II, Robbie K (2007) Nanomaterials and nanoparticles: sources and toxicity. Biointerphases 2:MR17–MR71. https://doi.org/10.1116/1.2815690

Bystrzejewski M, Pyrzyńska K, Huczko A, Lange H (2009) Carbon-encapsulated magnetic nanoparticles as separable and mobile sorbents of heavy metal ions from aqueous solutions. Carbon 47:1201–1204. https://doi.org/10.1016/j.carbon.2009.01.007

Çeribasi IH, Yetis U (2004) Biosorption of Ni (II) and Pb (II) by *Phanerochaete chrysosporium* from a binary metal system-kinetics. Water SA 27:15–20. https://doi.org/10.4314/wsa.v27i1.5004

Chen GQ, Zhang WJ, Zeng GM, Huang JH, Wang L, Shen GL (2011) Surface-modified *Phanerochaete chrysosporium* as a biosorbent for Cr (VI)-contaminated wastewater. J Hazard Mater 186:2138–2143. https://doi.org/10.1016/j.jhazmat.2010.12.123

Crini G (2005) Recent developments in polysaccharide-based materials used as adsorbents in wastewater treatment. Prog Polym Sci 30:38–70. https://doi.org/10.1016/j.progpolymsci.2004.11.002

Das J, Sarkar A, Sil PC (2015) Hexavalent chromium induces apoptosis in human liver (HepG2) cells via redox imbalance. Toxicol Rep 2:600–608. https://doi.org/10.1016/j.toxrep.2015.03.013

Dasgupta N, Ranjan S, Mundekkad D, Ramalingam C, Shanker R, Kumar A (2015) Nanotechnology in agro-food: from field to plate. Food Res Int 69:381–400

Dasgupta N, Ranjan S, Ramalingam C (2017) Applications of nanotechnology in agriculture and water quality management. Environ Chem Lett 15(4):591–605

Dean JG, Bosqui FL, Lanouette KH (1972) Removing heavy metals from wastewater. Environ Sci Technol 6:518–522. https://doi.org/10.1021/es60065a006

Dey S, Rao P, Bhattacharyya B, Bandyopadhyay M (1995) Sorption of heavy metals by four *Basidiomycetous* fungi. Bioprocess Eng 12:273–277. https://doi.org/10.1007/BF00369502

Di Z-C, Ding J, Peng X-J, Li Y-H, Luan Z-K, Liang J (2006) Chromium adsorption by aligned carbon nanotubes supported ceria nanoparticles. Chemosphere 62:861–865. https://doi.org/10.1016/j.chemosphere.2004.06.044

Dias M, Lacerda I, Pimentel P, De Castro H, Rosa C (2002) Removal of heavy metals by an *Aspergillus terreus* strain immobilized in a polyurethane matrix. Lett Appl Microbiol 34:46–50. https://doi.org/10.1046/j.1472-765x.2002.01040.x

Durali Tuzen M, Soylak M (2008) A biosorption system for metal ions on *Penicillium italicum*–loaded on Sepabeads SP 70 prior to flame atomic absorption spectrometric determinations. J Hazard Mater 152:1171–1178. https://doi.org/10.1016/j.jhazmat.2007.07.097

Dwivedi S, AlKhedhairy AA, Ahamed M, Musarrat J (2013) Biomimetic synthesis of selenium nanospheres by bacterial strain JS-11 and its role as a biosensor for nanotoxicity assessment: a novel Se-bioassay. PLoS One 8:e57404. https://doi.org/10.1371/journal.pone.0057404

El Saliby I, Shon H, Kandasamy J, Vigneswaran S (2008) Nanotechnology for wastewater treatment: in brief. Encyclopedia of Life Support System (EOLSS)

Espinosa-Ortiz EJ, Shakya M. Jain R, Rene ER, van Hullebusch ED, Lens PNL (2016) Sorption of zinc onto elemental selenium nanoparticles immobilized in *Phanerochaete chrysosporium* pellets. Environ Sci Poll Res 23: 21619–21630

Falih AM (1997) Influence of heavy-metals toxicity on the growth of *Phanerochaete chrysosporium*. Bioresour Technol 60:87–90. https://doi.org/10.1016/S0960-8524(96)00177-0

Fu F, Wang Q (2011) Removal of heavy metal ions from wastewaters: a review. J Environ Manag 92:407–418. https://doi.org/10.1016/j.jenvman.2010.11.011

Gao W, Majumder M, Alemany LB, Narayanan TN, Ibarra MA, Pradhan BK, Ajayan PM (2011) Engineered graphite oxide materials for application in water purification. ACS Appl Mater Interfaces 3:1821–1826. https://doi.org/10.1021/am200300u

Gautam RK, Mudhoo A, Lofrano G, Chattopadhyaya MC (2014) Biomass-derived biosorbents for metal ions sequestration: adsorbent modification and activation methods and adsorbent regeneration. J Environ Chem Eng 2:239–259. https://doi.org/10.1016/j.jece.2013.12.019

González PG, Pliego-Cuervo YB (2014) Adsorption of Cd(II), Hg(II) and Zn(II) from aqueous solution using mesoporous activated carbon produced from *Bambusa vulgaris* striata. Chem Eng Res Des 92:2715–2724. https://doi.org/10.1016/j.cherd.2014.02.013

Gopal M, Pakshirajan K, Swaminathan T (2002) Heavy metal removal by biosorption using *Phanerochaete chrysosporium*. Appl Biochem Biotechnol 102:227–237. https://doi.org/10.1385/ABAB:102-103:1-6:227

Grimm L, Kelly S, Krull R, Hempel D (2005) Morphology and productivity of filamentous fungi. Appl Microbiol Biotechnol 69:375–384. https://doi.org/10.1007/s00253-005-0213-5

Gupta VK, Rastogi A, Nayak A (2010) Biosorption of nickel onto treated algae (*Oedogonium hatei*): application of isotherm and kinetic models. J Colloid Interface Sci 342:533–539. https://doi.org/10.1016/j.jcis.2009.10.074

Han X, Wong YS, Tam NFY (2006) Surface complexation mechanism and modeling in Cr (III) biosorption by a microalgal isolate, *Chlorella miniata*. J Colloid Interface Sci 303:365–371. https://doi.org/10.1016/j.jcis.2006.08.028

Hu J, Chen G, Lo IM (2005a) Removal and recovery of Cr (VI) from wastewater by maghemite nanoparticles. Water Res 39:4528–4536. https://doi.org/10.1016/j.watres.2005.05.051

Hu J, Lo IM, Chen G (2005b) Fast removal and recovery of Cr (VI) using surface-modified jacobsite ($MnFe_2O_4$) nanoparticles. Langmuir 21:11173–11179. https://doi.org/10.1021/la051076h

Huang S-H, Chen D-H (2009) Rapid removal of heavy metal cations and anions from aqueous solutions by an amino-functionalized magnetic nano-adsorbent. J Hazard Mater 163:174–179. https://doi.org/10.1016/j.jhazmat.2008.06.075

Jain A, Ranjan S, Dasgupta N, Ramalingam C (2016) Nanomaterials in food and agriculture: an overview on their safety concerns and regulatory issues. Crit Rev Food Sci Nutr 15:1–21

Järup L (2003) Hazards of heavy metal contamination. Br Med Bull 68:167–182. https://doi.org/10.1093/bmb/ldg032

Kalfa OM, Yalçınkaya Ö, Türker AR (2009) Synthesis of nano B_2O_3/TiO_2 composite material as a new solid phase extractor and its application to preconcentration and separation of cadmium. J Hazard Mater 166:455–461. https://doi.org/10.1016/j.jhazmat.2008.11.112

Kanold JM, Wang J, Brümmer F, Šiller L (2016) Metallic nickel nanoparticles and their effect on the embryonic development of the sea urchin *Paracentrotus lividus*. Environ Pollut 212:224–229. https://doi.org/10.1016/j.envpol.2016.01.050

Karimi MA, Kafi M (2015) Removal, preconcentration and determination of Ni(II) from different environmental samples using modified magnetite nanoparticles prior to flame atomic absorption spectrometry. Arab J Chem 8:812–820. https://doi.org/10.1016/j.arabjc.2013.05.018

Kirman C, Hays S, Aylward L, Suh M, Harris M, Thompson C, Haws L, Proctor D (2012) Physiologically based pharmacokinetic model for rats and mice orally exposed to chromium. Chem Biol Interact 200:45–64. https://doi.org/10.1016/j.cbi.2012.08.016

Kitching M, Ramani M, Marsili E (2015) Fungal biosynthesis of gold nanoparticles: mechanism and scale up. Microb Biotechnol 8:904–917. https://doi.org/10.1111/1751-7915.12151

Kowshik M, Deshmukh N, Vogel W, Urban J, Kulkarni SK, Paknikar K (2002) Microbial synthesis of semiconductor CdS nanoparticles, their characterization, and their use in the fabrication of an ideal diode. Biotechnol Bioeng 78:583–588. https://doi.org/10.1002/bit.10233

Lee XJ, Lee LY, Foo LPY, Tan KW, Hassell DG (2012) Evaluation of carbon-based nanosorbents synthesised by ethylene decomposition on stainless steel substrates as potential sequestrating materials for nickel ions in aqueous solution. J Environ Sci 24:1559–1568. https://doi.org/10.1016/S1001-0742(11)60987-X

Lemos VA, Teixeira LSG, Bezerra MA, Costa ACS, Castro JT, Cardoso LAM, de Jesus DS, Santos ES, Baliza PX, Santos LN (2008) New materials for solid-phase extraction of trace elements. Appl Spectrosc Rev 43:303–334. https://doi.org/10.1080/05704920802031341

Li Q, Wu S, Liu G, Liao X, Deng X, Sun D, Hu Y, Huang Y (2004) Simultaneous biosorption of cadmium (II) and lead (II) ions by pretreated biomass of *Phanerochaete chrysosporium*. Sep Purif Technol 34:135–142. https://doi.org/10.1007/BF00167924

Li Y-H, Di Z, Ding J, Wu D, Luan Z, Zhu Y (2005) Adsorption thermodynamic, kinetic and desorption studies of Pb^{2+} on carbon nanotubes. Water Res 39:605–609. https://doi.org/10.1016/j.watres.2004.11.004

Liang P, Ding Q, Liu Y (2006) Speciation of chromium by selective separation and preconcentration of Cr (III) on an immobilized nanometer titanium dioxide microcolumn. J Sep Sci 29:242–247

Liu J-F, Z-s Z, G-b J (2008) Coating Fe_3O_4 magnetic nanoparticles with humic acid for high efficient removal of heavy metals in water. Environ Sci Technol 42:6949–6954. https://doi.org/10.1021/es800924c

Loukidou MX, Matis KA, Zouboulis AI, Liakopoulou-Kyriakidou M (2003) Removal of As (V) from wastewaters by chemically modified fungal biomass. Water Res 37:4544–4552. https://doi.org/10.1016/S0043-1354(03)00415-9

Lu C, Chiu H (2006) Adsorption of zinc (II) from water with purified carbon nanotubes. Chem Eng Sci 61:1138–1145. https://doi.org/10.1016/j.ces.2005.08.007

Luef E, Prey T, Kubicek CP (1991) Biosorption of zinc by fungal mycelial wastes. Appl Microbiol Biotechnol 34:688–692. https://doi.org/10.1007/BF00167924

Madaeni S, Mansourpanah Y (2003) COD removal from concentrated wastewater using membranes. Filtr Sep 40:40–46. https://doi.org/10.1016/S0015-1882(03)00635-9

Mahdavian AR, Mirrahimi MA (2010) Efficient separation of heavy metal cations by anchoring polyacrylic acid on superparamagnetic magnetite nanoparticles through surface modification. Chem Eng J 159:264–271. https://doi.org/10.1016/j.cej.2010.02.041

Moghadasali R, Mutsaers HAM, Azarnia M, Aghdami N, Baharvand H, Torensma R, Wilmer MJG, Masereeuw R (2013) Mesenchymal stem cell-conditioned medium accelerates regeneration of human renal proximal tubule epithelial cells after gentamicin toxicity. Exp Toxicol Pathol 65:595–600. https://doi.org/10.1016/j.etp.2012.06.002

Murray C, Norris DJ, Bawendi MG (1993) Synthesis and characterization of nearly monodisperse CdE (E= sulfur, selenium, tellurium) semiconductor nanocrystallites. J Am Chem Soc 115:8706–8715. https://doi.org/10.1021/ja00072a025

Narayanan KB, Sakthivel N (2010) Biological synthesis of metal nanoparticles by microbes. Adv Colloid Interf Sci 156:1–13. https://doi.org/10.1016/j.cis.2010.02.001

Nassar NN (2010) Rapid removal and recovery of Pb (II) from wastewater by magnetic nanoadsorbents. J Hazard Mater 184:538–546. https://doi.org/10.1016/j.jhazmat.2010.08.069

Nriagu JO, Pacyna JM (1988) Quantitative assessment of worldwide contamination of air, water and soils by trace metals. Nature 333:134–139. https://doi.org/10.1038/333134a0

Özlem Kocabaş-Ataklı Z, Yürüm Y (2013) Synthesis and characterization of anatase nanoadsorbent and application in removal of lead, copper and arsenic from water. Chem Eng J 225:625–635. https://doi.org/10.1016/j.cej.2013.03.106

Peng Q, Liu Y, Zeng G, Xu W, Yang C, Zhang J (2010) Biosorption of copper (II) by immobilizing *Saccharomyces cerevisiae* on the surface of chitosan-coated magnetic nanoparticles from aqueous solution. J Hazard Mater 177:676–682. https://doi.org/10.1016/j.jhazmat.2009.12.084

Perullini M, Jobbágy M, Mouso N, Forchiassin F, Bilmes SA (2010) Silica-alginate-fungi biocomposites for remediation of polluted water. J Mater Chem 20:6479–6483. https://doi.org/10.1039/C0JM01144D

Prasad K, Jha AK (2010) Biosynthesis of CdS nanoparticles: an improved green and rapid procedure. J Colloid Interface Sci 342:68–72. https://doi.org/10.1016/j.jcis.2009.10.003

Qin J-J, Wai M-N, Oo M-H, Wong F-S (2002) A feasibility study on the treatment and recycling of a wastewater from metal plating. J Membr Sci 208:213–221. https://doi.org/10.1016/S0376-7388(02)00263-6

Qu L, Peng X (2002) Control of photoluminescence properties of CdSe nanocrystals in growth. J Am Chem Soc 124:2049–2055. https://doi.org/10.1021/ja017002j

Ranjan S, Dasgupta N, Chakraborty AR, Melvin Samuel S, Ramalingam C, Shanker R, Kumar A (2014) Nanoscience and nanotechnologies in food industries: opportunities and research trends. J Nanopart Res 16(6):1–23

Rashed IGA-A, Afify HA, Ahmed AE-M, Ayoub MAE-S (2013) Optimization of chemical precipitation to improve the primary treatment of wastewater. Desalin Water Treat 51:7048–7056. https://doi.org/10.1080/19443994.2013.792147

Rengaraj S, Joo CK, Kim Y, Yi J (2003) Kinetics of removal of chromium from water and electronic process wastewater by ion exchange resins: 1200H, 1500H and IRN97H. J Hazard Mater 102:257–275. https://doi.org/10.1016/S0304-3894(03)00209-7

Romero-González ME, Williams CJ, Gardiner PH (2001) Study of the mechanisms of cadmium biosorption by dealginated seaweed waste. Environ Sci Technol 35:3025–3030. https://doi.org/10.1021/es991133r

Rubio J, Souza M, Smith R (2002) Overview of flotation as a wastewater treatment technique. Miner Eng 15:139–155. https://doi.org/10.1016/S0892-6875(01)00216-3

Sarı A, Tuzen M (2009) Kinetic and equilibrium studies of biosorption of Pb (II) and Cd (II) from aqueous solution by macrofungus (*Amanita rubescens*) biomass. J Hazard Mater 164:1004–1011. https://doi.org/10.1016/j.jhazmat.2008.09.002

Say R, Denizli A, Yakup Arıca M (2001) Biosorption of cadmium (II), lead (II) and copper (II) with the filamentous fungus *Phanerochaete chrysosporium*. Bioresour Technol 76:67–70. https://doi.org/10.1016/S0960-8524(00)00071-7

Sheet I, Kabbani A, Holail H (2014) Removal of heavy metals using nanostructured graphite oxide, silica nanoparticles and silica/ graphite oxide composite. Energy Procedia 50:130–138. https://doi.org/10.1016/j.egypro.2014.06.016

Silbernagel SM, Carpenter DO, Gilbert SG, Gochfeld M, Groth E, Hightower JM, Schiavone FM (2011) Recognizing and preventing overexposure to methylmercury from fish and seafood consumption: information for physicians. J Toxicol 2011. https://doi.org/10.1155/2011/983072

Soylak M, Tuzen M, Mendil D, Turkekul I (2006) Biosorption of heavy metals on *Aspergillus fumigatus* immobilized Diaion HP-2MG resin for their atomic absorption spectrometric determinations. Talanta 70:1129–1135. https://doi.org/10.1016/j.talanta.2006.02.027

Srivastava N, Majumder C (2008) Novel biofiltration methods for the treatment of heavy metals from industrial wastewater. J Hazard Mater 151:1–8. https://doi.org/10.1016/j.jhazmat.2007.09.101

Suresh AK (2014) Extracellular bio-production and characterization of small monodispersed CdSe quantum dot nanocrystallites. Spectrochim Acta A Mol Biomol Spectrosc 130:344–349. https://doi.org/10.1016/j.saa.2014.04.021

Syed A, Ahmad A (2013) Extracellular biosynthesis of CdTe quantum dots by the fungus *Fusarium oxysporum* and their anti-bacterial activity. Spectrochim Acta A Mol Biomol Spectrosc 106:41–47. https://doi.org/10.1016/j.saa.2013.01.002

Syed A, Saraswati S, Kundu GC, Ahmad A (2013) Biological synthesis of silver nanoparticles using the fungus *Humicola* sp. and evaluation of their cytoxicity using normal and cancer cell lines. Spectrochim Acta A Mol Biomol Spectrosc 114:144–147. https://doi.org/10.1016/j.saa.2013.05.030

Tan T, Cheng P (2003) Biosorption of metal ions with *Penicillium chrysogenum*. Appl Biochem Biotechnol 104:119–128. https://doi.org/10.1385/ABAB:104:2:119

Tuzen M, Saygi KO, Usta C, Soylak M (2008) *Pseudomonas aeruginosa* immobilized multiwalled carbon nanotubes as biosorbent for heavy metal ions. Bioresour Technol 99:1563–1570. https://doi.org/10.1016/j.biortech.2007.04.013

Venkateswarlu S, Himagirish Kumar S, Jyothi NVV (2015) Rapid removal of Ni(II) from aqueous solution using 3-Mercaptopropionic acid functionalized bio magnetite nanoparticles. Water Resour Ind 12:1–7. https://doi.org/10.1016/j.wri.2015.09.001

Vigneshwaran N, Ashtaputre N, Varadarajan P, Nachane R, Paralikar K, Balasubramanya R (2007) Biological synthesis of silver nanoparticles using the fungus *Aspergillus flavus*. Mater Lett 61:1413–1418. https://doi.org/10.1016/j.matlet.2006.07.042

Volesky B (1990) Biosorption of heavy metals. CRC press, Boca Raton

Wang J, Chen C (2006) Biosorption of heavy metals by *Saccharomyces cerevisiae*: a review. Biotechnol Adv 24:427–451. https://doi.org/10.1016/j.biotechadv.2006.03.001

Xu P, Zeng GM, Huang DL, Feng CL, Hu S, Zhao MH, Lai C, Wei Z, Huang C, Xie GX (2012a) Use of iron oxide nanomaterials in wastewater treatment: a review. Sci Total Environ 424:1–10. https://doi.org/10.1016/j.scitotenv.2012.02.023

Xu P, Zeng GM, Huang DL, Lai C, Zhao MH, Wei Z, Li NJ, Huang C, Xie GX (2012b) Adsorption of Pb (II) by iron oxide nanoparticles immobilized *Phanerochaete chrysosporium*: equilibrium, kinetic, thermodynamic and mechanisms analysis. Chem Eng J 203:423–431. https://doi.org/10.1016/j.cej.2012.07.048

Yalcinkaya O, M Kalfa O, R Turker A (2011) Preconcentration of trace copper, cobalt and lead from various samples by hybrid nano sorbent and determination by FAAS. Curr Anal Chem 7:225–234

Yan G, Viraraghavan T (2003) Heavy-metal removal from aqueous solution by fungus *Mucor rouxii*. Water Res 37:4486–4496. https://doi.org/10.1016/S0043-1354(03)00409-3

Zhang X, Yan S, Tyagi R, Surampalli R (2011) Synthesis of nanoparticles by microorganisms and their application in enhancing microbiological reaction rates. Chemosphere 82:489–494. https://doi.org/10.1016/j.chemosphere.2010.10.023

Chapter 8
Nanomaterials Reactivity and Applications for Wastewater Cleanup

Tamer Elbana and Mohamed Yousry

Abstract Treated wastewater is a reliable water resource for agriculture in arid and semiarid areas. Nanomaterials are promising to clean wastewater. Here we review nanomaterials characteristics, reactivity and potentiality to reduce or remove pollutants from wastewater. Characteristics include high reactivity of surface areas, quantum confinement effects, surface charge density and stability of nanophases. We discuss applications to remove from inorganic and organic contaminants, with focus on reaction kinetics, sorption and degradation. Remediation efficiency is also controlled by wastewater properties such as pH, ionic strength and water temperature.

The use of nanomaterials often allow a removal of more than 80% of most pollutants. Nonetheless, this review explains that the cost, the aggregate formation and the difficulty of recovering most applied nanomaterials are challenging. Alternatively, natural nanomaterials such as nano clay represent an inexpensive and environmental friendly substance for wastewater remediation.

Keywords Nanomaterials • Wastewater • Inorganic pollutant • Organic pollutant

Contents

T. Elbana (✉) • M. Yousry
Soils and Water Use Department, National Research Centre (NRC), Cairo, Egypt
e-mail: tamerelbana@yahoo.com; amryousry@hotmail.com

K M Gothandam et al. (eds.), *Nanotechnology, Food Security and Water Treatment*, Environmental Chemistry for a Sustainable World 11,
https://doi.org/10.1007/978-3-319-70166-0_8

Abbreviations

CNT	Carbon nanotubes
DOS	density of states
FAO	Food and Agriculture Organization
ISO	International Organization for Standardization
nZVI	nanoscale zero valent iron
POPs	persistent organic pollutants
PZC	point of zero charge
U.S EPA	U.S. Environmental Protection Agency
WHO	World Health Organization

8.1 Introduction

The reactivity of a material is strongly depending upon its particle size and surface area as well as its chemical properties. Thus, cutting down the bulk material into the nanoscale dimension increases surface area and enlarges the surface energy of this material. Since, the nanomaterials science has been attracted the attention of several researchers globally, the Nobel prize 2016 in chemistry was awarded for developing the world' smallest machine that executed nanomaterials. One of the most widely known definition of nanomaterial was introduced by the international organization for standardization (ISO 2015), as "a material with any external dimension in the nanoscale, length range approximately from 1 nm to 100 nm, or having internal structure or surface structure in the nanoscale". Comparison of the varied regulatory and advisory definitions of the term "nanomaterial" elucidates that the size is the only common aspect. While, numerous definitions disregard key factors such as the agglomerates and aggregates, distributional thresholds, novel properties, and solubility (Boverhof et al. 2015).

The nanomaterials are employed effectively in numerous modern daily life applications such as nanoelectronics (sensors and detectors), nanomedicine, nanobots and catalysis as well as various biological and environmental applications (Cao and Wang 2011). Murty et al. (2013) listed the applications of nanomaterials in environmental remediation as; (i) Nano-membranes and nano-clays for water filtration, (ii) Nanoparticle-activated wastewater reuse, and (iii) Nanosensors to monitor water quality. Additionally, the same authors considered the application of nanomaterials in green energy technologies. Practically, the nanomaterials are promising to remediate environmental pollution however rational manipulation and safety policies are essentially required for recycling the manufactured nanomaterials (Buzea et al. 2007). For instance, the nanoscale zero valent iron is commonly applied for remediating groundwater and contaminated soil (Kržišnik et al. 2014; O'Carroll et al. 2013; Ranjan et al. 2014; Dasgupta et al. 2015, 2017; Jain et al. 2016). Whereas, evaluation of the ecological impacts of the nanoscale zero valent iron applications is environmentally necessary due to its cytotoxicity and the adverse effects on the indigenous microbial communities (Araújo et al. 2016; Lefevre et al. 2016; Siripireddy et al. 2016, 2017; Maddinedi et al. 2015, 2017;

Tammina et al. 2017; Shukla et al. 2017; Walia et al. 2017; Ranjan and Ramalingam 2016; Dasgupta and Ramalingam 2016; Ranjan et al. 2016; Dasgupta et al. 2016a, b).

8.2 Wastewater: Contamination Source

Wastewater term is usually implemented to designate the generated water after its original use for domestic, commercial establishments, or industrial purposes. Based on the global water withdrawals data between 1950 and 2010, Flörke et al. (2013) estimated the global domestic water use as 390 km^3 which was increased by 3.7 folds during the past 60 years. Whereas, the water use in the industry sector increased by 2.9% to get 955 km^3 by 2010. The food and agriculture organization of the United Nations estimated the total global withdrawal by agricultural sector in 2010 as 2769 km^3 that represents 69% of the total withdrawal waters (FAO 2017). As the water sacristy is one of the most environmental problem that the world faces, so investigation a new alternatives for saving water resources is critically required. One of the promising alternatives is the reuse of treated wastewater in agriculture. Thus, treated wastewater is a vital resource of water especially in arid and semiarid regions to cope water scarcity issue in such area. Moreover, it is essential to apply a recommended treatment level before discharging it to the surrounding ecosystem for sustaining the performance of our environment.

In fact, wastewater pollutants are varied in corresponding to the source of water (e.g. agricultural, industrial, or domestic wastewaters). Generally, chemical, biological, and radiological pollutants can be found separately or collectively in wastewater. Chemical pollutants can be classified into organic (such as, chlorophenols) and inorganic contaminants (such as, trace elements). The U.S. environmental protection agency and the world health organization published a comprehensive guidelines for wastewater reuse that specified the maximum concentration and the allowable levels of organic and inorganic chemicals in the treated wastewater to be reused for agriculture, aquaculture and other ruse purposes (U.S. Environmental Protection Agency 2012; World Health Organization 2006a, b, c, d). Various domestic wastewaters contain microorganisms, organic materials, nutrients, and radioactive substances (Henze 2002). However, agricultural drainage water contains numerous fertilizers and pesticides residues at different concentrations. Specifically, quality of drainage water is identified by its salt contents and major ions (e.g. Na, Ca, Mg, CO_3, SO_4, Cl), toxic trace elements (e.g. Cd, Pd, Hg, As), pesticides (e.g. atrazine, carbofuran), as well as the concentrations of plant nutrients such as nitrate and phosphates (Tanji and Kielen 2002). On the other hand, industrial wastewater is usually classified as a point source pollution that is mainly contains a certain chemical composition. Industrial organic wastes can often be biologically or thermally degraded into carbon dioxide (CO_2) and water whereas, inorganic industrial wastes are not degradable and should be discharged after a treatment that can reduce the negative consequences on the environment and human health (Artiola 2006). However, the diffuse of pollutants from mining industries represents highly polluted wastewater which is hardly can be treated (UN-Water 2015).

Untreated wastewater is a source of contamination for air, fresh water, and soil (Chamtouri et al. 2008; Elbana et al. 2013; Karaouzas 2016). Environmental chemists are continuously seeking for affordable highly reactive materials that can safely remove pollutants from wastewater. Due to the high surface area of the nanomaterials, these materials exhibit reactive chemical characteristics and high potential for cleaning up the polluted aquatic solutions. The applications of nanomaterials for water and wastewater treatment principally depend upon adsorption, oxidation (Photocatalysis), and disinfection as well as separation processes (Qu et al. 2013). In this chapter nanomaterials reactivity and potentiality to reduce or eliminate pollutants concentrations from wastewater will be discussed. Specifically, the main objectives of this text are; (1) introducing the importance of nanomaterials and their reactivity and (2) emphasizing the application of nanomaterials in cleaning up the wastewater from inorganic and organic contaminates.

8.3 Nanomaterials Reactivity

The reactivity of any material is highly related to its surface area. In order to explain the nanomaterials reactivity, there is a need to illustrate how those materials get their huge surface area. Theoretically, if there is one cube of iron with 0.2 μm edge dimension, then this cube is segmented into eight identical cubes. Repetitively, if each new cube is divided up equally into another eight cubes and keep dividing of each new cub to smaller identical cubes for another time. Finally, there will be 512 similar size cubes in total (1, 8, 64, 512 cubes for the consecutive segmentation, see Fig. 8.1). As the edge length of the original cub is 200 nm with total surface area of 0.24 μm^2. However, with sequential splitting up of a cube for only three times, the total area of the 512 cubes will be 1.92 μm^2 as the edge length of each smallest cube is 25 nm. Accordingly, the surface area increased by eight folds when the cube is divided three times. This increasing in the surface area can increase the edge energies and chemical reactivity of the materials in nano-size compared with the case of the normal size. Additionally, this hypothetical example illustrates how nanoparticles have an extremely large surface area per unit mass or volume "specific surface area".

The reactivity of nanomaterials can be ascribed to the surface and the quantum confinement effects (Roduner 2006). Precisely, atoms on the surface of nanoparticles have less bonding and coordination number compared with that ones in the materials bulk. Additionally, the nanomaterials are characterized by a set of discrete energy levels due to the change of the density of states. Moreover, the reactivity of nanomaterials depends on their surface curvature, as high reactivity is expected to associate with the high curvature (small radius). Agrawal (2013) emphasized the influence of surface curvature on nanomaterial reactivity and explained that the chemical potential of atoms on convex surface is higher than its chemical potential on concave surface. In accord, Solveyra and Szleifer (2016)

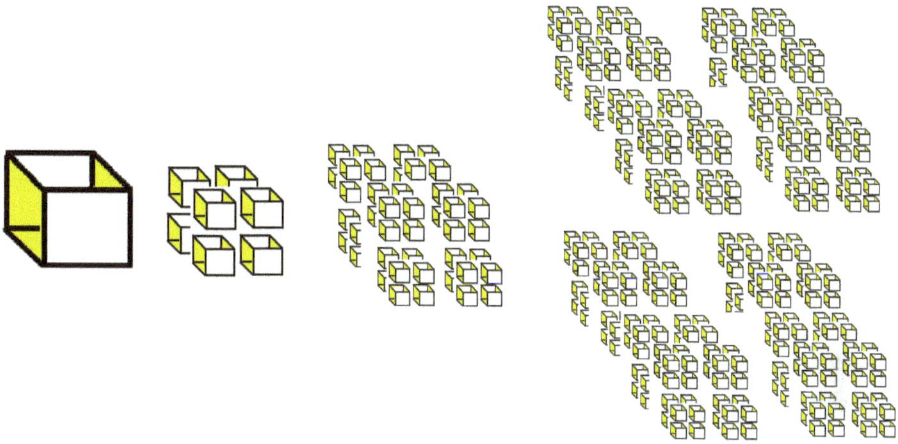

Fig. 8.1 Schematic of the segmentation of one cube into identical eight ones for three times consecutively

justified the substantial role of nanomaterial's confinement and curvature in the sorption process of the molecules from the contiguous environment.

Additional key characteristic of nanomaterials reactivity is the surface charge density. The surface charge is known to be dependent on the surrounding media properties such as pH, ion concentration, and temperature. The electrical charge density (E) can be quantified by Nernst equation:

$$E = E_o + \frac{R\,T}{F\,C_i} \ln\, a_i \tag{8.1}$$

Where E_o is the standard electrode potential; C_i is the ion concentration; a_i is the activity of ions; R is the gas constant; T is the temperature, F is the Faraday's constant. The material surface is positively charged when the surrounding media has a pH less than the material's point of zero charge and vice versa, it is negatively charged when the surrounding media has a pH value higher than the material's point of zero charge. Clavier et al. 2015 showed that in addition to the acid–base properties, the surface charge variations of nano-metal oxide were controlled by the site distribution and dielectric constant. Furthermore, Barisik et al. (2014) elucidated the increase of the surface charge density magnitude for silica nanoparticles with the increases of pH and salt concentration. Additionally, Barisik et al. (2014) clarified that the magnitude of the surface charge is particle size dependent till a critical diameter of the silica nanoparticles. In other words, the surface charge density magnitude was decreased by increasing particle size till the critical diameter. The authors attributed this variation to the change in the surface concentration of H^+ ions.

Stability of nanophases is an additional important characteristic that is affecting nanomaterials reactivity. Wang et al. (2011) demonstrated that the particle size and surface charge are key factors for the stability of nanoparticle suspension. He and Alexandridis (2015) found that the stability of nanoparticle dispersions in ionic liquids affected by temperature, pressure, and water concentration as well as the intermolecular interactions and the physicochemical properties of ionic liquids and nanoparticles. Brant et al. (2005) examined the effect of ionic stress on the stability of fullerene nanoparticles (nC60) suspension, they found that 0.001 *M* of NaCl was able to destabilize the nC60 by forming large aggregates whereas; nC60 was stable overtime with the absence of the electrolytes. Unquestionable, preserving the stability of nanophases and avoiding the growing of grains is critical to maintain the nanostructure and the reactivity of the nanomaterials (Andrievski 2014). In another example of instability, the bara nanomaterial of iron oxide is oxidized in low pH and consequently forming large aggregates. This instability can be resolved by functionalized or coating nanoparticles in acidic media (Gautam and Chattopadhyaya 2016).

Quantifying the nanomaterial reactions is a prerequisite for contaminant removal from wastewater. Exploring the reaction kinetic provides a comprehensive knowledge on nanomaterials behavior in the environment. Removal of contaminants from different solutions can be described by quantifying the rate of sorption, degradation, decay, or transformation of the toxic chemical. The rate of reaction (r) can be calculated by:

$$r = -\frac{dC_{cont}}{dt} \tag{8.2}$$

Where, dC_{cont} is the change of contaminant concentration during a certain period of time (dt). Based on Eq. 8.2, a high absolute value of "r" is an indication of the speedy removal of contaminant from the solution. This simple calculation provides a measuring way to select the appropriated materials or treatment design for removing contaminant. Besides, calculating the half-life of a reaction is an important measure for wastewater treatment design that is calculating the required time to reduce contaminant concentration to the half of its initial one. Further calculations will be needed to simulate the reaction and identifying its order. First, second, and n th order kinetic models can be considered for understanding the chemical behavior of the applied material for cleaning up wastewater. For instance, Fu et al. (2014) elucidated the kinetic of the oxidation reaction between nanosized zinc sulfide and oxygen. Authors explained that reaction constant and reaction order were increased by decreasing the particle size of zinc sulfide (from 71.9 to 17.9 nm) and increasing temperature (from 473 K to 553 K).

Kinetic studies can progressively help in improving and understanding the nanomaterials catalysts size, shape and surface-composition (Mondloch et al. 2012). Moreover, adsorption kinetic information can provide a manner to evaluate nanoparticles reactivates and its potentialities to remove pollutants. For example, Ahmadi et al. (2015) studied the removal kinetic of Congo red dye by nickel-zinc sulfide nanoparticle and palladium nanoparticles which were loaded on activated carbon. This study indicated the fast removing of the dye with contact time (less than 26 min) and the pseudo second order kinetic successfully simulated experimental data whereas, the pseudo first order kinetic model failed to simulate the experimental data.

8.4 Nanomaterials for Wastewater Cleanup

One of the main applications of nanomaterials is the water treatment through filtration by nano-membranes and nano-clay, nanoparticle-activated wastewater reuse systems, and nanosensors to monitor water quality (Murty et al. 2013). Several efforts have been accomplished to remediate wastewater and to improve quality of water for achieving particular water quality criteria. Cleanup can be done by reducing contaminant concentration to a certain permissible limit, transforming a toxic chemical to a nonhazardous compound, and/or to eliminate pathogens and biological threats. Hasemzadeh et al. (2014) emphasized the wide range of nanomaterials uses for wastewater treatment; they considered treating of inorganic polluted water, dye wastewater, papermaking wastewater, pesticide wastewater and oily wastewater through adsorption of chemicals and removal/inactivation of pathogens. For example, the advanced oxidation process using nanoscale zero valent iron as a catalyst is appropriate for the toxic or non-biodegradable materials; that can be ascribed to the positive impact of nanoscale zero valent iron on the production of hydroxyl radicals (Rosales et al. 2017). Additionally, graphene-based nanocomposites exhibit a robust potential as reactive adsorbents for removing inorganic and organic contaminates from wastewaters whereas, further experimental examination is needed on large scale applications (Gautam and Chattopadhyaya 2016).

Recently, Zhang et al. (2016) reviewed the most commonly categories of nanomaterials that are used in water treatment specially carbon based nanomaterials (graphene based nanomaterials, carbon nanotubes), metal and metal oxides nanoparticles (nanoscale zero valent iron, nanosized iron oxides, nanosized titanium oxides), noble metal nanoparticles (such as gold (Au) and silver (Ag)); authors concluded that these materials are promising candidate for the development of next generation water treatment technology (Zhang et al. 2016). For example, carbon nanotubes have a great performance for removing contaminants. That's owing to its high adsorption capacity, its high speed reaction kinetic, and carbon nanotubes were considered as efficient catalysts as well as its high ability to remove bacteria and viruses (Liu et al. 2013; Ren et al. 2011). Actually, it was expected to have the current rapidly development for large scale applications of nanotechnology for wastewater treatment (Hasemzadeh et al. 2014).

Despite of nanomaterials benefits for wastewater treatment, challenges and obstacles should be taken into consideration. Zhang et al. (2016) pointed out that; (i) the instability of nanomaterials due to aggregates formation; (ii) the difficulty of recycling nanomaterials except for magnetic nanoparticles; (iii) the long-term fate of nanomaterials is unclear yet. Liu et al. (2013) emphasized the engineering and economical challenges for carbon nanotubes such as uniform dispersion and manufacturing cost, respectively. Concurring with that, Savage and Diallo (2005) explained that however the great functionality of nanomaterials for water purification, there is a necessity for synthesizing cost-effective and environmental friendly nanomaterials. Recently, Adeleye et al. (2016) highlighted the importance of practicing risk assessment of nanotechnology to gain the promising implementation of it in the pollution cleanup, especially for affordable nanotechnologies which exhibits economical comparable to the conventional methods. Understanding nanomaterials toxicology and safety practices of its application will expand the advantages of those materials implementations (Dasgupta et al. 2016a, b).

8.5 Nanomaterials for Inorganic Contaminants Removal

The high affinity of the inorganic chemical to be sorbed on nanomaterials makes the sorption process one of the widespread approach for wastewater cleanup. Actually, ion-exchange, adsorption and membrane separation are commonly considered for removing of heavy metals from wastewater (Fu and Wang 2011). The nanoadsorbents are considered as practical and simple process to remove heavy metals and radioactive elements from wastewater (Ray and Shipley 2015).

Generally, sorption term is applied to describe the attachment of contaminant (adsorbate) to a reactive particle (adsorbent) without referring to the bonding mechanism. Freundlich equation is a popular equation that is applied to mathematically quantify the sorbed amount:

$$S = K_F\, C^b \tag{8.3}$$

where S is the sorbed amount on a specific adsorbent; C is the concentration in surrounding solution; b is a dimensionless reaction order and K_F is the Freundlich distribution coefficient. Moreover, Langmuir equation is commonly applied to quantify maximum sorption capacity and to simulate the relation between the sorbed amount (S) and the contaminant concentration (C) at the apparent equilibrium using the following formula:

$$S = \frac{S_{max} K_L C}{1 + K_L C} \tag{8.4}$$

Where S_{max} is the maximum sorption capacity; K_L is a Langmuir coefficient.

Table 8.1 shows some examples of nanomaterial applications for trace elements removal from wastewater. Data in Table 8.1 reveals the various uses of nanomaterials to remove potential toxic elements from contaminated water. The pH of wastewater, temperature and contact time should be considered to obtain high sorption capacities and consequently to perform a high removal efficiency. Data in Table 8.1 indicates the importance of quantifying the maximum sorption capacity for comparison purpose. For example, zeolite nano-particles sorbed approximately seven times higher of Pb (682 mg/g) in comparing with nano-alumina that sorbed 100 mg/g. Whereas, carbon nanotubes exhibited the highest sorption capacity for Ni of 3900 mg/g (See Table 8.1).

Table 8.1 Selected examples of nanomaterial applications for trace element removal from wastewater

Nanomaterial (adsorbent)	Inorganic contaminant	Sorption capacity/ removal potential	Note	References
Nanoscale zero-valent iron (nZVI)	As (III)	Sorption capacity of 1.80, 2.47, and 1.56 mg of As(III)/g of nZVI at 25. 35, and 45 °C, respectively	The maximum adsorption capacity was calculated based on Langmuir equation for sorption isotherm data.	Kanel et al. (2005)
	Zn(II)	Removal of 5 mg Zn^{2+} L^{-1} and 25 mg Zn^{2+} L^{-1} using 0.4 g Fe L^{-1} during a contact time of 1 h and 24 h, respectively	Removal at pH of 7.	Kržišnik et al. (2014)
	U (VI)	The maximum removal capacity of 350.47 mg/g was realized at pH of 5.5	Using nZVI- polyaniline-graphene composite, Uranium removal is increased by temperature increasing.	Chen et al. (2017)
Nano-alumina	Pb (II), Cd (II), Cr (III), Co (II), Ni (II), and Mn (II).	Sorption capacity of 100, 83.3, 100, 41.7, 18.2, and 6.3 mg/g for Pb, Cd, Cr, Co, Ni, and Mn, respectively	Nano-alumina was modified with 2,4-dinitrophenylhydrazine, Simultaneous removal, Langmuir equation was applied	Afkhami et al. (2010)
Carbon nanotubes	Ni (II), Zn (II), As (III), and Co (II)	Sorption capacity of 3900, 3650, 3500, and 3800 mg/g for Ni, Zn, As, and Co, respectively	PAMAM/CNT nanocomposite, Langmuir equation was applied.	Hayati et al. (2016)

(continued)

Table 8.1 (continued)

Nanomaterial (adsorbent)	Inorganic contaminant	Sorption capacity/ removal potential	Note	References
Zeolite nano-particles	Pb (II) and Ni (II)	Sorption capacity of 682 and 122 mg/g respectively for Pb and Ni	Implementation of adsorption and the filtration processes, After 60 min of filtration with 1 bar of transmembrane pressure.	Yurekli (2016)
Mercaptoamine-functionalised silica-coated magnetic nano-adsorbents	Hg (II) and Pb (II)	Sorption capacity of 355 and 292 mg/g respectively for Hg and Pb	Langmuir equation was applied, The maximum adsorptions were occurred at pH 5–6 and 6–7 for Hg and Pb, respectively.	Bao et al. (2017)
PVA/TiO_2 nanohybrid adsorbent	Cd (II), Ni (II) and U (VI)	Sorption capacity of 49.0, 13.1 and 36.1 mg g^{-1} for Cd, Ni and U, respectively	Sorption at pH of 5.5, 5 and 4.5, for Cd, Ni and U, respectively.	Abbasizadeh et al. (2014)

Removing toxic elements from wastewater is an environmental demand. Kumar and Chawla (2014) provided a comprehensive review on the application of nano-metal oxides as adsorbents to remove cadmium (Cd) from polluted water, where the sorption capacity varied between 15.2 to 625 mg/g for different nano-metal oxides. According to their review, the selection of an appropriated nanosorbent for Cd removal from wastewater depends on the concentration of Cd in the water, pH, temperature, cost and toxicity as well as adsorption capacity. For another example, removal of selenium (Se) from wastewater is receiving a great attention nowadays. Holmes and Gu (2016) compared the reported adsorption capacities of varied nanomaterials to explore Se adsorption which diverged between 0.05 to 178 mg/g. Definitely, the initial concentration and the pH of wastewater are important key factors that affect Se affinity to nanomaterials.

Furthermore, application of the natural nanoscale materials is promising as adsorbents for inorganic removal too. Zhang et al. (2015b) emphasized the importance of nano-clay as an adsorbent material for heavy metals removal from wastewater and authors provided the properties of nano-Kaolinite, nano-Montmorillonite, and nano-layered double hydroxides (anionic clay) that control the removal of heavy metals from aqueous solutions. Based on the available literature, nano-clay provides affordable materials for wastewater cleanup.

Removal of nitrate (NO_3) from wastewater is very critical to sustain environmental resources and safely reuse treated wastewater. Similar to the removal of heavy metals, the initial concentration, wastewater pH, contact time, and temperature are significantly affect NO_3 removal. In a study to assess the use of nano-alumina to remove NO_3 from contaminated water, the sorption capacity of 4.0 mg/g was reported at pH of 4.4 and 25 °C. However, the NO_3 adsorption found to be strongly pH dependent and sharp sorption decrease was observed with increasing the pH above 5 due to the change of surface charge (Bhatnagar et al. 2010). Application of nanoscale zero valent iron for reducing NO_3 concentration in water is commonly practiced. The use of nanoscale zero valent iron -based permeable reactive barriers for NO_3 reduction is effective technique and can remove up to 100% (Araújo et al. 2016). One advantage of using the nano-alumina (sorption process) compared with using of nanoscale zero valent iron (reduction technique) is that using nano-alumina is avoiding the production of NH_4.

8.6 Nanomaterials for Organic Pollutant Removal

Various uses of nanomaterials are implemented to decontaminate wastewater by removing or reducing the organic contaminate concentrations. Along with sorption, contaminant degradation process is commonly practiced for organic contaminant removal. Table 8.2 shows examples of recently applications for cleanup wastewaters from petroleum, dairy, and textile industrial sectors. Application of various kinetic calculations can provides a numerical approach to understand the fate of the targeted organic contaminants. For example, the nonlinear degradation kinetics can be simulated by calculating the change of the relative contaminant concentration (C/C_o) using Eq. 8.5 (McDonald et al. 2013):

$$\frac{d\left(\frac{C}{C_o}\right)}{dt} = -K_d\left(\frac{C}{C_o}\right)^m \tag{8.5}$$

Where, C is the contaminant concentration at specific time (t), C_o is the initial concentration (at time 0), K_d is the degradation rate, and m is the degradation order. Experimental data can be fitted to the integrated form of the Eq. 8.5 that is shown in Eq. 8.6:

$$\frac{C}{C_o} = (1 + [m - 1]K_d t)^{-1/(m-1)} \tag{8.6}$$

Moreover, the calculation of the half-life of the organic contaminant due to the use of certain nanomaterial is useful to design an appropriated treatment. For the first order degradation, half-life time ($t_{0.5}$) can be calculated as $t_{0.5} = 0.693/K_d$ (for further details on the modeling the degradation kinetics see Levenspiel 1999; Méndez et al. 2010).

Literature review of utilization of nanomaterials proved the high potential of nanomaterials to remove and reduce pesticides, dyes, antibiotics and among other

Table 8.2 Selected examples of nanomaterial applications for organic contaminant removal from wastewater

Nanomaterial	contaminant	Removal/degradation	Note	Reference
Nano-TiO_2 supported on Fe-ZSM-5 zeolite	Petroleum refinery wastewater	Photodegradation efficiencies of 63% and 67.05% were obtained at pH 4 and 8, respectively	High degradation at pH of 4, 45 °C, and UV irradiation for 120 min, The degradation followed the first order kinetics.	Ghasemi et al. (2016)
TiO_2-Ag nanofibers	Dairy effluent	60% degradation after 180 minutes	Photocatalytic degradation	Kanjwal et al. (2016)
Amino acid proline based polymer nanocomposite	Textile dyes	Sorption capacities of 24.4 to 28.3 mg/g were achieved within 180 minutes.	Dyes: Reactive blue 222, reactive red 195, and reactive Yellow145, Increasing of temperature, pH, and salt contents reduced the sorption efficiency.	Raghunath et al. (2016)
Zinc nanoparticles (ZnNPs)	Trypan blue dye	The maximum sorption capacity of 129.87 was observed at pH 7 within 30 min	Langmuir isotherm model was applied, Pseudo-second order kinetic model provide a successful simulation of kinetic data.	Nadaroglu et al. (2017)
Palladium/hydroxyapatite/Fe_3O_4 nanocatalyst (Pd/HAP/Fe_3O_4)	Azo dyes (methyl red, methyl orange and methyl yellow)	A complete decomposition of azo dyes was observed after 80 min	Catalytic activity of Pd/HAP/Fe_3O_4 enhanced with the low pH, The pseudo-first-order provided a good-fitting of the data.	Safavi and Momeni (2012)
Cu-nano zeolite	2-chlorophenol	Maximum adsorption capacity of 204.68 mg/g	The highest removal at pH 6 during 150 minutes.	Huong et al. (2016)
Nano zeolite	Meta, ortho and para-nitrophenols isomers	Sorption capacities of 121.7, 143.8, and 156.6 mg/g for meta, ortho and para-nitrophenols, respectively	With adsorbent dose of 0.6 g at pH 6.0 within 150 minutes.	Pham et al. (2016)

organic contaminants from contaminated water (Ghasemi et al. 2016; Pham et al. 2016; Raghunath et al. 2016; Ramya et al. 2016). Also, natural nanomaterial such as nano zeolite can offer an inexpensive remediation material for wastewater treatment. Pham et al. (2016) evaluated the use of nano zeolite to remove nitrophenol from contaminated water and showed that nano zeolite economically removed more than 70% of the nitrophenol effectively with 46.6% cost reduction compared with the use of activated carbon.

Occurrence of antibiotics in wastewater represents serious environmental threats. Karthikeyan and Meyer (2006) surveyed 21 antibiotic compounds in wastewater in Wisconsin, USA and emphasized the importance of monitoring of at least six compounds namely: sulfamethazine, sulfamethoxazole, tetracycline, ciprofloxacin, erythromycin-H_2O, and trimethoprim. Recently, Meng et al. (2016) showed that nano-MgO/diatomite membrane was able to remove >95% of tetracycline for a wide range of pH (6–10.2) with water flux of 120.5 L/m^2 h. Here is another example of removing antibiotic compound by nanomaterial, Fang et al. (2011) completely removed metronidazole (80 mg/L) using a dosage of 0.1 g/L nanoscale zero valent iron within 5 minutes.

Organic pollutants in wastewater represent challenge for reusing the treated wastewater or even to discharge it to the environment safely. Positively charged nanomaterials are needed to sorb anionic organic pollutants. Cai and Larese-Casanova (2016) functionalized graphene oxide with ethylenediamine to remove anionic ibuprofen in a batch reactor experiment. Their results proved the enhancing of the sorptive capacity for removing ibuprofen (maximum sorption capacity of 95.2 mg/g) compared with the case of using activated carbon and regular graphene oxide. Organofluorine compounds exhibits a harmful impact on the environment; Stahl et al. (2011) provided a toxicological evaluation of perfluorinated compounds that showing their contamination potential to the environmental resources and adverse impacts on human health. Zhang et al. (2016) used nano-ZnO coated electrodes for degradation of perfluorinated compounds in wastewater treatment and achieved removal efficiency of 39% to 66% within 40 minutes of electrochemical oxidation process. This removal efficiency is higher than that one of using stainless steel plates as anode and cathode electrode where the reported efficiency was 39% to 66% (Zhang et al. 2015a). Actually, organic contaminants represent a real threatens to environmental resources particularly due to its various categories that occasionally need specific treatments. In other words, organic contaminants could be cationic, anionic, volatile, short or long chains, persistent organic pollutants and others classes, thus the elimination of them from wastewater is a problematic issue that need integrated remediation approach. Nanomaterials provide an effective solution to cleanup wastewater from those tenacious contaminants.

The high reactivity of applied nanomaterials for cleaning up wastewater can modify the conventional treatment approaches. The expected consequences of using nanomaterials in wastewater treatment plants includes reducing the required time and energy of the treatment, types and quantities of the applied chemicals as well as improving the quality of the treated wastewater. For instance, implementation of nanofiltration technology in wastewater treatment shows an effective improvement of the treated water quality (Bunani et al. 2013, 2014; Li et al. 2016). In a pilot-scale experiment for cleaning up a hospital wastewater, nanofiltration equipped with membrane bioreactor was effectively applied (Kootenaei and Rad 2013). The results of the experiment after 20 days revealed that the average removal efficiencies were 94%, 88%, > 80%, 87%, and 46 to 68% for chemical oxygen demand, ammonia, nitrite, and nitrate, and phosphorus, respectively. On the other hand, chemicals incompatibility, limited temperature range, and fouling can restrict the application of nanofiltration for cleaning up wastewater (Shahmansouri and Bellona 2015).

According to the current literature review, designing of wastewater treatment based upon the use of nanomaterials should consider the application scale, time and cost of treatment as well as the existing type of contaminants. Prior of applying such nanotechnology the following assessments are suggested:

- Assessing the nanomaterials potentiality for reduce the concentration to acceptable level or remove the targeted contaminants.
- Check the required optimization of operating conditions and reactor configuration.
- Evaluate the cost-effectiveness of the nanotechnology implementation on the large scale.

Additionally, integrating of nanomaterials-based treatment with the current conventional techniques for wastewater cleaning up is simply compatible (Gehrke et al. 2015). In the following schematic diagram we suggested different contribution of implementing nanomaterials-based treatment unit for cleaning up wastewater (Fig. 8.2). Specifically, three cases are suggested: Case "A" represents the

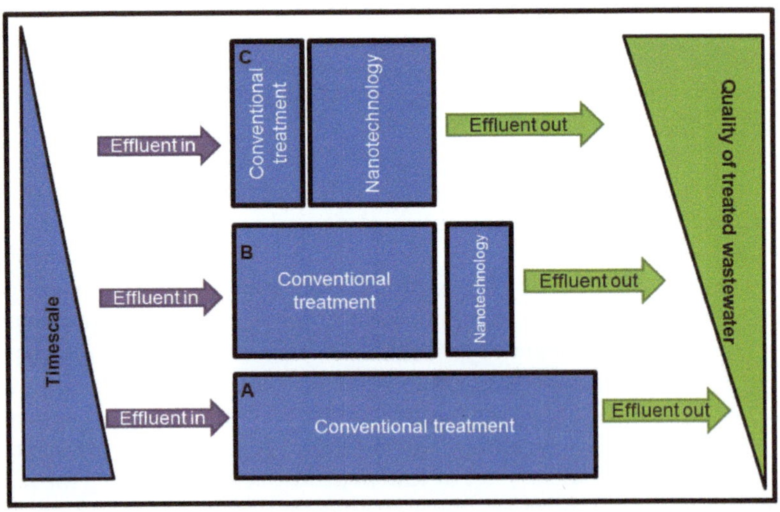

Fig. 8.2 Schematic diagram of the consequences of combining nanotechnologies with the conventional wastewater treatment on timescale and the quality of the treated wastewater

conventional wastewater treatment; Case "B" represents the situation of using the nanomaterials in one main process such as applying nano-membranes technology; Case "C" represents the use of more than two nanotechnologies (such as nano-membranes, nanomaterials for oxidation and sorption processes) during the wastewater treatment.

Generally, the conventional wastewater treatment incudes different processes and units of operation such as screening, sedimentation, flotation, chemical oxidation, membrane filtration, sorption, ion exchange, ultraviolet, and others. Figure 8.2 reveals the influence of combining nanotechnology with the conventional treatment on the required time of treatment and on the quality of treated wastewater. The diagram shows the decrease of treatment time on the left side and the improvement in the treated wastewater quality on the right side as a result of implementing nanomaterials-based treatment. For example, the nanoscale zero-valent iron particles are largely applied in wastewater treatment. As a result of simple oxidation process, the surface of zero-valent iron is covered by iron (hydr)oxides that provides a surface for contaminants sorption and reduction reactions. One can clearly imagine the high rate of reaction in case of using the nanoscale zero-valent iron in reference to the high reactivity of nanomaterials that has been discussed in this chapter.

8.7 Summary

In summary, cleanup of wastewater using nanomaterial is a potentially promising approach that has been used recently. Nanomaterials are applied for sorption, oxidation, reduction, disinfection and degradation of wastewater contaminants. The effectiveness of nanomaterials to eliminate contaminant concentration in wastewater is highly depending on the physical and chemical characteristics of the nanomaterial, the wastewater, and the contaminant. Reactivity of nanomaterial varies based upon its surface features such as surface area, surface confinement effect, surface charge density, energy, and surface curvature, as well as the stability of nanophase system in the wastewater. Selectivity of the appropriated nanomaterials for wastewater cleanup should consider the reactivity of the nanomaterial (sorption capacity, removal efficiency, etc) as well as the rate of reaction that is a critical factor sometimes in the treatment design. The use of the simulation models provides a powerful tool for such purpose. Although, the developing of nanoscience is exponentially grown, the main challenge of using nanomaterials could be attributed to the absence of the full understanding of its fate and behavior in the environment. Developing the application of natural nanomaterials such as nano-clay is highly promising technique in reducing the contamination of wastewater, simultaneously, it can reduce the expenses of using this advanced technology.

References

Abbasizadeh S, Keshtkar AR, Mousavian MA (2014) Sorption of heavy metal ions from aqueous solution by a novel cast PVA/TiO_2 nanohybrid adsorbent functionalized with amine groups. J Ind Eng Chem 20(4):1656–1664. https://doi.org/10.1016/j.jiec.2013.08.013

Adeleye AS, Conway JR, Garner K, Huang Y, Su Y, Keller AA (2016) Engineered nanomaterials for water treatment and remediation: costs, benefits, and applicability. Chem Eng J 286:640–662. https://doi.org/10.1016/j.cej.2015.10.105

Afkhami A, Saber-Tehrani M, Bagheri H (2010) Simultaneous removal of heavy-metal ions in wastewater samples using nano-alumina modified with 2,4-dinitrophenylhydrazine. J Hazard Mater 181(1–3):836–844. https://doi.org/10.1016/j.jhazmat.2010.05.089

Agrawal DC (2013) Introduction to nanoscience and nanomaterials. World Scientific Publishing Co. Pte. Ltd, Singapore

Ahmadi K, Ghaedi M, Ansari A (2015) Comparison of nickel doped zinc sulfide and/or palladium nanoparticle loaded on activated carbon as efficient adsorbents for kinetic and equilibrium study of removal of Congo red dye. Spectrochim Acta A 136:1441–1449. https://doi.org/10.1016/j.saa.2014.10.034

Andrievski RA (2014) Review of thermal stability of nanomaterials. J Mater Sci 49:1449–1460. https://doi.org/10.1007/s10853-013-7836-1

Araújo R, Castro ACM, Baptista JS, Fiúza A (2016) Nanosized iron based permeable reactive barriers for nitrate removal-systematic review. Phys Chem Earth Pt A/B/C 94:29–34. https://doi.org/10.1016/j.pce.2015.11.007

Artiola JF (2006) Industrial and municipal solid waste treatment and disposal. In: Pepper IL, Gerba CP, Brusseau ML (eds) Environmental & pollution science, 2nd edn. Elsevier Inc, California, pp 415–428

Bao S, Li K, Ning P, Peng J, Jin X, Tang L (2017) Highly effective removal of mercury and lead ions from wastewater by mercaptoamine-functionalised silica-coated magnetic nano-adsorbents: behaviours and mechanisms. Appl Surf Sci 393:457–466. https://doi.org/10.1016/j.apsusc.2016.09.098

Barisik M, Atalay S, Beskok A, Qian S (2014) Size dependent surface charge properties of silica nanoparticles. J Phys Chem C 118:1836–1842. https://doi.org/10.1021/jp410536n

Bhatnagar A, Kumar E, Sillanpää M (2010) Nitrate removal from water by nano-alumina: characterization and sorption studies. Chem Eng J 163:317–323. https://doi.org/10.1016/j.cej.2010.08.008

Boverhof DR, Bramante CM, Butala JH, Clancy SF, Lafranconi M, West J, Gordon SC (2015) Comparative assessment of nanomaterial definitions and safety evaluation considerations. Regul Toxicol Pharm 73:137–150. https://doi.org/10.1016/j.yrtph.2015.06.001

Brant J, Lecaotnet H, Wiessner MR (2005) Aggregation and deposition characteristics of fullerene nanoparticles in aqueous systems. J Nanopart Res 7:545–553. https://doi.org/10.1007/s11051-005-4884-8

Bunani S, Yörükoğlu E, Sert G, Yüksel Ü, Yüksel M, Kabay N (2013) Application of nanofiltration for reuse of municipal wastewater and quality analysis of product water. Desalination 315:33–36. https://doi.org/10.1016/j.desal.2012.11.015

Bunani S, Yörükoğlu E, Yüksel Ü, Kabay N, Yüksel M (2014) Application of nanofiltration for reuse of wastewater. Int J Global Warming 6(2/3):325–338. https://doi.org/10.1504/IJGW.2014.061028

Buzea C, Blandino IIP, Robbie K (2007) Nanomaterials and nanoparticles: sources and toxicity. Biointerphases 2(4):MR17–MR172. https://doi.org/10.1116/1.2815690

Cai N, Larese-Casanova P (2016) Application of positively-charged ethylenediamine-functionalized graphene for the sorption of anionic organic contaminants from water. J Environ Chem Eng 4:2941–2951. https://doi.org/10.1016/j.jece.2016.06.004

Cao G, Wang Y (2011) Nanostructures and nanomaterials: synthesis, properties, and applications, World scientific series in nanoscience and nanotechnology: volume 2, 2nd edn. World Scientific Publishing Co. Pte. Ltd, Hackensack, p 581

Chamtouri I, Abida H, Khanfir H, Bouri S (2008) Impacts of at-site wastewater disposal systems on the groundwater aquifer in arid regions: case of Sfax City, southern Tunisia. Environ Geol 55:1123–1133. https://doi.org/10.1007/s00254-007-1060-8

Chen L, Feng S, Zhao D, Chen S, Li F, Chen C (2017) Efficient sorption and reduction of U(VI) on zero-valent iron-polyaniline-graphene aerogel ternary composite. J Colloid Interf Sci 490:197–206. https://doi.org/10.1016/j.jcis.2016.11.050

Clavier A, Seijo M, Carnal F, Stoll S (2015) Surface charging behavior of nanoparticles by considering site distribution and density, dielectric constant and pH changes – a Monte Carlo approach. Phys Chem Chem Phys 17:4346–4353. https://doi.org/10.1039/c4cp04733h

Dasgupta N, Ramalingam C (2016) Silver nanoparticle antimicrobial activity explained by membrane rupture and reactive oxygen generation. Environ Chem Lett 14(4):477–485

Dasgupta N, Ranjan S, Mundekkad D, Ramalingam C, Shanker R, Kumar A (2015) Nanotechnology in agro-food: from field to plate. Food Res Int 69:381–400

Dasgupta N, Ranjan S, Rajendran B, Manickam V, Ramalingam C, Avadhani GS, Kumar A (2016a) Thermal co-reduction approach to vary size of silver nanoparticle: its microbial and cellular toxicology. Environ Sci Pollut Res 23(5):4149–4163

Dasgupta N, Ranjan S, Chakraborty AR, Ramalingam C, Shanker R, Kumar A (2016b) Nanoagriculture and water quality management. Pp 1-42. In: Ranjan S, Dasgupta N, Lichtfouse E (eds) Nanoscience in food and agriculture 1, sustainable agriculture reviews (20). Springer, Cham

Dasgupta N, Ranjan S, Ramalingam C (2017) Applications of nanotechnology in agriculture and water quality management. Environ Chem Lett 15(4):591–605

Elbana TA, Ramadan MA, Gaber HM, Bahnassy MH, Kishk FM, Selim HM (2013) Heavy metals accumulation and spatial distribution in long term wastewater irrigated soils. J Environ Chem Eng 1:925–933. https://doi.org/10.1016/j.jece.2013.08.005

Fang Z, Chen J, Qiu X, Qiu X, Cheng W, Zhu L (2011) Effective removal of antibiotic metronidazole from water by nanoscale zero-valent iron particles. Desalination 268:60–67. https://doi.org/10.1016/j.desal.2010.09.051

FAO (Food and Agriculture Organization of the United Nations) (2017) AQUASTAT. Available at: http://www.fao.org/nr/water/aquastat/water_use/index.stm. Accessed 28 Jan 2017

Flörke M, Kynast E, Bärlund I, Eisner S, Wimmer F, Alcamo J (2013) Domestic and industrial water uses of the past 60 years as a mirror of socio-economic development: a global simulation study. Glob Environ Chang 23(1):144–156. https://doi.org/10.1016/j.gloenvcha.2012.10.018

Fu F, Wang Q (2011) Removal of heavy metal ions from wastewaters: a review. J Environ Manag 92:407–418. https://doi.org/10.1016/j.jenvman.2010.11.011

Fu Q, Xue Y, Cui Z, Wang M (2014) Study on the size-dependent oxidation reaction kinetics of nanosized zinc sulfide. J Nanomat. Article ID 856489,8 pages. doi:https://doi.org/10.1155/2014/856489

Gautam RK, Chattopadhyaya MC (2016) Nanotechnology for water cleanup. In: Gautam RK, Chattopadhyaya MC (eds) Nanomaterials for wastewater remediation. Butterworth-Heinemann, Boston, pp 1–18

Gehrke I, Geiser A, Somborn-Schulz A (2015) Innovations in nanotechnology for water treatment. Nanotechnol Sci Appl 8:1–17. https://doi.org/10.2147/NSA.S43773

Ghasemi Z, Younesi H, Zinatizadeh AA (2016) Kinetics and thermodynamics of photocatalytic degradation of organic pollutants in petroleum refinery wastewater over nano-TiO_2 supported on Fe-ZSM-5. J Taiwan Inst Chem E 65:357–366. https://doi.org/10.1016/j.jtice.2016.05.039

Hasemzadeh G, Momenpour M, Omidi F (2014) Applications of nanomaterials in water treatment and environmental remediation. Front Environ Sci En 8(4):471–482. https://doi.org/10.1007/s11783-014-0654-0

Hayati B, Maleki A, Najafi F, Daraei H, Gharibi F, McKay G (2016) Synthesis and characterization of PAMAM/CNT nanocomposite as a super-capacity adsorbent for heavy metal (Ni^{2+}, Zn^{2+}, As^{3+}, Co^{2+}) removal from wastewater. J Mol Liq 224:1032–1040. https://doi.org/10.1016/j.molliq.2016.10.053

He Z, Alexandridis P (2015) Nanoparticles in ionic liquids: interactions and organization. Phys Chem Chem Phys 17:18238–18261. https://doi.org/10.1039/C5CP01620G

Henze M (2002) Wastewater, volumes and composition. In: Henze M, Harremoës P, JlC J, Arvin E (eds) Wastewater treatment, Biological and chemical processes, 3rd edn. Springer, Berlin/Heidelberg, pp 11–42

Holmes AB, Gu FX (2016) Emerging nanomaterials for the application of selenium removal for wastewater treatment. Environ Sci Nano 3:982–996. https://doi.org/10.1039/C6EN00144K

Huong P-T, Lee B-K, Kim J (2016) Improved removal of 2-chlorophenol by a synthesized Cu-nano zeolite. Process Saf Environ Prot 100:272–280. https://doi.org/10.1016/j.psep.2016.02.002

ISO (2015) ISO/TS 80004-1:2015. Nanotechnologies – vocabulary – part 1: core terms. International Organization for Standardization

Jain A, Ranjan S, Dasgupta N, Ramalingam C (2016) Nanomaterials in food and agriculture: an overview on their safety concerns and regulatory issues. Crit Rev Food Sci Nutr 15:1–21

Janardan S, Suman P, Ragul G, Anjaneyulu U, Shivendu R, Dasgupta N, Ramalingam C, Swamiappan S, Vijayakrishna K, Sivaramakrishna A (2016) Assessment on the antibacterial activity of nanosized silica derived from hypercoordinated silicon(iv) precursors. RSC Adv 6 (71):66394–66406

Kanel SR, Manning B, Charlet L, Choi H (2005) Removal of arsenic(III) from groundwater by nanoscale zero-valent iron. Environ Sci Technol 39:1291–1298. https://doi.org/10.1021/es048991u

Kanjwal MA, Alm M, Thomsen P, Barakat NAM (2016) Hybrid matrices of TiO_2 and TiO_2–Ag nanofibers with silicone for high water flux photocatalytic degradation of dairy effluent. J Ind Eng Chem 33:142–149. https://doi.org/10.1016/j.jiec.2015.09.026

Karaouzas I (2016) Agro-industrial wastewater pollution in Greek river ecosystems. In: The handbook of environmental chemistry. Springer, Berlin/Heidelberg. https://doi.org/10.1007/698_2016_453.

Karthikeyan KG, Meyer MT (2006) Occurrence of antibiotics in wastewater treatment facilities in Wisconsin, USA. Sci Total Environ 361:196–207. https://doi.org/10.1016/j.scitotenv.2005.06.030

Kootenaei FG, Rad HA (2013) Treatment of hospital wastewater by novel nano-filtration membrane bioreactor (NF-MBR). Iran J Energ Env 4(1):60–67. https://doi.org/10.5829/idosi.ijee.2013.04.01.10

Kržišnik N, Mladenovič A, Škapin AS, Škrlep L, Ščančar J, Milačič R (2014) Nanoscale zero-valent iron for the removal of Zn^{2+}, Zn(II)–EDTA and Zn(II)–citrate from aqueous solutions. Sci Total Environ 476-477:20–28. https://doi.org/10.1016/j.scitotenv.2013.12.113

Kumar R, Chawla J (2014) Removal of cadmium ion from water/ wastewater by nano-metal oxides: a review. Water Qual Expo Health 5:215–226. https://doi.org/10.1007/s12403-013-0100-8

Lefevre E, Bossa N, Wiesner MR, Gunsch CK (2016) A review of the environmental implications of in situ remediation by nanoscale zero valent iron (nZVI): behavior, transport and impacts on microbial communities. Sci Total Environ 565:889–901. https://doi.org/10.1016/j.scitotenv.2016.02.003

Levenspiel O (1999) Chemical reaction engineering, 3rd edn. Wiley, New York, p 668

Li K, Wang J, Liu J, Wei Y, Chen M (2016) Advanced treatment of municipal wastewater by nanofiltration: operational optimization and membrane fouling analysis. J Environ Sci 43:106–117. https://doi.org/10.1016/j.jes.2015.09.007

Liu X, Wang M, Zhang S, Pan B (2013) Application potential of carbon nanotubes in water treatment: a review. J Environ Sci 25(7):1263–1280. https://doi.org/10.1016/S1001-0742(12)60161-2

Maddinedi SB, Mandal BK, Ranjan S, Dasgupta N (2015) Diastase assisted green synthesis of size-controllable gold nanoparticles. RSC Adv 5(34):26727–26733

Maddinedi SB, Mandal BK, Patil SH, Andhalkar VV, Ranjan S, Dasgupta N (2017) Diastase induced green synthesis of bilayered reduced graphene oxide and its decoration with gold nanoparticles. J Photochem Photobiol B Biol 166:252–258

McDonald J, Gaston L, Elbana T, Kevin A, Eileen C (2013) Dimoxystrobin sorption and degradation in sandy loam soil: impact of different landscape positions. Soil Sci 178:662–670. https://doi.org/10.1097/SS.0000000000000030

Méndez V, Fedotov S, Horsthemke W (2010) Reaction-transport systems, mesoscopic foundations, fronts, and spatial instabilities. Springer-Verlag, Berlin/Heidelberg, p 454

Meng X, Liu Z, Deng C, Zhu M, Wang D, Li K, Deng Y, Jiang M (2016) Microporous nano-MgO/diatomite ceramic membrane with high positive surface charge for tetracycline removal. J Hazard Mater 320:495–503. https://doi.org/10.1016/j.jhazmat.2016.08.068

Mondloch JE, Bayram E, Finke RG (2012) A review of the kinetics and mechanisms of formation of supported-nanoparticle heterogeneous catalysts. J Mol Catal A-Chem 355:1–38. https://doi.org/10.1016/j.molcata.2011.11.011

Murty BS, Shankar P, Raj B, Rath BB, Murday J (2013) Applications of nanomaterials. In: Textbook of nanoscience and nanotechnology. Springer, Berlin/Heidelberg, pp 107–148

Nadaroglu H, Cicek S, Gungor AA (2017) Removing Trypan blue dye using nano-Zn modified Luffa sponge. Spectrochim Acta A 172:2–8. https://doi.org/10.1016/j.saa.2016.08.052

O'Carroll D, Sleep B, Krol M, Boparai H, Kocur C (2013) Nanoscale zero valent iron and bimetallic particles for contaminated site remediation. Adv Water Resour 51:104–122. https://doi.org/10.1016/j.advwatres.2012.02.005

Pham T, Lee B, Kim J (2016) Improved adsorption properties of a nano zeolite adsorbent toward toxic nitrophenols. Process Saf Environ 104(322):314. https://doi.org/10.1016/j.psep.2016.08.018

Qu X, Alvarez PJJ, Li Q (2013) Applications of nanotechnology in water and wastewater treatment. Water Res 47:3931–3946. https://doi.org/10.1016/j.watres.2012.09.058

Raghunath S, Anand K, Gengan RM, Nayunigari MK, Maity A (2016) Sorption isotherms, kinetic and optimization process of amino acid proline based polymer nanocomposite for the removal of selected textile dyes from industrial wastewater. J Photoch Photobio B 165:189–201. https://doi.org/10.1016/j.jphotobiol.2016.10.012

Ramya T, Anbazhagi M, Muthukumar M (2016) Electrochemical oxidation of fipronil contaminated wastewater by $RuO_2/IrO_2/TaO_2$ coated titanium anodes and sorbent nano hydroxyapatite. Mater Today Proc 3(6):2509–2517. https://doi.org/10.1016/j.matpr.2016.04.169

Ranjan S, Ramalingam C (2016) Titanium dioxide nanoparticles induce bacterial membrane rupture by reactive oxygen species generation. Environ Chem Lett 14(4):487–494

Ranjan S, Dasgupta N, Chakraborty AR, Melvin Samuel S, Ramalingam C, Shanker R, Kumar A (2014) Nanoscience and nanotechnologies in food industries: opportunities and research trends. J Nanopart Res 16(6):1–23

Ranjan S, Dasgupta N, Rajendran B, Avadhani GS, Ramalingam C, Kumar A (2016) Microwave-irradiation-assisted hybrid chemical approach for titanium dioxide nanoparticle synthesis: microbial and cytotoxicological evaluation. Environ Sci Pollut Res 23(12):12287–12302

Ray PZ, Shipley HJ (2015) Inorganic nano-adsorbents for the removal of heavy metals and arsenic: a review. RSC Adv 5:29885–29907. https://doi.org/10.1039/C5RA02714D

Ren X, Chen C, Nagatsu M, Wang X (2011) Carbon nanotubes as adsorbents in environmental pollution management: a review. Chem Eng J 170:395–410. https://doi.org/10.1016/j.cej.2010.08.045

Roduner E (2006) Size matters: why nanomaterials are different. Chem Soc Rev 35:583–592. https://doi.org/10.1039/B502142C

Rosales E, Pazos M, Sanromán MÁ (2017) Advanced oxidation processes using nanomaterials. In: Gautam RK, Chattopadhyaya MC (eds) Advanced nanomaterials for wastewater remediation. Taylor & Francis Group, LLC, Boca Raton, pp 109–132

Safavi A, Momeni S (2012) Highly efficient degradation of azo dyes by palladium/hydroxyapatite/ Fe_3O_4 nanocatalyst. J Hazard Mater 201–202:125–131. https://doi.org/10.1016/j.jhazmat.2011.11.048

Savage N, Diallo MS (2005) Nanomaterials and water purification: opportunities and challenges. J Nanopart Res 7:331–342. https://doi.org/10.1007/s11051-005-7523-5

Shahmansouri A, Bellona C (2015) Nanofiltration technology in water treatment and reuse: applications and costs. Water Sci Technol 71(3):309–319. https://doi.org/10.2166/wst.2015.015

Shukla A, Dasgupta N, Ranjan S, Singh S, Chidambram R (2017) Nanotechnology towards prevention of anaemia and osteoporosis: from concept to market. Biotechnol Biotechnol Equip 31(5):863–879

Siripireddy B, Mandal BK, Ranjan S, Dasgupta N, Ramalingam C (2017) Nano-zirconia – evaluation of its antioxidant and anticancer activity. J Photochem Photobiol B Biol 170:125–133

Solveyra ES, Szleifer I (2016) What is the role of curvature on the properties of nanomaterials for biomedical applications? WIREs Nanomed Nanobi 8:334–354. https://doi.org/10.1002/wnan.1365

Stahl T, Mattern D, Brunn H (2011) Toxicology of perfluorinated compounds. Environ Sci Eur 23:38–52. https://doi.org/10.1186/2190-4715-23-38

Tammina SK, Mandal BK, Ranjan S, Dasgupta N (2017) Cytotoxicity study of Piper nigrum seed mediated synthesized SnO2 nanoparticles towards colorectal (HCT116) and lung cancer (A549) cell lines. J Photochem Photobiol B Biol 166:158–168

Tanji KK, Kielen NC (2002) Agricultural drainage water management in arid and semi-arid areas, FAO irrigation and drainage paper 61. Food and Agriculture Organization of the United Nations, Rome

U.S. Environmental Protection Agency (2012) Guidelines for water reuse, 600/R-12/618. U.S. EPA, Washington, DC

UN-Water (2015) Wastewater management - a UN-water analytical brief. United Nations, New York. Note. http://www.unwater.org/publications/publications-detail/en/c/275896/. Accessed 18 Nov 2016

Walia N, Dasgupta N, Ranjan S, Chen L, Ramalingam C (2017) Fish oil based vitamin D nanoencapsulation by ultrasonication and bioaccessibility analysis in simulated gastro-intestinal tract. Ultrason Sonochem 39:623–635

Wang Y, Gao H, Xu H (2011) Nanogeochemistry: nanostructures and their reactivity in natural systems. In: Parker A, Russell H (eds) Frontiers in geochemistry. Wiley-Blackwell, Chichester, pp 200–220

WHO, World Health Organization (2006a) Guidelines for the safe use of wastewater, excreta and greywater, volume 1: policy and regulatory aspects. World Health Organization, Geneva, pp 1–114

WHO, World Health Organization (2006b) Guidelines for the safe use of wastewater, excreta and greywater, volume 2: wastewater use in agriculture. World Health Organization, Geneva, pp 1–196

WHO, World Health Organization (2006c) Guidelines for the safe use of wastewater, excreta and greywater, volume 3: wastewater and excreta use in aquaculture. World Health Organization, Geneva, pp 1–140

WHO, World Health Organization (2006d) Guidelines for the safe use of wastewater, excreta and greywater, volume 4: excreta and greywater use in agriculture. World Health Organization, Geneva, pp 1–182

Yurekli Y (2016) Removal of heavy metals in wastewater by using zeolite nano-particles impregnated polysulfone membranes. J Hazard Mater 309:53–64. https://doi.org/10.1016/j.jhazmat.2016.01.064

Zhang M, Gao B, Jin J, Chen H, Yao Y, Fang J, Creamer AE (2015a) Use of nanotechnology against heavy metals present in water. In: Sharma SK (ed) Heavy metals in water: presence, removal and safety. The Royal Society of Chemistry, Cambridge, pp 177–192

Zhang C, Wang L, Li J, Su P, Peng C (2015b) Removal of perfluorinated compounds in wastewater treatment plant effluents by electrochemical oxidation. Water Sci Technol 71(12):1783–1789. https://doi.org/10.2166/wst.2015.160

Zhang Y, Wu B, Xu H, Liu H, Wang M, He Y, Pan B (2016) Nanomaterials-enabled water and wastewater treatment. NanoImpact 3-4:22–39. https://doi.org/10.1016/j.impact.2016.09.004

Chapter 9
Bioremediation of Heavy Metals

Anamika Das and Jabez William Osborne

Abstract Human activities and industrial processes have led to worldwide heavy metal pollution. Several strategies have been developped for metal remediation. The conventional strategies are expensive, usually low in efficiency and may alter the soil nature. Here we review bioremediation using plants, microbes, e.g. bacteria, fungi, and actinobacteria, earthworms, and algae for metal removal. Bioaugmentation of microbes using plants, earthworms and algae is used to enhance the bioremediation efficiency. We discuss the importance of metagenomics, metabolomics and proteomics approach to assess the response of the living organisms under stress and how they can contribute to the improvement of the already existing strategies.

Keywords Heavy metals • Bioremediation • Biosystems • Bioaugmentation • Metagenomics • Metabolomics • Proteomics

Contents

A. Das • J.W. Osborne (✉)
School of Bio Sciences and Technology, VIT University, Vellore, Tamil Nadu, India
e-mail: jabezosborne@vit.ac.in

K M Gothandam et al. (eds.), *Nanotechnology, Food Security and Water Treatment*, Environmental Chemistry for a Sustainable World 11,
https://doi.org/10.1007/978-3-319-70166-0_9

9.1 Introduction

Environmental pollution occurs when the natural environment cannot destroy an element without creating harm or damage to itself (Wijnhoven et al. 2007). The elements involved are not produced by nature, and the destroying process can vary from a few days to thousands of years. Current research has documented elemental pollutants as "emerging contaminants" (Yu et al. 2014). Intense industrialization, modern agricultural practices, increased anthropogenic activities, and unauthorized disposal methods have increased the concentrations of elemental pollutants in the environment, creating adverse effects to all the living organisms (Wijnhoven et al. 2007). Heavy metals are one of the major pollutants which has been the chief concern in past decade. They can enter the environment in a single high-level exposure or the cumulative effect of repeated high or low-level exposures but when introduced into an environment, it can stay there in toxic form for a long period of time.

A number of physical, chemical and biological techniques can be used to remediate metal contaminated soils. Physico-chemical methods are, however, not appreciated as they generate a large amount of sludge and result in more contamination (Ahluwalia and Goyal 2007). Thus, bioremediation provides the best answer. Many reports have established the bioremoval of heavy metals by the use of either plants, earthworms or microbes (Wang et al. 2015; Rodriguez-Campos et al. 2014; Dharni et al. 2014; Ma et al. 2015). But recent reports have studied the uptake studies by using more than one living organism and have come out with more efficient and improved results (Emenike et al. 2016; Wood et al. 2016; Lemtiri et al. 2016). Thus, they have opened the gate of exploring the more diverse flora and fauna for achieving the best result in bioremediation. Scientists have also developed and studied the three main 'omics' approach for understanding the response of the organism under the stressed condition, i.e., metagenomics, metabolomics and proteomics (Gillan et al. 2015; Tomanek 2014). The integrated 'omics' analysis can be a powerful technique to identify the vast microbial communities which are unculturable but still possess the ability of bioremediation and the various metabolites released under stress along with their function. This approach has brought a revolution in the field of bioremediation. Figure 9.1 summarizes the bioremediation technologies described in this review.

This review emphases on the utilization of different tactics of bioremediation using plants, bacteria (rhizobacteria, actinobacteria), earthworms, algae, fungi and highlights the advantages of the integrated approach of using multi-biosystem for

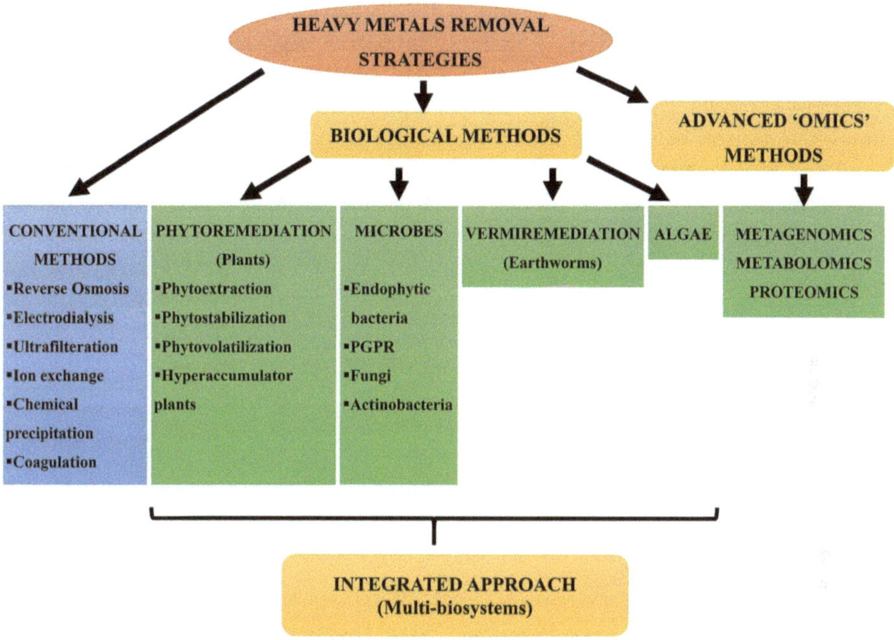

Fig. 9.1 An outline of the remedial strategies applied for bioremoval of heavy metals. The conventional methods are ineffective or expensive when the concentration of heavy metals is very low and produces a large amount of derivatives of contaminants. Alternately, biological methods with the usage of living biosystems has proven efficient in heavy metals bioremediation. The 'omics' approach also enhanced the understanding of the living biosystems under stressed condition (*PGPR* Plant growth promoting rhizobacteria)

the bioremediation of Heavy metals. To support the statement, many evidence has been provided representing different case studies along with their mechanism and limitations. In this context, the scope of 'omics' tool to enhance the overall bioremediation process has also been discussed.

9.2 Heavy Metals

Heavy metals represent a class of metallic element present abundantly in the earth's crust (Yu et al. 2014). They are defined as the metals possessing density greater than 5 gm/cm^3 (Das et al. 2014). Different from other organic pollutants, heavy metals are harder to be chemically or biologically degraded. Irrespective of the origin of the metals in the soil, excessive levels of many metals can result in the deprivation of soil quality, crop yield and agricultural products and can be significantly hazardous to human, animal and ecosystem health (Das et al. 2014). The metals or metalloids including arsenic (As), cadmium (Cd), chromium (Cr), copper (Cu), lead (Pb), mercury (Hg), nickel (Ni), selenium (Se), silver (Ag), and zinc (Zn) can be of severe threat to human and animal health due to its intensified long-term persistence in the environment (Gisbert et al. 2003). Toxic heavy metals are also known as cumulative poison because of it persistence in nature and the ability to get transferred and accumulated in various tropic levels causing DNA damage and carcinogenic effects by their mutagenic ability (Knasmüller et al. 1998). Therefore an alarm has been triggered for the researchers to conserve the environment from toxic heavy metals. The Agency for Toxic Substances and Disease Registry (ATSDR) in Atlanta, Georgia, (a part of the U.S. Department of Health and Human Services) compiled a Priority List called the "Top 20 Hazardous Substances." The heavy metals arsenic, lead, mercury, and cadmium appear on this list (ATSDR 2011).

9.3 Conventional Strategies for Detoxification of Heavy Metals

In order to make the environment healthier, contaminated water bodies and land need to be remedied to make them free from heavy metals and trace elements. There are several conventional techniques to remove these heavy metals, including chemical precipitation, oxidation or reduction, filtration, ion-exchange, reverse osmosis, membrane technology, evaporation and electrochemical treatment. But most of these techniques become ineffective when the concentrations of heavy metals are less than 100 mg/L (Ahluwalia and Goyal 2007). Additionally, physico-chemical methods are ineffective or expensive. Some of the techniques are mentioned in Table 9.1 with their drawbacks.

Biological methods for removal of heavy metals has become an attractive alternative to physico-chemical methods. Bioremediation has proved to be an innovative and promising technology available for removal of heavy metals and recovery of the heavy metals in polluted water and lands.

9.4 Bioremediation of Heavy Metals

According to Environmental Protection Agency (EPA), bioremediation is a technique that uses naturally occurring organisms to break down hazardous substances into less toxic or nontoxic substances (Agouborde and Navia 2009). Various living biosystems can be utilized for the bioremoval of heavy metals. The biomass-based systems are more satisfactory compared to the conventional treatment methods as it is cost effective with high efficiency of detoxification of dilute effluents and reducing the quantity of sludge disposal in the environment. There are many reports about biodegradation and bioremediation strategies being utilized by bacteria or

Table 9.1 Conventional techniques for heavy metals removal and their drawbacks

Technique	Application to heavy metal	Drawback	References
Reverse osmosis- A semi permeable membrane is used to separate the heavy metal at a pressure greater than the osmotic pressure	Cu^{2+}, Ni^{2+}, Zn^{2+}	High power consumption due to the pumping pressures, and the restoration of the membranes	Fu and Wang (2011)
Electrodialysis-Ion selective semi permeable membrane are used to separate heavy metals by applying electrical potential between two electrodes	Cr(III), Cu, Fe	The separation percentage decreased with an increasing flow rate	Sadrzadeh et al. (2009)
Ultrafiltration- A porous membrane is used to remove heavy metals by applying pressure.	Cd^{2+}, Cu^{2+}, Ni^{2+}, Pb^{2+} and Zn^{2+}	If the surfactant and heavy metals are not disposed of, it lead to secondary pollution by generating sludge	Landaburu-Aguirre et al. (2009)
Ion exchange- From the dilute solution containing heavy metal, the metal ion gets exchanged to the exchange resin by the ions held by electrostatic force	Ce^{2+}, Fe^{2+} and Pb^{2+}	It can be used only with low concentrated metal solution and is highly sensitive with the pH of the aqueous phase.	Gunatilake (2015)
Chemical precipitation- Chemicals react with heavy metal ions to form insoluble precipitates	Cu2+, Cd^{2+} and Pb^{2+}	Generates large volumes of low density sludge, which can cause disposal problems	Kongsricharoern and Polprasert (1995)
Coagulation- Removal of heavy metals by charge neutralization of particles	Ni^{2+}	Unable to treat the heavy metal wastewater completely	Chang and Wang (2007)

plant species (Wang et al. 2015; Ma et al. 2016; Glick 2010) but so far very few investigations have been carried out using other living biosystems such as earthworms, algae, fungi and their integrated approach.

9.4.1 Phytoremediation of Heavy Metals

The word "phytoremediation" is derived from Greek word phyto (mean plant) and Latin word Remedium (to remove an evil). Phytoremediation utilizes a variety of plant processes and the physical characteristics of plants to aid in remediation of contaminated sites. It is an *in situ* remediation technology driven by solar energy. Remediation of metals using plants seems an effective approach in the present scenario since plants are the primary recipients of heavy metals (Ali et al. 2013;

Wang et al. 2015). Phytoremediation technique includes processes such as phytoextraction, phytostabilization, phytovolatilization (Alkorta et al. 2004).

9.4.1.1 Phytoextraction

It is the process of uptake of contaminants from soil or water by plant roots and their accumulation in biomass, *i.e.*, shoots (Seth 2012). Generally shoot metal concentration and shoot biomass mainly determine a suitable plant species for phytoextraction of metals. Depending upon these parameters, two different phytoextraction approaches have been used, i.e., use of hyperaccumulator plants with relatively low biomass production and use of plants with relatively higher above ground biomass production but lesser metal accumulation such as *Brassica juncea* (Robinson et al. 1998; Ali et al. 2013). A recent report by Ma et al. (2016) suggested that the highly developed root system of Napier grass makes it an ideal candidate for phytoextraction process by absorbing, transporting and storing both contaminants and nutrients into the plant tissue.

9.4.1.2 Phytostabilization

Phytostabilisation is a method where the plants are used to immobilise metals in the rhizosphere and reduce the above ground wind and water erosion (Gil-Loaiza et al. 2016). There are two main factors which are considered when determining the aptness of plants with a large biomass for phytostabilisation: root accumulation and rhizosphere immobilisation (Sun et al. 2016). The plants selected must be able to develop abundant root systems, and translocate metals from roots to shoots at as low concentrations as possible (Mendez and Maier 2008). Giant reed (*Arundo donax*) and silvergrass (*Miscanthus sinensis*) genotypes are bioenergy crops well suited for the phytostabilisation of metal(-loid)-contamination of dry land (Barbosa et al. 2015). But phytostabilization is not a permanent solution as heavy metals remains in the soil as it is; only with restricted movement and needs to monitor regularly.

9.4.1.3 Phytovolatilization

This approach involves conversion of heavy metals into volatile forms by plants and subsequently released into the atmosphere. This process has been used for removal of some volatile heavy metals like Hg and Se from polluted soils (Karami and Shamsuddin 2010). However, this is limited by the fact that it does not remove the metals completely but rather transfers them from one medium (soil or water) to another (atmosphere) from which they can re-enter soil and water.

9.4.1.4 Hyperaccumulator Plants

Recently, removal of heavy metals through hyperaccumulators to degrade the contaminants, has received wide attention due to its efficacy and cost efficiency (Ahemad 2014). Hyperaccumulators have been found to exhibit higher heavy metal tolerance and accumulating abilities compared to other plants (Prasad and Freitas 2003). Many reports are provided for hyperaccumulators being utilized such as *Arabidopsis halleri* and *Solanum nigrum* L. for uptake of Cd (Dahmani-Muller et al. 2000; Wei et al. 2005), *Zea mays* for uptake of Pb, Cd and Zn (Meers et al. 2010), *Brassica juncea*, *Astragalus bisulcatus* for uptake of Se (Bitther et al. 2012). However, the disadvantages that limit the use of hyperaccumulators include difficulty in finding heavy metal hyperaccumulators, slow growth and lower biomass yield. This makes the process quite time-consuming and therefore not feasible for rapidly contaminated sites or sewage treatments (Xiao et al. 2010).

9.4.1.5 Mechanism of Heavy Metals Phytoremediation

The uptake of heavy metals by plants depends mainly on the bioavailability of the heavy metals in the soil as well as the plant nutrients. The heavy metals either gets accumulated in the root tissues or get translocated to the aerial regions of the plants through xylem vessels by symplastic and/or apoplastic pathways (Sarwar et al. 2016). The tolerance against heavy metals is a prerequisite for phytoremediation process to minimize the adverse effects on the plants. The tolerance potential of the plant depends on mechanisms like cell wall metal binding, active transport of metal ion into the vacuoles, chelation of metal ions with proteins and peptides and complex formation (Memon and Schroder 2009).

9.4.1.6 Challenges in Phytoremediation

Phytoremediation, no doubt, is an attractive process for heavy metals uptake but the researchers have confronted several limitations when only plants were used for the bioremediation (Karami and Shamsuddin 2010; Naees et al. 2011; Ramamurthy and Memarian 2012) which has been summarized in Fig. 9.2.

9.4.2 Microbial Remediation of Heavy Metals

Microorganisms as metal accumulators possess an inherent novel remediation property for toxic metals in the soil with increased crop productivity. Many researchers have studied the close interactions among plants-microorganisms heavy metals in rhizosphere soils to enhance phytoremediation process (Glick

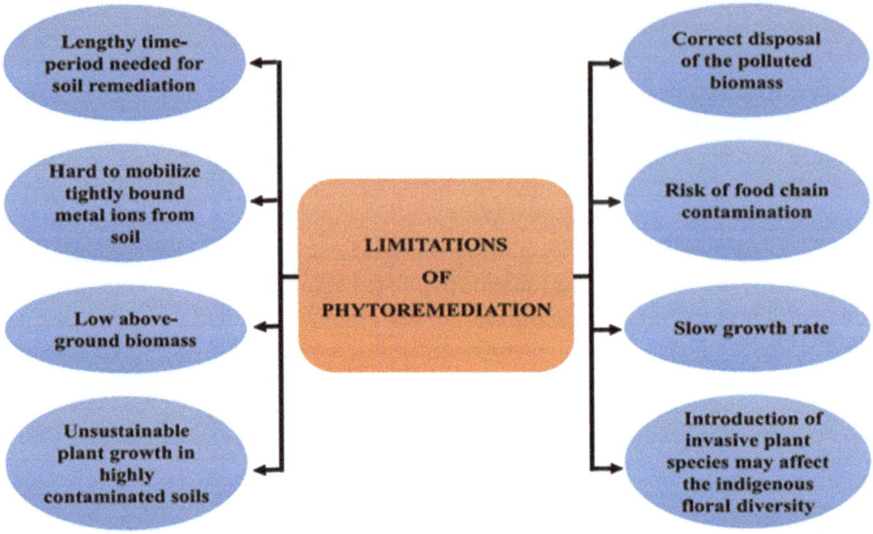

Fig. 9.2 An overview of the common problems faced by the plants in bioremediation. These limitations inhibit the application of the traditional phytoremediation techniques on large scale applications. The limitations can be overcome by synergistic integration of the plants with other living organism for bioremediation by advanced bioremediation research

2010; Dharni et al. 2014; Ma et al. 2015). Inoculation of plants with selected and acclimatized microbes (bioaugmentation) has attained prominence for phytoremediation of metal polluted soils (Lebeau et al. 2008; Glick 2010; Ma et al. 2011). Some microorganisms live in association with plant roots while others are free living. A recent report by Abd-Elnaby et al. (2016) identified three marine Psychrobacter strains which were able to resist and accumulate several metals (Pb^{2+}, Cu^{2+} and Cd^{2+}) with variable degrees, depending on bacterial strains and metal ion species. There are few bacterial species such as Alphaproteobacteria and *P. aeruginosa* which are isolated from sponge genera such as *Sarcotragus* sp. *Suberites clavatus* and *Crella cyathophora* and have been recognized as a source for secondary metabolites having the potential for heavy metal bioremediation (Saurav et al. 2016a, b).

9.4.2.1 Endophytic Bacteria

Endophytes mostly lives under the epidermal cells of plant tissues and colonize (Schulz and Boyle 2006). The extensive co-evolution of plants and endophytes has developed an intimate ecosystem which helps the plants to survive in stressed conditions and helps in enhanced bioremoval of Heavy metals (Ryan et al. 2008).

Bioaugmentation with such endophytic bacteria can diminish the metal phytotoxicity and alter the phytoavailability of heavy metals in contaminated soils, making them ideal for microbial assisted phytoremediation studies (Weyens et al. 2009; Ma et al. 2011). The hyperaccumulator plants constitute a complex and specialized endophytic bacterial flora such as *Pseudomonas koreensis*, *Bacillus* sp., *Rahnella* sp. with high levels of resistance to heavy metals such as Pb, Mn and Cd (Babu et al. 2015; Luo et al. 2012; Yuan et al. 2014).

9.4.2.2 Plant Growth Promoting Rhizobacteria

Plant growth promoting rhizobacteria (PGPR) are a group of microbial community which can improve the growth of the host plant in heavy metal contaminated soils by mitigating toxic effects of heavy metals on the plants (Seth 2012). These may be free- living bacteria, in symbiotic associations, or endophytic bacteria (Glick 2012). Some important genera of PGP bacteria include *Bacillus, Pseudomonas, Enterobacter, Erwinia, Klebsiella, Flavobacterium* and *Gluconacetobacter* (Dardanelli et al. 2010; Nadeem et al. 2010). PGPR improve plant growth and effect heavy metals mobility by atmospheric nitrogen fixation, production of phytohormones and siderophores and solubilisation of insoluble phosphate (Ullah et al. 2015). A wide range of PGPR has been identified which aid in uptake of Heavy metals (Glick 2010). A report by Jing et al. (2014) showed enhanced accumulation of Cd, Pb, Zn in *Brassica napus* when inoculated with PGPR strains such as *Enterobacter* sp. and *Klebsiella* sp..

9.4.2.3 Fungi

Fungi have been chiefly ignored as constituents of the host microbiota and their role in bioremediation (Moyes and Naglik 2012). Fungi have emerged as potential biocatalysts to access heavy metals and transform them into less toxic compounds. They possess metal sequestration and chelation systems to increase their tolerance to heavy metals. Moreover, their high biomass makes them suitable for bioremediation of Heavy metals (Aly et al. 2011). Some fungi such as, *Allescheriella* sp., *Stachybotrys* sp., *Phlebia* sp. *Pleurotus pulmonarius*, have metal binding potential (D'Annibale et al. 2007). Fungi of the genera *Penicillium*, *Aspergillus* and *Rhizopus* have been studied extensively as potential microbial agents for the removal of heavy metals from aqueous solutions (Volesky and Holan 1995; Huang and Huang 1996). Pb (II) contaminated soils can be biodegraded by fungal species like *Aspergillus parasitica* and *Cephalosporium aphidicola* with biosorption process (Tunali et al. 2006; Akar et al. 2007). Recent reports identifies 20 fangal taxa in which *Alternaria,* and *Peyronellaea* are the dominant genera and shows excellent uptake of Pb^{2+} and Zn^{2+} (Li et al. 2012).

9.4.2.4 Actinobacteria

Actinobacteria are a group of bacteria which play an important role in recycling substances, since they are able to metabolize complex organic matter (Kieser et al. 2000). They prove to be an important ecological agent by possessing the ability to remove Heavy metals (Albarracín et al. 2005; Polti et al. 2009). Several reports signifies *Corynebacterium* strain tolerant to heavy metals such as Cd(II), Co(II), Cr (VI), Hg(II), and Ni(II) (Oyetibo et al. 2010). Other reports by Mangold et al. (2012) demonstrates the strain *Acidimicrobium ferrooxidans* tolerant to higher concentrations of Zn(II) and adapting to the adverse environment. Although the bioremediation skills of the genera such as *Streptomyces, Rhodococcus,* and *Amycolatopsis* were extensively studied but the lack of information to enhance the bioremediation process of actinobacteria through pathway engineering techniques did not supported their further use (Alvarez et al. 2017).

9.4.2.5 Mechanism of Bioremediation by Microbes

We know microorganisms are omnipresent and reside in heavy metal contaminated soil. The bioremediation strategy for Heavy metals depends on the active metabolizing capabilities of microorganisms. The microbes mineralize the organic contaminants to end-products such as carbon dioxide and water which are used as substrates for cell growth. The production of degradative enzymes by the microbes for the target pollutants is one way to resist against Heavy metals. Microbes are capable of dissolving metals and reducing or oxidizing transition metals. A short summary of microbial mechanism for Heavy metals tolerance with some examples are provided in Table 9.2.

9.4.2.6 Challenges in Microbial Bioremediation

The lack of information on the cellular responses of microbes towards utilization and interaction with trace heavy metal pollutants restricts their successful execution (Boopathy 2000). Large-scale application of microbes is limited because of their requirements for extra nutrients which in turn increases the biological oxygen demand in the waste (Dixit et al. 2015). Few challenges in bioremediation by microbes has been summarized in Fig. 9.3.

9.4.3 Bioremediation of Heavy Metals Using Earthworms

As one of the most important species in soil fauna, earthworms play a major role in the functioning of the soil ecosystem (van Gestel et al. 2009). They have been

Table 9.2 Summary of microbial bioremediation mechanisms

Microorganisms	Mechanism of bioremediation	References
Endophytic bacteria	Bioremoval of Heavy metals in metal amended medium; Increased biomass, chlorophyll content, nodule number and metal accumulation	Babu et al. (2013)
Endophytic bacteria	Increased root elongation of plant; Reduced metal phytotoxicity and increase metal accumulation	Shin et al. (2012)
Endophytic bacteria	Improved heavy metal availability in soil, shoot dry biomass and uptake of Heavy metals	Chen et al. (2014)
PGPR	Produce metal chelating agents termed siderophores, which are able to bind metals and thus enhance their bioavailability in the rhizosphere through a complexation reaction	Rajkumar et al. (2013)
PGPR	Decrease the level of ethylene in plants, which increases plant growth. This attributed to ACC deaminase, which hydrolyzes ACC, the biosynthetic precursor for ethylene in plants, into ammonia and α ketobutyrate	Ullah et al. (2015)
PGPR	Phosphate solubilization and nitrogen fixation which affect heavy metals mobility and availability to the plant	Gadd (2010)
Fungi	Extracellular metal sequestration and precipitation, metal binding to the fungal cell walls, intracellular sequestration and complexation, compartmentation, and volatilization	Fomina et al. (2005)
Fungi	Fungi can compete with roots and other microorganisms for water and metal uptake, protect the roots from direct interaction with the metals and impeded metal transport through increased soil hydrophobicity	Wenzel (2009)
Fungi	Fungal endophytes possess chelation systems to increase the tolerance of host plants to heavy metals	Aly et al. (2011)
Actinobacteria	Upregulation of genes to antioxidant proteins like superoxide dismutase, alkyl hydroperoxide reductase and mycothiol reductase,	Costa et al. (2012)
Actinobacteria	Use of immobilized microbial cells provides high degradation efficiency and good operational stability	Ahamad and Kunhi (2011)
Actinobacteria	Production of 'Surface active compounds' which form complexes with pollutants attached to soil matrix and promote their desorption	Shafiei et al. (2014)

Heavy metals Heavy metals, *PGPR* Plant Growth Promoting Rhizobacteria, *ACC* 1-Aminocyclopropane-1-Carboxylate

described as the soil ecosystem engineers with physical, chemical and biological effects on plants and the environment (Lavelle et al. 2006). The potential use of worms in so-called vermiremediation process was recently reviewed (Rodriguez-Campos et al. 2014). Indeed, earthworms can be exploited in the process of remediation of contaminated soils due to their ability to enhance the removal of some heavy metal trace pollutants. Earthworms can survive in heavy-metal contaminated soils, can accumulate efficiently high tissue metal concentrations such as Pb, Cd, and Zn using a variety of sequestration mechanisms (Sinha et al. 2008; Andre et al. 2009). They may expose to heavy metals through their intestine and

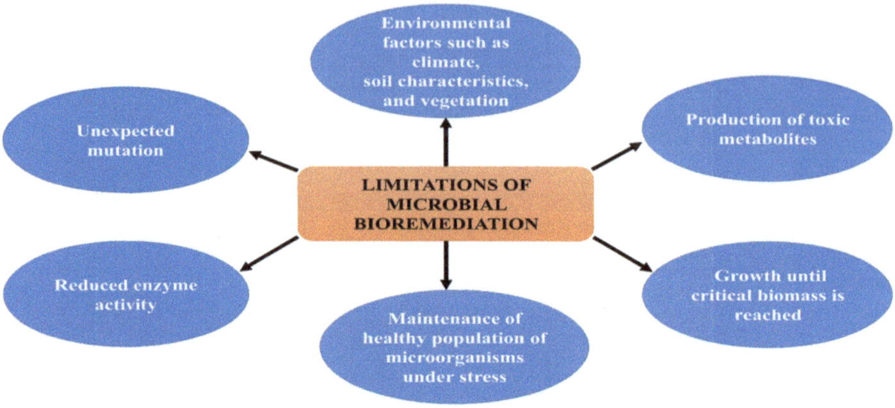

Fig. 9.3 Limitations of the microbial remediation. It is difficult to maintain the healthy condition of microbes in contaminated soil throughout as it is exposed to various environmental factors which inhibit the bacterial growth. Unexpected mutation in microbes can lead to loss of their enzymatic activity which will affect their heavy metal degrading property

skin via alimentary and dermal uptake routes (Homa et al. 2010). There are reports which indicates that earthworms reduced the concentrations of Cr, Cu, Pb and Zn in the vermicomposted sludge below the limits set by the USEPA in 60 days (Contreras-Ramos et al. 2006). Earthworms collected from the roadsides and mining sites show higher amounts of heavy metals than those from the other sites and hence can be a 'bioindicator' of heavy metal contamination in soil. The choice of the right species of earthworm and proper selection of earthworm for vermicomposting is the prime step as it affects the rate of waste stabilization. For eg. a recent report by Sizmur et al. (2011) showed that *Lumbricus terrestris* decreased water soluble Cu and As but increased the solubility of Pb and Zn in soil but at the same time, Natal-da Luz et al. (2009) did not observed an influence of *Dendrobaena veneta* on the solubility of Cr, Cu, Ni, and Zn in soil. A brief report on successful bioremediation cases of Heavy metals by earthworms is provided in Table 9.3.

Earthworm, no doubt, is beneficial candidate for bioremediation as they easily available, easy to handle and to measure the toxic parameters such as growth, reproduction and biochemical responses but taking into account the indicator role of earthworms in contaminated environments is a topic of limited practicality.

9.4.3.1 Mechanism of Vermiremediation

Earthworms ingests a large amount of different substrates and thus, concentrates Heavy metals in their body through their skin and intestine (Mohee and Soobhany 2014). Thus, vermicomposting can be used to breakdown the toxic metals into its non-toxic forms. Dia et al. (2004) suggested that bioaccumulation of metals in

Table 9.3 A report on bioremoval of heavy metals by earthworms

Earthworm species	Heavy metals uptake	References
Eisenia andrei	Body accumulation factor exceeded 1 only for Cd (17.4 4). BAFs calculated for all analyzed metals can be ranked as follows: Cd > Cu > Zn > Ni > Cr > Pb	Rorat et al. (2017)
Eisenia fetida	A slight reduction of total Pb in a binary biological system was observed with an adverse impact of Pb on the morphological parameters of the earthworms	Liu et al. (2017)
Eisenia fetida and *Metaphire guillelmi*	*M. guillelmi* accumulated more Cd than *E. fetida* but at higher doses of Cd, inverse results were obtained. This behavioural response indicates higher bioaccumulation at low-dose exposure and to the lower detoxification ability of *M. guillelmi*	Chen et al. (2017)
Eisenia fetida	Co uptake was higher than Hg which proves that Hg is more toxic to earthworms as it effects coccon production, coelomocytes, body weight and length also	Jatwani et al. (2016)
Eudrilus eugeniae	An increased concentration of Cd, Co and Ni were obtained in the tissue of the earthworms after the vermicomposting processes which showed that vermicomposting can efficiently remove heavy metals	Soobhany et al. (2015)
Metaphire posthuma and Eisenia fetida	The removal efficiency of *M. posthuma* was positive for Zn but it was negative in *E. fetida*	Sahariah et al. (2015)
Eisenia fetida	Indicated a reduction in As mobility and bioavailability in all matured composts and vermicomposts.	Maňáková et al. (2014)
Lumbricus rubellus	The heavy metals Cr, Cd and Pb contained in vermicompost of sewage sludge were lower than initial concentrations, with 90–98.7% removal	Azizi et al. (2013)
Eisenia fetida	Cu and Zn appear to be less toxic to earthworms than Cd and Pb referring to Cytochrome P450 monooxygenase activity.	Cao et al. (2012)
Eisenia fetida	Bioaccumulation of Cu and Zn within 10 weeks of experiment	Malley et al. (2006)

BAF Bioaccumulation factor

earthworms is their ability to eliminate the excess of metals. Sizmur and Hodson (2009) suggested four prime mechanisms of metal bioremoval by earthworms (Fig. 9.4).

Few reports on mechanism of vermiremediation suggested by various scientists are given below:

(a) The heavy metal accumulation in the tissue of earthworms is the result of their detritivorous lifestyle coupled with their highly permeable body walls and Chloragosomes (phosphate- sulphur rich stuctures) which function as metal sequestering organelles (Morgan et al. 2002).
(b) Some metals are taken up by earthworms and bound by a protein called 'metallothioneins (MT)' which have the capacity to bind metals. Stürzenbaum et al. (2004) found that Cd detoxification in *E. fetida* was due to

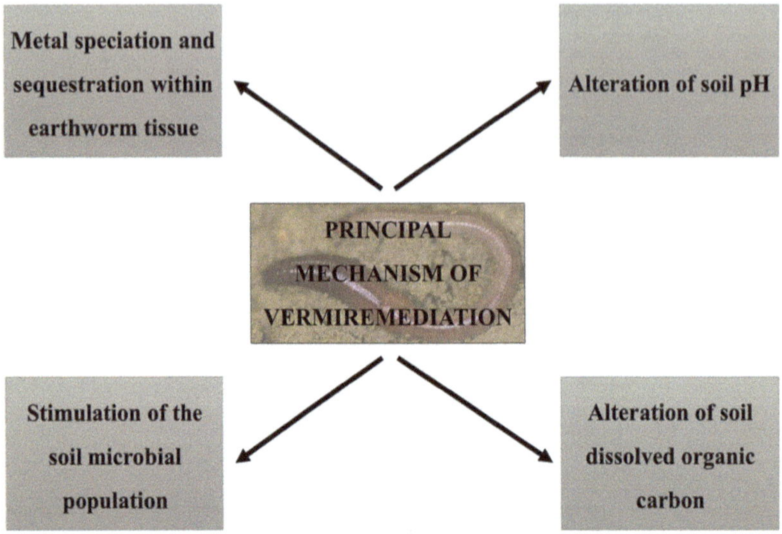

Fig. 9.4 Principal mechanism of vermiremediation. Vermiremediation is very cost-effective, environmentally sustainable way to treat heavy metals polluted soil. It lead to significant improvement in the quality of soil

compartmentalization of the metal by binding it to metallothioneins in the form of Cd-metallothioneins

(c) Wang et al. 2014 studied the relative contribution of the dermal and the gut exposure route to the uptake of heavy metals in earthworms. Using a modeling approach, it was estimated that the dermal exposure route accounted for more than 96% of the total uptake of Cd and Cu in the *E. Andrei*.

(d) Malonylaldehyde (MDA) is an important indicator of lipid peroxidation whose level increases reactive oxygen species stress (produced in response to oxidative stress). Sinhorin et al. (2014) measured decreased malonylaldehyde content in *E. fetida* under Cd stress which may be one of the mechanism to resist against heavy metal.

(e) Earthworms either bio-transform' or 'biodegrade' the contaminants rendering them harmless in their bodies. The process takes place in their gut followed metabolization, complexation and sequesteration in tissues or vacuoles (Gu et al. 2016).

(f) The worm's digestive system is capable of detaching heavy metal ions from the complex aggregates between these ions and humic substances in the waste as it rots. Various enzyme-driven process accumulate the metal ions in the worms' tissues rather than being released back into the environment. The separation of dead worms from compost is a relatively straight forward process allowing the heavy metal to be removed from the organic waste (Jatwani et al. 2016).

9.4.3.2 Challenges in Vermiremediation

Although earthworms are capable of accumulating heavy metal from the soil, it is not considered worldwide as a practical approach of enriching contaminated sludge or soils since there are evidences which indicates that progressive mineralization tends to increase the total metal concentration of metals in the substrates (Sizmur et al. 2011). Moreover, the application of metal-containing vermicomposts, to any contaminated site will inevitably introduce heavy metal into terrestrial food chain by earthworms which are significant prey organisms (Roodbergen et al. 2008). The general concept is that when earthworms are available for their predators with high concentrations of heavy metals in their tissues, the heavy metal should not get transfer to higher trophic levels and lead to biomagnification of heavy metal. There are few reports which exhibits such predator-prey phenomenon along with transfer of metals in terrestrial and aquatic food chains (DeForest et al. 2007). There are no recent cases reported on biomagnification of heavy metal from one trophic level to another via earthworms but that does not mean that the bioaccumulation of heavy metal by earthworms during vermicomposting, or during field exposure has no potentially serious ecotoxicological impacts on consumer species since earthworms can transfer metal fractions both from internal cellular compartments and alimentary canal. Future research is needed to better understand the interaction mechanism between heavy metal exposure and soil macroorganism in polluted soil.

9.4.4 Bioremediation of Heavy Metals by Algae

Accumulation of heavy metal by algae has received attention only in recent years because of its potential for application in environmental protection and recovery of some important metals (Zeraatkar et al. 2016; Malik 2004). The algal biomass may serve as an ecologically safer, cheaper and efficient means to remove heavy metal ions from waste water by biosorption process (Pohl and Schimmack 2006). The metal content of the indigenous algae can be used for biomonitoring metal pollution in a water body since the amount of metal accumulated by algae is related with the concentration of metal in water (De Filippis and Pallaghy 1994). The heavy metal uptake may depend upon the specificity of the algal strain used in the process for interaction. For eg. Monteiro et al. (2010) investigated removal of Cd ions using two strains of *Desmodesmus pleiomorphus* cells and found 25% difference between them for cadmium biosorption. Romera et al. (2007) introduced brown algae as a very good candidate for biosorbents of heavy metal ions. Alginate is one of the main constituents of the cell wall of brown algae and it is well recognized to be involved in metal accumulation (Davis et al. 2003). The phenomenon of remediation by algae can be broadly categorized in two different sets. (i) Bioaccumulation of heavy metal by living cells and biosorption by non-living cells, (ii) Macroalgae and microalgae.

9.4.4.1 Live vs. Non-living Biomass for Heavy Metal Biosorption

Heavy metal ions can be removed from wastewaters by either live cells or dead cells by the usage of inactive biomass. Lamaia et al. (2005) reported the limited sortion capacity of heavy metal ions by live cells as they were harmed by the increased heavy metal ions. The live cells are affected by many environmental factors which influence their sorption capacity. The absorption mechanism of the live cells are more complex as the intracellular heavy metal uptake occur at the growth phase where adsorption occurs whereas in dead algal cells, the heavy metal are absorbed on the surface of the cell and it is an extracellular process (Godlewska-Zyłkiewicz 2001). The non-living algal biomass is an assemblage of polymers such as sugars, cellulose, pectins, etc. which are capable of binding heavy metal ions (Volesky 2007; Arief et al. 2008). Moreover, they do not require a nutrient supply and therefore can be used for multiple sorption desorption cycles (Areco et al. 2012).

9.4.4.2 Macro Algae vs Micro Algae

The green macroalgae (seaweed) exhibit high affinity for many metal ions (Mani and Kumar 2014). The adsorption capacity of the macroalgae is directly related to the alginate content, availability and its specific macromolecular conformation. Lee and Chang (2011) tested the bioremoval capacity of two macroalgae *Spirogyra* and *Cladophora* for Pb(II) and Cu(II) and found that although the functional groups of these two genera of algae were similar but the sorption capacity of *Spirogyra* was superior to *Cladophora*.

Microalgae has gained more demand due to the development of innovative mass-production and more efficient biosorption of heavy metal ions. Minimal growth requirements (solar light and CO_2) make them suitable for bioremediation of heavy metal. Microalgae have developed an extensive spectrum of mechanisms (extracellular and intracellular) to cope with heavy metal toxicity (Kumar et al. 2015). *Spirulina* spp. and *Planothidium lanceolatum* are reported to remediate Ni and Zn (Doshi et al. 2008; Sbihi et al. 2012).

9.4.4.3 Mechanism of Algal Bioremediation

The accumulation of heavy metal ions in algae occurs in two phases (Monteiro et al. 2012). The first is a rapid passive biosorption where the metal ions adsorb onto the cell surface within a short span of time, and the process is metabolism independent. The second phase is a slower active sorption of heavy metal ions into the cytoplasm of algal cells. This phase is metabolism-dependent (Talebi et al. 2013).

The biosorption capacity for heavy metal ions has been attributed to presence of various functional groups on the algal cell surface such as hydroxyl (OH), phosphoryl (PO_3O_2), amino (NH_2), carboxyl (COOH), sulphydryl (SH), etc., which confer negative charge to the cell surface (Kaplan 2013). Since heavy metal ions are in the cationic form in water, they get adsorbed onto the algal cell surface. The functional groups are associated with various cell wall components such as peptidoglycan, teichoic acids, polysaccharides and proteins which provide metal binding sites (Kuyucak and Volesky 1988).

Other mechanisms have also been reported like complexation which is important in metal sorption by algae (Davis et al. 2003). Adhiya et al. (2002) reported that Cd biosorption to *Chlamydomonas reinhardtti* involves complexation with carboxylic groups. Electrostatic attraction and covalent binding, respectively, mediate Ni and Zn adsorption on *Chaetophora elegans* (Andrade et al. 2005). Aluminum sorption onto algal cells involves a different kind of mechanism. Aluminium (Al) ions bind to biomass in the form of polynuclear Al species and thus prevents other heavy metal ions from accessing the binding site (Bottero et al. 1980).

9.4.4.4 Challenges in Algal Bioremediation

Use of algae for biosorption of heavy metal ions from wastewaters has shown promising results but an efficient and commercially viable algal technology still need to be developed. There is a need to develop a thorough understanding of the mechanism of metal sorption. Still there are many freshwater and marine algae which has not been explored for their metal binding capacity. Therefore, screening of algae is a necessary step for selection of the best algal species with high affinity for a particular metal. The algal biomass has to be immobilized before passing wastewater through it. For this purpose, alginate is used which is an expensive chemical and thus not feasible for metal removal from wastewater always.

Although the use of inactivated algal biomass has been preferred, there are some limitations to it as well. Dead cells cannot be used where biological alteration in valency of a metal is sought. Moreover, there is no scope for biosorption improvement through mutant isolation. On the other hand, use of live cells also carries some demerits. The metal recovery might be limited since it is bound intracellularly and the metabolic extracellular products may interact with metals and retain them within the solution. However, to achieve the highest removal efficiency, interaction between algal strains, dead or live cells and pollutants should be optimized.

9.5 Integrated Approach Using Multi-biosystems for Remediation of Heavy Metals

There are many cases of heavy metal bioremediation reported using single biosystem but very few reports on biological approaches using multi-biosystems. When compared bioremediation strategies applied to polluted soils between combined and single process, it can be easily concluded that combined multiple bioremediation approaches removed much more heavy metal from the soil and highly efficient hydrophobic than each single process alone.

Bacterial consortia have gained interest of environmentalists where the ultimate aim of the bacterial mixtures system is to deliver benefits environmental applications of cleaning up the contaminants (Emenike et al. 2016). Compared with single strain, the bacterial mixtures showed higher growth rate and a considerably higher heavy metal bioremediation which might due to higher bacterial cell density at high levels of heavy metals (Kang et al. 2016).

Phytoremediation alone sometimes may not be sufficient to bring out the best result and may cause toxic effects to the plants at higher concentrations of heavy metal. Inoculation of the plant rhizosphere with microorganisms is an established route to improving phytoextraction efficiency. The plants are benefited from synergistic effects with rhizobacteria that improve plant growth and metal accumulation, mitigating the toxic effects on plants and increasing their tolerance to heavy metals (Wood et al. 2016; Sumi et al. 2015). PGPBs-legumes associations represent an alternative procedure for phytostabilisation of heavy metals polluted soils mainly generated by industrial and agricultural practices (Hao et al. 2014).

We know microorganisms are responsible for the biodegradation of heavy metal but the combination of earthworms and microbes have shown better results. Tomar and Suthar (2011) have reported a successful treatment of waswater by microbial-earthworm ecofilters as a promising economical process. The concept behind the approach is that microorganisms perform biochemical degradation of waste material while earthworms regulate microbial biomass and activity by directly or/and indirectly grazing on microorganisms (Liu et al. 2012). Earthworms have a complex digestive system in which the earthworm and microbes in the gut are mutually benefited from each other and lead to the degradation of ingested contaminants (Brown et al. 2000). However, it is difficult to differentiate between the metabolism of earthworms microorganisms which contribute to the bioremediation of heavy metal.

Algae and bacteria have coexisted ever since the early stages of evolution. They synergistically affect each other's physiology and metabolism. Many studies have dealt with algae-bacteria consortium for metal bioremediation (Boivin et al. 2007). Higher concentrations of heavy metal can cause toxic effects in algae but the consortia of algae and bacteria overcomes it and they mutually detoxify and assimilate metals from metal rich environments.

Generally, fungi are more tolerant to metals than bacteria (Kidd et al. 2009). They can proficiently explore the soil microbes which are not accessible for plant roots due to their small diameters. Fungi can compete with roots and other microorganisms for water and metal uptake, protect the roots from direct interaction with the metals and inhibit metal transport through increased soil hydrophobicity (Wenzel 2009). The endophytic fungi could increase resistance of the host plant to multimetal contamination. They can also reduce the level of growth-inhibiting stress ethylene within the plants and also provide the plants with iron from the soil. Thus, they prove to be a suitable candidate for remediation of heavy metal in combination with plants with reduced toxicity of plants under stressed condition.

Remediation of contaminated soils using earthworms and plants appears to be cost-effective and environmentally friendly technology. Wang and Li (2006) observed higher uptake of heavy metal by plants under earthworm inoculation which was probably due to the increase in dry matter production stimulated by earthworms. However, further research is needed to optimize the species combinations for suitable heavy metal uptake. A brief summary of the remediation of the heavy metal by integrated approach is demonstrated in Table 9.4.

9.6 Metagenomics

To bioremediate the heavy metal contaminated site, various biosystems are used. But very often, remediation techniques fail because of the difficulty to control and expand key biodegradative processes from bench to full scale (Fantroussi and Agathos 2005; Paerl and Steppe 2003). To get better results, a better understanding of the ecology of microbial communities inhabiting contaminated sites is needed, as well as of their interactions with the environment (Rittmann et al. 2006). But, the complete study of the microbial communities of the environment is challenging as most of them are recalcitrant to conventional cultivation (Stewart 2012). The proper management of microbial resources needs a comprehensive characterization of their genetic pool to measure the fate of contaminants and enhance bioremediation processes (Gillan et al. 2015). The emergence of metagenomics has the potential to revolutionize the overall bioremediation process as it gives direct access to microbial communities inhabiting polluted environments independently of their culturability (Bouhajja et al. 2016).

There are few main metagenomic approaches:

9.6.1 Library-Based Targeted Metagenomics

The environmental DNA is isolated from the environmental samples and cloned inside suitable host (usually *Escherichia coli*), then the clones of interest are selected based on their expression of biodegradative functions or sequence homology with probes and primers, thus establishing a metagenomic library. As host, *Escherichia coli* has been extensively used in metagenomic studies (Gabor et al. 2004) but use of multiple-host systems and broad-host-range vectors can be used to overcome the limitations of gene expression machinery or toxicity of some gene products in a single host (Cheng et al. 2014; Ekkers et al. 2012).

Table 9.4 Some examples of bioremediation of heavy metal by integrated approach

Biosystems	Reports on bioremediation on heavy metal	References
Bacterial consortia- *Bacillus* sp., *Lysinibacillus* sp. and *Rhodococcus* sp	Optimal removal of Pb, Mn and Cu in leachate-polluted soil of a land fill environment. Enhanced metabolic activity due to bioaugmentation of the microcosm using bacterial inoculums	Emenike et al. (2016)
Bacterial consortia- *Viridibacillus arenosi* B-21, *Sporosarcina soli* B-22, *Enterobacter cloacae* KJ-46, and *E. cloacae* KJ-47	Compared with single strain cultures, the bacterial mixtures demonstrated greater resistance and efficiency for the remediation of heavy metals such as Cd, Pb, Cu	Kang et al. (2016)
Plant and bacteria-*Sedum alfredii* and *Burkholderia cepacia*	Increase in the plant biomass and leading to enhanced Zn and Cd uptake	Li et al. (2007)
Plant and bacteria-*Brassica juncea* and *Bacillus* spp.	Increase in the plant dry weight with an increase in Cd uptake	Jeong et al. (2013)
Plant and bacteria-*Vicia faba*, *Lens culinaris* and *Sulla coronaria* co-inoculated with *Enterobacter cloacae, Pseudomonas sp.* and *Rhizobium sullea*	Inoculations decreased heavy metals (Cu and Pb) availability in the soil indicating a positive effect of co-inoculation of legumes by appropriate heavy metals resistant bacteria for the phytostabilisation of mine tailings	Saadani et al. (2016)
Plant and bacteria-*Lepidium sativum* and *Azotobacter*	Stimulate the plant growth and enhance its tolerance to Cr(VI) and Cd(II), to ultimately provide a reliable phytoremediation system.	Sobariu et al. (2016)
Fungi and plant-*Trichoderma atroviride* and *Brassica juncea*	Significantly alleviates the cellular toxicity of Cdand Ni from contaminated soil	Cao et al. (2008)
Fungi and plant- *Cryptococcus sp.* (yeast), *Rhodotorula sp.* and *B. chinensis*	Fungi helps in plant growth in multi-metal contaminated soils and give resistance to Cd, Pb, Zn, and Cu	Deng et al. (2012) and Wang et al. (2013)
Fungi and plant-*Microsphaeropsis sp.* and *Solanum nigrum*	Shows enhanced Cd biosorption capacity	Xiao et al. (2010)
Fungi consortia-*Mucor sp.* and *Fusarium sp.*	Increased metal concentrations in the canola (Cd, Pb, and Zn), elevated the extractable metal amount, and increased metal translocation from roots to shoots	Deng et al. (2014)
Plant and Earthworm-*Vicia faba, Zea mays*, and *Eisenia fetida*	Earthworms and plants increased the uptake of metals (Pb, Cd and Zn) from contaminated soils. The earthworm-plant-soil interaction influence both the health of the plant and the uptake of heavy metals by plants	Lemtiri et al. (2016)

(continued)

Table 9.4 (continued)

Biosystems	Reports on bioremediation on heavy metal	References
Plant and Earthworm-*Lantana camara* and *Pontoscolex corethrurus*	Interaction between earthworm and plant have a positive effect on Pb-phytoextraction yield and was significantly correlated with the increase in total microbial activity and richness index of the fungal community	Jusselme et al. (2015)

9.6.2 Direct Sequencing of Metagenomes

It does not involve a cloning step and has been more often applied to polluted environments for characterization of the taxonomic and functional composition of microbial communities and their dynamics. The analysis has focused on 16S rRNA genes and marker genes of biodegradation.

9.7 Next Generation Sequencing

Next Generation Sequencing (NGS) was introduced in 2005 (Margulies et al. 2005).There has been a remarkable increase in metagenomic studies based on NGS. It includes immense parallel sequencing of clonally amplified or single DNA molecules spatially separated in a flow cell (van Dijk et al. 2014).

Lastly, the huge amount of data generated by metagenomic studies is analyzed using bioinformatic tools to predict the microbial diversity, enhance the discovery and characterization of unknown bacterial and fungal metabolic pathways involved in the degradation of hazardous pollutants. Even though metagenomics is having some technical and computational challenges, the positive claims of it can be used to efficiently monitor the clean-up process of the environment and mitigate the effects of the pollutants on the eco-system.

9.8 Metabolomics

The main challenge faced by plants growing under heavy metal stressed condition is biomass reduction, nutrient deficiency aided with increased toxicity of heavy metal. Plants are considered to biosynthesize specialized (primary and secondary) metabolites to adapt to the environmental stresses (Auge et al. 2014).

Metabolomics is a newly emerging discipline which can serve to analyze the whole set of small molecular weight chemical compounds (<1000 Da) in organism (Ji et al. 2015; Watanabe et al. 2015). It provides a glimpse of dynamic changes in metabolic pathways in the host plant regulated by microbial population and their response to highly dynamic environmental conditions in their unique ecological niches. This field is coupled with functional genomics to understand biochemical phenotypes across a range of biological systems. Metabolomics measures all metabolites at a specific time point, reflecting a snapshot of all the regulatory events responding to the external environmental conditions (Kumar et al. 2016). The metabolites reflect the true integration of gene regulation and protein expression incorporating the impact of the environment and other organisms. The metabolites fate can be employed as bioindicators to monitor the biological effects of the pollutants on living organism and help in better understanding of the environment (Tomanek 2014).

Recent developments in analytical instrumentation and bioinformatics tools has led to evaluate numerous plant metabolites, metabolic changes and finally elucidate metabolic pathways responsible for heavy metal tolerance to plants (Obata and Fernie 2012). Current studies are mostly restricted to targeted metabolomics, which focuses on amino acid and/or lipid metabolism (Kumari et al. 2015; Melo et al. 2015).

9.8.1 Various Metabolomic Platforms to Identify Metabolites

The main strategies engaged to analyse the metabolome of plants include (i) metabolite profiling; (ii) targeted analysis; and (iii) metabolic fingerprinting (Hill and Roessner 2013). Metabolite profiling is a semi-quantitative which allows for detection of a large set of both known and unknown metabolites. Target analysis is an absolute quantitative approach which detects metabolites involved in a particular pathway by utilizing specialized protocols and detection techniques. Finally, metabolic fingerprinting is the highest throughput procedure and generates fingerprints characterizing a specific metabolic state of a sample by non-specific and rapid analysis of crude metabolite mixtures.

Without adequate knowledge of the metabolites under stressful conditions, a targeted metabolomic approach possess a high risk of missing significant changes in the metabolome. In order to achieve desired results, there is a need to expand beyond the known targets that can only be accomplished with non-targeted, unbiased metabolomics also known as global metabolomics (Kueger et al. 2012). Global metabolomics provides a panoramic view covering both primary (including sugars, amino acids and tricarboxylic acids involved in primary metabolic processes such as respiration and photosynthesis) and secondary metabolites (including alkaloids, phenolics, steroids, lignins, tannins, etc.) in a single run and has advantages of uncovering many novel compounds.

9.8.2 Analytical Platform to Analyze the Metabolites

A range of analytical platforms have been established which includes nuclear magnetic resonance (NMR), Fourier transform ion cyclotron resonance mass spectrometry (FT-ICRMS) and mass spectrometry (MS). MS-based metabolomics combines chromatographic separation with mass spectra and are available in multiple forms such as liquid chromatography (LC –MS), gas chromatography (GC –MS), capillary electrophoresis (CE –MS) and matrix-assisted laser desorption/ionization (MALDI-MS). However, due to high grade of molecular weight and structural diversity between primary and secondary metabolites, a single platform is not sufficient to indentify and quantify the metabolites (Kueger et al. 2012). Therefore, a combination of different techniques will reveal a vast metabolite profile. However, investigations have demonstrated 1H NMR as efficient approach for detection of the metabolites released in responses to metal pollutants whereas, MS-based analytical approaches are preferred to investigate plant responses to environmental cues due to its sensitivity to low abundant molecules and the flexibility for detecting multiple classes of molecules (Hill and Roessner 2013).

9.8.3 Bioinformatics Tools

The vast amount of metabolic data generated need to be archived, managed and integrated for metabolic analysis. So, various bioinformatics tools are designed for processing of raw data, mining, statistical analysis, management and mathematical modelling of metabolomic networks. A range of bioinformatics tools for effective insilico data pre- processing have been designed for this purpose including Analyzer Pro, Automated Mass Spectral Deconvolution and Identification system, and many more (Fukushima and Kusano 2013).

Though metabolomics is a relatively new approach in plant biology, it can be combined with other 'omics' disciplines turning out to be a major tool in revealing new knowledge on diverse metabolites produced by plants to heavy metal contaminants, and also on their metabolomic reprogramming for acclimation to extreme perturbations.

9.9 Proteomics

It is very important to understand why a particular metal at a certain concentration can alter from non-toxic to toxic form for other species at a slightly higher concentration (Ge et al. 2009; Vido et al. 2001). In the past years, substantial improvements in protein separation and identification techniques have opened the application of proteomic methods to answer the biological questions along with metagenomics and metabolomics methods (Isaacson et al. 2006). The heavy metal uptake process across a number of unrelated plant species appears to be associated with proteins involved in energy metabolism, the oxidative stress response and abiotic and biotic stress (Visioli and Marmiroli 2013). Examining the toxic effects of heavy metals on protein expression can be useful for gaining insight into the biomolecular mechanisms of toxicity and for identifying potential candidate metal-specific protein markers of exposure and response (Luque-Garcia et al. 2011). Proteomics, an important omic approach facilitates both identification and quantification of differentially expressed proteins. Moreover, the identification of post-transcriptionally regulated array of functionally diverse genes playing a key role in conferring resistance towards stress has also been advanced (Zargar et al. 2017). Proteomic data supplement the huge genomic and transcriptomic data sets in providing a clear picture of the process and thus helps in determination of major genetic determinants of the hyperaccumulation phenomenon (Visioli and Marmiroli 2013). Research analysis has depicted that proteomics in union with bioinformatic tools, can facilitate the discovery of new and better biomarkers of heavy metal toxicity (Zhai et al. 2005).

The current state of knowledge regarding the proteomics of hyperaccumulation is inadequate to understand the role of the large number of proteins involved and the level of cross-talk between different pathways (Visioli and Marmiroli 2013). Few proteomic methodologies appropriate for the identification of key regulators of hyperaccumulation are as follows.

9.9.1 Gel and Non-gel Approaches

For most of the plant proteomics studies, pre-fractionation of the sample prior to mass spectometry (MS) analysis is carried out which can be achieved by gel electrophoresis or by certain gel-free techniques. 2D-Gel electrophoresis (2D-GE) has become the optimum choice for separating complex protein mixtures with respect to achievable resolution and reproducibility (Rose et al. 2004). However, there are certain drawbacks such as limited capacity to fractionate hydrophobic proteins and glycoproteins successfully, detection of small peptide molecules and the risk of quantification (Visioli and Marmiroli 2013). Generally, the reproducibility of LC-based separation is better than that achieved by 2D-GE which is an important advantage for comparative proteomics (Lambert et al. 2005; Pirondini et al. 2006). There are various statistical packages which facilitate semi-quantitative proteomics such as Progenesis (Nonlinear Dynamics), ImageMaster 2D Platinum (Ge Healthcare, Amersham Biosciences) and PDQuest (Bio-Rad).

9.9.2 Mass Spectometry-Based Quantification

The introduction of MS technology has widely enhanced the throughput of proteomic data compared to electrophoretic or chromatographic methods and provides a more reliable characterization of all the protein species along with identification of post-translational modifications such as phosphorylation and acetylation, which are important in cell signalling and various epigenetic phenomena (Bantscheff et al. 2012).

9.10 Conclusion

Heavy metal contamination has taken a serious turn leading to devastating effects on environment and human health. Compared to the complexity and time consumption involved in the conventional methods for remediation of soil, bioremediation techniques has proven to be the best alternative techniques where in addition to bioremoval of heavy metal, it also replenishes the site and maintain the ecological balance of the environment. Plants are the most widely accepted bio-tool for remediation of soil. But the traditional phytoremediation approaches are less economical because the hyperaccumulators are generally slow growing and have less biomass production. Earthworms, being the soil organism, leads to significant improvement in the quality of soil and assist in heavy metal bioremoval in their biomass but higher concentrations of heavy metal produce toxic effects in earthworms. The bioremediation capacity of the algae and fungi have been studied extensively and has been effective remediators in many cases. Use of microbes has arisen as the savior for bioremediation. Recently, the integrated approach of using more than one organism for bioremediation has gained popularity as it helps to overcome the drawback of a single biosystem. Moreover, the symbiosis relation between has resulted in high performances such as more metal accumulation, high biomass production and well adapted to variety of climatic conditions, therefore driving us towards a sustainable environment. A successful bioremediation strategy require a detailed understanding of the functioning of degradative microbial communities which is quite a challenge for microbiologists. Thus, metagenomics, metabolomics and proteomics have come into play and has become the major tool for identification of all the unexplored microbial communities possessing the ability to degrade heavy metal and identification of the diverse metabolites produced by organism to tolerate under stress conditions. Thus, coupling both the 'omics' will give a comprehensive understanding of the microbial communities and their biodegradation pathways.

Moreover, in order to achieve even better results for bioremediation, certain points have to be considered as follows:

(a) The exploitation of the floral diversity should be extended for obtaining the effective hyperaccumulator plant which can maintain effective rate of heavy metal uptake throughout. Prior comprehensive risk assessment studies should be carried out to protect the local plant diversity.
(b) Studies need to be conducted to have a better understanding of the interactions between heavy metal, soil, microbe, earthworm and plant roots to comprehend the fate of metal ions in the soil.
(c) More research is needed to obtain effective and environmentally safe chemicals which can increase the metal solubility in soil and thus, enhanced the bioavailability of metals to the plant roots.
(d) In spite of all the advances, most of the research is still limited to laboratory scale studies. Long-term in-situ field trails are actually required for to prove the efficacy of the strategy in real-contaminated area.
(e) More sophisticated bioinformatics tools should be developed to reconstruct full length metabolic and catabolic pathways. More studies have to carried out opting an integrated approach using the 'omics' tool together for better insights. Thus, it can provide a practical implementation of large-scale application of bioremediation.

References

Abd-Elnaby HM, Abou-Elela GM, Ghozlan HA, Hussein H, Sabry SA (2016) Characterization and bioremediation potential of marine Psychrobacter species. Egypt J Aquat Res 42 (2):193–203

Adhiya J, Cai X, Sayre RT, Traina SJ (2002) Binding of aqueous cadmium by the lyophilized biomass of Chlamydomonas reinhardtii. Coll Surf A 210(1):1–11

Agouborde L, Navia R (2009) Heavy metals retention capacity of a non-conventional sorbent developed from a mixture of industrial and agricultural wastes. J Hazard Mater 167 (1):536–544

Ahamad PYA, Kunhi AAM (2011) Enhanced degradation of phenol by Pseudomonas sp. CP4 entrapped in agar and calcium alginate beads in batch and continuous processes. Biodegradation 22:253–265

Ahemad M (2014) Remediation of metalliferous soils through the heavy metal resistant plant growth promoting bacteria: paradigms and prospects. Arab J Chem. https://doi.org/10.1016/j.arabjc.2014.11.020

Ahluwalia SS, Goyal D (2007) Microbial and plant derived biomass for removal of heavy metals from wastewater. Bioresour Technol 98(12):2243–2257

Akar T, Tunali S, Cabuk A (2007) Study on the characterization of lead (II) biosorption by fungus Aspergillus parasiticus. Appl Biochem Biotech 136:389–406

Albarracín VH, Amoroso MJ, Abate CM (2005) Isolation and characterization of indigenous copper-resistant actinomycete strains. Chem Erde Geochem 65:145–156

Ali H, Khan E, Sajad MA (2013) Phytoremediation of heavy metals-concepts and applications. Chemosphere 91(7):869–881

Alkorta I, Hernández-Allica J, Becerril JM, Amezaga I, Albizu I, Garbisu C (2004) Recent findings on the phytoremediation of soils contaminated with environmentally toxic heavy metals and metalloids such as zinc, cadmium, lead, and arsenic. Rev Environ Sci Biotechnol 3(1):71–90

Alvarez A, Saez JM, Costa JSD, Colin VL, Fuentes MS, Cuozzo SA, Benimeli CS, Polti MA, Amoroso MJ (2017) Actinobacteria: Current research and perspectives for bioremediation of pesticides and heavy metals. Chemosphere 166:41–62

Aly AH, Debbab A, Proksch P (2011) Fungal endophytes: unique plant inhabitants with great promises. Appl Microbiol Biotechnol 90:1829–1845

Andrade AD, Rollemberg MCE, Nobrega JA (2005) Proton and metal binding capacity of the green freshwater alga Chaetophora elegans. Process Biochem 40(5):1931–1936

Andre J, Charnock J, Stürzenbaum SR, Kille P, Morgan AJ, Hodoson ME (2009) Accumulated metal speciation in earthworm populations with multigenerational exposure to metalliferous soils: cell fractionation and high-energy synchrotron analyses. Environ Sci Technol 43:6822–6829

Areco MM, Hanela S, Duran J, Afonso Mdos S (2012) Biosorption of Cu(II), Zn(II), Cd(II) and Pb (II) by dead biomasses of green alga Ulva lactuca and the development of a sustainable matrix for adsorption implementation. J Hazard Mater 213–214:123–132

Arief VO, Trilestari K, Sunarso J, Indraswati N, Ismadji S (2008) Recent progress on biosorption of heavy metals from liquids using low cost biosorbents: characterization, biosorption parameters and mechanism studies. Clean 36(12):937–962

ATSDR. Agency for toxic substances and disease registry (2011) toxicological profile for uranium- an update. U.S. department of health and human services. Public Health Service, Atlanta. Accessed at http://www.atsdr.cdc.gov/toxprofiles/tp150.html. Aug 2011

Auge RM, Toler HD, Saxton AM (2014) Arbuscular mycorrhizal symbiosis and osmotic adjustment in response to NaCl stress: a meta-analysis. Front Plant Sci 5

Azizi AB, Lim MPM, Noor ZM, Noorlidah A (2013) Vermiremoval of heavy metal in sewage sludge by utilising Lumbricus rubellus. Ecotoxicol Environ Saf 90:13–20

Babu AG, Kim JD, Oh BT (2013) Enhancement of heavy metal phytoremediation by Alnus firma with endophytic Bacillus thuringiensis GDB-1. J Hazard Mater 250–251:477–483

Babu AG, Shea PJ, Sudhakar D, Jung IB, BT O (2015) Potential use of Pseudomona koreensis AGB-1 in association with Miscanthus sinensis to remediate heavy metal(loid)-contaminated mining site soil. J Environ Manag 151:160–166

Bantscheff M, Lemeer S, Savitski MM, Kuster B (2012) Quantitative mass spectrome try in proteomic s: critical review upd ate from 2007 to the present. Anal Bioanal Chem 404:939–965

Barbosa B, Boléo S, Sidella S, Costa J, Duarte MP, Mendes B, Cosentino SL, Fernando AL (2015) Phytoremediation of heavy metal-contaminated soils using the perennial energy crops Miscanthus, spp. and Arundo donax, L. Bioenergy Res 8:1500–1511

Bitther OP, Pilon-Smits EAH, Meagher RB, Doty S (2012) Biotechnological approaches for phytoremediation. In: Arie Altman A, Hasegawa PM (eds) Plant biotechnology and agriculture. Academic, Oxford, pp 309–328

Boivin MEY, Greve GD, Garcia-Meza JV, Massieux B, Sprenger W, Kraak MHS, Breure AM, Rutgers M, Admiraal W (2007) Algal–bacterial interactions in metal contaminated floodplain sediments. Environ Pollut 145(3):884–894

Boopathy R (2000) Factors limiting bioremediation technologies. Bioresour Technol 74(1):63–67

Bottero JY, Cases JM, Fiessinger F, Poirier JE (1980) Studies of hydrolyzed aluminum chloride solutions. I. Nature of aluminum species and composition of aqueous solutions. J Phys Chem 84:2933

Bouhajja E, Agathos SN, George IF (2016) Metagenomics: probing pollutant fate in natural and engineered ecosystems. Biotechnol Adv 34(8):1413–1426

Brown GG, Barois I, Lavelle P (2000) Regulation of soil organic matter dynamics and microbial activityin the drilosphere and the role of interactionswith other edaphic functional domains. Eur J Soil Biol 36(3):177–198

Cao L, Jiang M, Zeng Z, Liu Y (2008) Trichoderma atroviride F6 improves phytoextraction efficiency of mustard (Brassia juncea (L) Coss. var. foliosa Bailey) in Cd, Ni contaminated soils. Chemosphere 71:1769–1773

Cao X, Song Y, Kai J, Yang X, Ji P (2012) Evaluation of EROD and CYP3A4 activities in earthworm Eisenia fetida as biomarkers for soil heavy metal contamination. J Hazard Mater 243:146–151
Chang Q, Wang G (2007) Study on the macromolecular coagulant PEX which traps heavy metals. Chem Eng Sci 62(17):4636–4643
Chen L, Luo SL, Li XJ, Wan Y, Chen JL, Liu CB (2014) Interaction of Cdhyperaccumulator Solanum nigrum L. and functional endophyte Pseudomonas sp. Lk9 on soil heavy metals uptake. Soil Biol Biochem 68:300–308
Chen X, Wang X, Gu X, Jiang Y, Ji R (2017) Oxidative stress responses and insights into the sensitivity of the earthworms Metaphire guillelmi and Eisenia fetida to soil cadmium. Sci Total Environ 574:300–306
Cheng J, Pinnell L, Engel K, Neufeld JD, Charles TC (2014) Versatile broad-host-range cosmids for construction of high quality metagenomic libraries. J Microbiol Methods 99:27–34
Contreras-Ramos SM, Alvarez-Bernal D, Dendooven L (2006) Eisenia fetida increased removal of polycyclic aromatic hydrocarbons from soil. Environ Pollut 141(3):396–401
Costa JSD, Kothe E, Abate CM, Amoroso MJ (2012) Unraveling the Amycolatopsis tucumanensis copper-resistome. Biometals 25(5):905–917
D'Annibale A, Leonardi V, Federici E, Baldi F, Zecchini F, Petruccioli M (2007) Leaching and microbial treatment of a soil contaminated by sulphide ore ashes and aromatic hydrocarbons. Appl Microbiol Biotechnol 74:1135–1144
Dahmani-Muller H, Van Oort F, Gelie B, Balabane M (2000) Strategies of heavy metal uptake by three plant species growing near a metal smelter. Environ Pollut 109(2):231–238
Dardanelli MS, Manyani H, González-Barroso S, Rodríguez-Carvajal MA, Gil-Serrano AM, Espuny MR, López-Baena FJ, Bellogín RA, Megías M, Ollero FJ (2010) Effect of the presence of the plant growth promoting rhizobacterium (PGPR) Chryseobacterium balustinum Aur9 and salt stress in the pattern of flavonoids exuded by soybean roots. Plant Soil 328(1–2):483–493
Das S, Raj R, Mangwani N, Dash HR, Chakraborty J (2014) 2-Heavy metals and hydrocarbons: adverse effects and mechanism of toxicity. Microb Biodegrad Biorem. Elsevier, Oxford, pp 23–54. https://doi.org/10.1016/B978-0-12-800021-2.00002-9
Davis TA, Volesky B, Mucci A (2003) A review of the biochemistry of heavy metal biosorption by brown algae. Water Res 37:4311–4330
De Filippis LF, Pallaghy CK (1994) Heavy metals: sources and biological effects. In: Rai LC, Gaur JP, Soeder CJ (eds) Advances in limnology series: algae and water pollution. E. Scheizerbartsche Press, Stuttgart, pp 31–77
DeForest DK, Brix KV, Adams WJ (2007) Assessing metal bioaccumulation in aquatic environments: the inverse relationship between bioaccumulation factors, trophic transfer factors and exposure concentration. Aquat Toxicol 84(2):236–246
Deng Z, Wang W, Tan H, Cao L, (2012). Characterization of heavy metal-resistant endophytic yeast Cryptococcus sp. CBSB78 from rapes (Brassica chinensis) and its potential in promoting the growth of Brassica spp. in metal-contaminated soils. Water Air Soil Pollut. 223, 5321–5329
Deng Z, Cao L, Zhang R, Shi Y, Hu L, Tan H, Cao L (2014) Enhanced phytoremediation of multimetal contaminated soils by interspecific fusion between the protoplasts of endophytic Mucor sp.CBRF59 and Fusarium sp. CBRF14. Soil Biol Biochem 77:31–40
Dharni S, Maurya A, Samad A, Srivastava SK, Sharma A, Patra DD (2014) Purification, characterization, and in vitro activity of 2, 4-di-tert-butylphenol from Pseudomonas monteilii PsF84: conformational and molecular docking studies. J Agric Food Chem 62(26):6138–6146
Dia J, Becquer T, Rouiller JH, Reversat G, Bernhard-Reversat F, Nahmani J, Lavelle P (2004) Heavy metal accumulation by two earthworm species and its relationship to total and DTPA extractable metals in soils. Soil Biol Biochem 36:91–98
Dixit R, Wasiullah MD, Pandiyan K, Singh UB, Sahu A, Shukla R, Singh BP, Rai JP, Sharma PK, Lade H, Paul D (2015) Bioremediation of heavy metals from soil and aquatic environment: an overview of principles and criteria of fundamental processes. Sustainability 7:2189–2212. https://doi.org/10.3390/su7022189

Doshi H, Seth C, Ray A, Kothari IL (2008) Bioaccumulation of heavy metals by green algae. Curr Microbiol 56:246–255

Ekkers DM, Cretoiu MS, Kielak AM, Elsas JD (2012) The great screen anomaly-a new frontier in product discovery through functional metagenomics. Appl Microbiol Biotechnol 93:1005–1020

El Fantroussi S, Agathos SN (2005) Is bioaugmentation a feasible strategy for pollutant removal and site remediation? Curr Opin Microbiol 8:268–275

Emenike CU, Agamuthu P, Fauziah SH (2016) Blending Bacillus sp., Lysinibacillus sp. and Rhodococcus sp. for optimal reduction of heavy metals in leachate contaminated soil. Environ Earth Sci 75(1):26

Fomina MA, Alexander IJ, Colpaert JV, Gadd GM (2005) Solubilization of toxic metal minerals and metal tolerance of mycorrhizal fungi. Soil Biol Biochem 37:851–866

Fu F, Wang Q (2011) Removal of heavy metal ions from wastewaters: a review. J Environ Manag 92(3):407–418

Fukushima A, Kusano M (2013) Recent progress in the development of metabolome databases for plant systems biology. Front Plant Sci 4:73

Gabor E, Alkema W, Janssen D (2004) Quantifying the accessibility of the metagenome by random expression cloning techniques. Environ Microbiol 6:879–886

Gadd GM (2010) Metals, minerals and microbes: geomicrobiology and bioremediation. Microbiology 156(3):609–643

Ge F, Lu XP, Zeng HL, He QY, Xiong S, Jin L, He QY (2009) Proteomic and functional analyses reveal a dual molecular mechanism underlying arsenic-induced apoptosis in human multiple myeloma cells. J Proteome Res 8(6):3006–3019

Gillan DC, Roosa S, Kunath B, Billon G, Wattiez R (2015) The long-term adaptation of bacterial communities in metal-contaminated sediments: a metaproteogenomic study. Environ Microbiol 17:1991–2005

Gil-Loaiza J, White SA, Root RA, Solís-Dominguez FA, Hammond CM, Chorover J, Maier RM (2016) Phytostabilization of mine tailings using compost-assisted direct planting: translating greenhouse results to the field. Sci Total Environ 565:451–461

Gisbert C, Ros R, De Haro A, Walker DJ, Bernal MP, Serrano R, Navarro-Aviñó J (2003) A plant genetically modified that accumulates Pb is especially promising for phytoremediation. Biochem Biophys Res Commun 303(2):440–445. https://doi.org/10.1016/s0006-291x(03)00349-8

Glick BR (2010) Using soil bacteria to facilitate phytoremediation. Biotechnol Adv 28(3):367–374

Glick BR (2012) Plant growth-promoting bacteria: mechanisms and applications. Scientifica 2012

Godlewska-Zyłkiewicz B (2001) Analytical applications of living organisms for preconcentration of trace metals and their speciation. Crit Rev Anal Chem 31(3):175–189

Gu JQ, Zhou WQ, Jiang BQ, Wang LH, Ma YN, Guo HY, Schulin R, Ji R, Evangelou MWH (2016) Effects of biochar on the transformation and earthworm bioaccumulation of organic pollutants in soil. Chemosphere 145:431–437

Gunatilake SK (2015) Methods of removing heavy metals from industrial wastewater. Methods 1(1)

Hao X, Taghavi S, Xie P, Orbach MJ, Alwathnani HA, Rensing C, Wei G (2014) Phytoremediation of heavy and transition metals aided by legume-rhizobia symbiosis. Int J Phytoremediat 16:179–202. https://doi.org/10.1080/15226514.2013.773273

Hill CB, Roessner U (2013) Metabolic pro filing of plants by GC–MS. In: Weckwerth W, Kahl G (eds) The handbook of plant metabolomics. Wiely-VCH Verlag GmbH and Co. KGaA, Weinheim, pp 1–23

Homa J, Klimek M, Kruk J, Cocquerelle C, Vandenbulcke F, Plytycz B (2010) Metal-specific effects on metallothionein gene induction and riboflavin content in coelomocytes of Allolobophora chlorotica. Ecotox Environ Saf 73(8):1937–1943

Huang C, Huang CP (1996) Application of Aspergillus oryzae and Rhizopus oryzae for Cu (II) removal. Water Res 30:1985–1990

Isaacson T, Damasceno CM, Saravanan RS, He Y, Catalá C, Saladié M, Rose JK (2006) Sample extraction techniques for enhanced proteomic analysis of plant tissues. Nat Protoc 1(2):769

Jatwani C, Gupta RK, Rai R, Bansal N (2016) Effects of Hg/Co toxicity in soil on biomolecules of earthworm, eisenia fetida. Procedia Environ Sci 35:450–455

Jeong S, Moon HS, Shin D, Nam K (2013) Survival of introduced phosphatesolubilizing bacteria (PSB) and their impact on microbial community structure during the phytoextraction of Cd-contaminated soil. J Hazard Mater 263:441e449

Ji C, Cao L, Li F (2015) Toxicological evaluation of two pedigrees of clam Ruditapes philippinarum as bioindicators of heavy metal contaminants using metabolomics. Environ Toxicol Pharmacol 39:545–554. https://doi.org/10.1016/j.etap.2015.01.004

Jing YX, Yan JL, He HD, Yang DJ, Xiao L, Zhong T, Yuan M, Cai XD, Li SB (2014) Characterization of bacteria in the rhizosphere soils of polygonum pubescens and their potential in promoting growth and Cd, Pb, Zn uptake by Brassica napus. Int J Phytoremediation 16:321–333

Jusselme MD, Poly F, Lebeau T, Rouland-lefèvre C, Miambi E (2015) Effects of earthworms on the fungal community and microbial activity in root-adhering soil of Lantana camara during phytoextraction of lead. Appl Soil Ecol 96:151–158

Kang CH, Kwon YJ, So JS (2016) Bioremediation of heavy metals by using bacterial mixtures. Ecol Eng 89:64–69

Kaplan D (2013) Absorption and adsorption of heavy metals by microalgae. In: Richmond A, Hu Q (eds) Handbook of microalgal culture: applied phycology and biotechnology. Blackwell Publishing, pp 439–447

Karami A, Shamsuddin ZH (2010) Phytoremediation of heavy metals with several efficiency enhancer methods. Afr J Biotechnol 9(25):3689–3698

Kidd P, Barceló J, Bernal MP, Navari-Izzo F, Poschenrieder C, Shilev S, Clemente R, Monterroso C (2009) Trace element behaviour at the root–soil interface: implications in phytoremediation. Environ Exp Bot 67(1):243–259

Kieser T, Bibb MJ, Buttner MJ, Chater KF, Hopwood DA (2000) Practical streptomyces genetics. JIF, Norwich

Knasmüller S, Parzefall W, Sanyal R, Ecker S, Schwab C, Uhl M,... & Darroudi F (1998) Use of metabolically competent human hepatoma cells for the detection of mutagens and antimutagens. Mutat Res Fundam Mol Mech Mutagen, 402(1), 185–202

Kongsricharoern N, Polprasert C (1995) Electrochemical precipitation of chromium (Cr^{6+}) from an electroplating wastewater. Water Sci Technol 31(9):109–117

Kueger S, Steinhauser D, Willmitzer L, Giavalisco P (2012) High-resolution plant metabolomics: from mass spectral features to metabolites and from whole-cell analysis to subcellular metabolite distributions. Plant J 70(1):39–50

Kumar KS, Daheavy metals HU, Won EJ, Lee JS, Shin KH (2015) Microalgae – a promising tool for heavy metal remediation. Ecotoxicol Environ Saf 113:329–352

Kumar M, Kuzhiumparambil U, Pernice M, Jiang Z, Ralph PJ (2016) Metabolomics: an emerging frontier of systems biology in marine macrophytes. Algal Res 16:76–92

Kumari P, Reddy CRK, Jha B (2015) Methyl jasmonate-induced lipidomic and biochemical alterations in the intertidal macroalga Gracilaria dura (Gracilariaceae, Rhodophyta). Plant Cell Physiol 56(10):1877–1889

Kuyucak N, Volesky B (1988) Biosorbents for recovery of metals from industrial solutions. Biotechnol Lett 10(2):137–142

Lamaia C, Kruatrachuea M, Pokethitiyooka P, Upathamb ES, Soonthornsarathoola V (2005) Toxicity and accumulation of lead and cadmium in the filamentous green alga Cladophora fracta (OF Muller ex Vahl) Kutzing: a laboratory study. Sci Asia 31(2):121–127

Lambert JP, Ethier M, Smith JC, Figeys D (2005) Proteomics: from gel based to gel free. Anal Chem 77:3771–3788

Landaburu-Aguirre J, García V, Pongrácz E, Keiski RL (2009) The removal of zinc from synthetic wastewaters by micellar-enhanced ultrafiltration: statistical design of experiments. Desalination 240(1–3):262–269
Lavelle P, Decaens T, Aubert M, Barot S, Blouin M, Bureau F, Margerie P, Mora P, Rossi JP (2006) Soil invertebrates and ecosystem services. Eur J Soil Biol 42:3–15
Lebeau T, Braud A, Jézéquel K (2008) Performance of bioaugmentation-assisted phytoextraction applied to metal contaminated soils: a review. Environ Pollut 153(3):497–522
Lee YC, Chang SP (2011) The biosorption of heavy metals from aqueous solution by Spirogyra and Cladophora filamentous macroalgae. Bioresour Technol 102(9):5297–5304
Lemtiri A, Liénard A, Alabi T, Brostaux Y, Cluzeau D, Francis F, Colinet G (2016) Earthworms Eisenia fetida affect the uptake of heavy metals by plants Vicia faba and Zea mays in metal-contaminated soils. Appl Soil Ecol 104:67–78
Li WC, Ye ZH, Wong MH (2007) Effects of bacteria on enhanced metal uptake of the Cd/Zn-hyperaccumulating plant. Sedum Alfredii J of Exp Bot 58:4173–4182
Li H, Wei D, He C, Shen M, Zhou Z, Mei T, Xu H-M (2012) Diversity and heavy metal tolerance of endophytic fungi from six dominant plant species in a Pb-Zn mine wasteland in China. Fungal Ecol 5:309–315
Liu J, Lu Z, Yang J, Xing M, Yu F, Guo M (2012) Effect of earthworms on the performance and microbial communities of excess sludge treatment process in vermifilter. Bioresour Technol 117:214–221
Liu G, Ling S, Zhan X, Lin Z, Zhang W, Lin K (2017) Interaction effects and mechanism of Pb pollution and soil microorganism in the presence of earthworm. Chemosphere
Luo S, Xu T, Chen L, Chen J, Rao C, Xiao X, Wan Y, Zeng G, Long F, Liu C, Liu Y (2012) Endophyte-assisted promotion of biomass production and metaluptake of energy crop sweet sorghum by plant-growth-promoting endophyte Bacillus sp. SLS18. Appl Microbiol Biotechnol 93(4):1745–1753
Luque-Garcia JL, Cabezas-Sanchez P, Camara C (2011) Proteomics as a tool for examining the toxicity of heavy metals. TrAC Trends Anal Chem 30(5):703–716
Ma Y, Prasad MNV, Rajkumar M, Freitas H (2011) Plant growth promoting rhizobacteria and endophytes accelerate phytoremediation of metalliferous soils. Biotechnol Adv 29:248–258. https://doi.org/10.1016/j.biotechadv.2010.12.001.
Ma Y, Oliveira RS, Nai F, Rajkumar M, Luo Y, Rocha I, Freitas H (2015) The hyperaccumulator Sedum plumbizincicola harbors metal-resistant endophytic bacteria that improve its phytoextraction capacity in multi-metal contaminated soil. J Environ Manag 156:62–69
Ma C, Ming H, Lin C, Naidu R, Bolan N (2016) Phytoextraction of heavy metal from tailing waste using Napier grass. Catena 136:74–83
Malik A (2004) Metal bioremediation through growing cells. Environ Int 30(2):261–278
Malley C, Nair J, Ho G (2006) Impact of heavy metals on enzymatic activity of substrate and on composting worms Eisenia fetida. Bioresour Technol 97:1498–1502
Maňáková B, Kuta J, Svobodová M, Hofman J (2014) Effects of combined composting and vermicomposting of waste sludge on arsenic fate and bioavailability. J Hazard Mater 280:544–551
Mangold S, Potrykus J, Bjorn E, Lovgren L, Dopson M (2012) Extreme zinc tolerance in acidophilic microorganisms from the bacterial and archaeal domains. Extremophiles 17:75–85
Mani D, Kumar C (2014) Biotechnological advances in bioremediation of heavy metals contaminated ecosystems: an overview with special reference to phytoremediation. Int J Environ Sci Technol 11(3):843–872
Margulies M, Egholm M, Altman W, Attiya S, Bader J, Bemben L (2005) Genome sequencing in microfabricated high-density picolitre reactors. Nature 437:376–380
Meers E, Van Slycken S, Adriaensen K, Ruttens A, Vangronsveld J, Du Laing G, Witters N, Thewys T, Tack FM (2010) The use of bio-energy crops (Zea mays) for 'phytoattenuation'of heavy metals on moderately contaminated soils: a field experiment. Chemosphere 78(1):35–41

Melo T, Alves E, Azevedo V, Martins AS, Neves B, Domingues P, Calado R, Abreu MH, Domingues MR (2015) Lipidomics as a new approach for the bioprospecting of marine macroalgae-Unraveling the polar lipid and fatty acid composition of Chondrus crispus. Algal Res 8:181–191

Memon AR, Schroder P (2009) Implications of metal accumulation mechanisms to phytoremediation. Environ Sci Pollut Res 16:162–175

Mendez MO, Maier RM (2008) Phytostabilization of mine tailings in arid and semiarid environments-an emerging remediation technology. Environ Health Perspect 116(3):278

Mohee R, Soobhany N (2014) Comparison of heavy metals content in compost against vermicompost of organic solid waste: past and present. Resour Conserv Recycl 92:206–213

Monteiro CM, Castro PML, Malcata FX (2010) Cadmium removal by two strains of Desmodesmus pleiomorphus cells. Water Air Soil Pollut 208:17–27

Monteiro CM, Castro PML, Malcata FX (2012) Metal uptake by microalgae: underlying mechanisms and practical applications. Biotechnol Prog 28(2):299–311

Morgan AJ, Turner MP, Morgan JE (2002) Morphological plasticity in metal-sequestering earthworm chloragocytes: morphometric electron microscopy provides a biomarker of exposure in field populations. Environ Toxicol Chem 21(3):610–618

Moyes DL, Naglik JR (2012) The mycobiome: influencing IBD severity. Cell Host Microb 11:551–552

Nadeem SM, Zahir ZA, Naveed M, Asghar HN, Arshad M (2010) Rhizobacteria capable of producing ACC-deaminase may mitigate salt stress in wheat. Soil Sci Soc Am J 74:533–542

Naees M, Ali Q, Shahbaz M, Ali F (2011) Role of rhizobacteria in phytoremediation of heavy metals: an overview. Int Res J Plant Sci 2:220–232

Natal-da Luz T, Tidona S, Jesus B, Morais PV, Sousa JP (2009) The use of sewage sludge as soil amendment. The need of an ecotoxicological evaluation. J Soils Sediments 9:246–260

Obata T, Fernie AR (2012) The use of metabolomics to dissect plant responses to abiotic stresses. Cell Mol Life Sci 69:3225–3243

Oyetibo GO, Ilori MO, Adebusoye SA, Obayori OS, Amund OO (2010) Bacteria with dual resistance to elevated concentrations of heavy metals and antibiotics in Nigerian contaminated systems. Environ Monit Assess 168:305–314

Paerl HW, Steppe TF (2003) Scaling up: The next challenge in environmental microbiology. Environ Microbiol 5:1025–1038

Pirondini A, Visioli G, Malcevschi A, Marmiroli N (2006) A 2-D liquid – phase chromatogra phy analysis in plant tissues. J Chromatogr B 833:91–100

Pohl P, Schimmack W (2006) Adsorption of radionuclides (134Cs, 85Sr, 226Ra, 241Am) by extracted biomasses of cyanobacteria (Nostoc Carneum, N. Insulare, Oscillatoria Geminata and Spirulina Laxis-sima) and phaeophyceae (Laminaria digitata and L. Japonica; waste products from alginate production) at different pH. J Appl Phycol 18(2):135e143

Polti MA, García RO, Amoroso MJ, Abate CM (2009) Bioremediation of chromium (VI) contaminated soil by Streptomyces sp. MC1. J Basic Microbiol 49:285–292

Prasad MNV, Freitas HMDO (2003) Metal hyperaccumulation in plants: biodiversity prospecting for phytoremediation technology. Electro J Biotechnol 6(3):285–321

Rajkumar M, Ma Y, Freitas H (2013) Improvement of Ni phytostabilization by inoculation of Ni resistant bacillus megaterium sr28c. J Environ Manag 128:973–980

Ramamurthy AS, Memarian R (2012) Phytoremediation of mixed soil contaminants. Water Air Soil Pollut 223:511–518

Rittmann BE, Hausner M, Loffler F et al (2006) A vista for microbial ecology and environmental biotechnology. Environ SciTechnol 40:1096–1103

Robinson BH, Leblanc M, Petit D, Brooks RR, Kirkman JH, Gregg PE (1998) The potential of Thlaspi caerulescens for phytoremediation of contaminated soils. Plant Soil 203(1):47–56

Rodriguez-Campos S, Smith NH, Boniotti MB, Aranaz A (2014) Overview and phylogeny of Mycobacterium tuberculosis complex organisms: implications for diagnostics and legislation of bovine tuberculosis. Res Vet Sci 97:S5–S19

Romera E, González F, Ballester A, Blázquez ML, Muñoz JA (2007) Comparative study of biosorption of heavy metals using different types of algae. Bioresour Technol 98:3344–3353
Roodbergen M, Klok C, van der Hout A (2008) Transfer of heavy metals in the food chain earthworm Black-tailed godwit (Limosa limosa): comparison of a polluted and a reference site in The Netherlands. Sci Total Environ 406(3):407–412
Rorat A, Wloka D, Grobelak A, Grosser A, Sosnecka A, Milczarek M, Jelonek P, Vandenbulcke F, Kacprzak M (2017) Vermiremediation of polycyclic aromatic hydrocarbons and heavy metals in sewage sludge composting process. J Environ Manag 187:347–353
Rose JKC, Bashir S, Giovannoni JJ, Jahn MM, Saravanan RS (2004) Tackling the plant proteome: practical approaches, hurdles and experimental tools. Plant J 39:715–733
Ryan RP, Germaine K, Franks A, Ryan DJ, Dowling DN (2008) Bacterial endophytes: recent developments and applications. FEMS Microbiol Lett 278:1–9
Saadani O, Fatnassi IC, Chiboub M, Abdelkrim S, Barhoumi F, Jebara M, Jebara SH (2016) In situ phytostabilisation capacity of three legumes and their associated Plant Growth Promoting Bacteria (PGPBs) in mine tailings of northern Tunisia. Ecotox Environ Saf 130:263–269
Sadrzadeh M, Mohammadi T, Ivakpour J, Kasiri N (2009) Neural network modeling of Pb^{2+} removal from wastewater using electrodialysis. Chem Eng Process 48(8):1371–1381
Sahariah B, Goswami L, Kim KH, Bhattacharyya P, Bhattacharya SS (2015) Metal remediation and biodegradation potential of earthworm species on municipal solid waste: a parallel analysis between Metaphire posthuma and Eisenia fetida. Bioresour Technol 180:230–236
Sarwar N, Imran M, Shaheen MR, Ishaq W, Kamran A, Matloob A, Rehim A, Hussain S (2016) Phytoremediation strategies for soils contaminated with heavy metals: modifications and future perspectives. Chemosphere 171:710–721. https://doi.org/10.1016/j.chemosphere.2016.12.116
Saurav K, Bar-Shalom R, Haber M, Burgsdorf I, Oliviero G, Costantino V, Morgenstern D, Steindler L (2016a) In search of alternative antibiotic drugs: quorum-quenching activity in sponges and their bacterial isolates. Front Microbiol 7
Saurav K, Burgsdorf I, Teta R, Esposito G, Bar-Shalom R, Costantino V, Steindler L (2016b) Isolation of marine Paracoccus sp. Ss63 from the sponge sarcotragus sp. and characterization of its quorum-sensing chemical-signaling molecules by LC-MS/MS analysis. Isr J Chem 56 (5):330–340
Sbihi K, Cherifi O, El Gharmali A, Oudra B, Aziz F (2012) Accumulation and toxicological effects of cadmium, copper and zinc on the growth and photosynthesis of the freshwater diatom Planothidium lanceolatum (Brébisson) Lange-Bertalot: a laboratory study. J Mater Environ Sci 3(3):497–506
Schulz B, Boyle C (2006) What are endophytes? In: Schulz BJE, Boyle CIC, Sieber TN (eds) Microbial root endophytes. Springer, Berlin, pp 1–13
Seth CS (2012) A review on mechanisms of plant tolerance and role of transgenic plants in environmental clean-up. Bot Rev 78(1):32–62
Shafiei Z, Hamid AA, Fooladi T, Yusoff WMW (2014) Surface active components: review. Curr Res J Biol Sci 6(2):89–95
Shin M, Shim J, You Y, Myung H, Bang KS, Cho M, Kamala-Kannan S, Oh BT (2012) Characterization of lead resistant endophytic Bacillus sp. MN3-4 and its potential for promoting lead accumulation in metal hyperaccumulator Alnus firma. J Hazard Mater 199–200:314–320
Sinha RK, Bharambe G, Chaudhari U (2008) Sewage treatment by vermifiltration with synchronous treatment of sludge by earthworms: a low-cost sustainable technology over conventional systems with potential for decentralization. Environmentalist 28(4):409–420
Sinhorin VDG, Sinhorin AP, Teixeira JMDS, Miléski KML, Hansen PC, PSA M, Kawashita NH, Baviera AM, Loro VL (2014) Effects of the acute exposition to glyphosate-based herbicide on oxidative stress parameters and antioxidant responses in a hybrid Amazon fish surubim (Pseudoplatystoma sp). Ecotoxicol Environ Saf 106:181–187
Sizmur T, Hodson ME (2009) Do earthworms impact metal mobility and availability in soil? – a review. Environ Pollut 157(7):1981–1989

Sizmur T, Tilston EL, Charnock J, Palumbo-Roe B, Watts MJ, Hodson ME (2011) Impacts of epigeic, anecic and endogeic earthworms on metal and metalloid mobility and availability. J Environ Monit 13(2):266–273
Sobariu DL, Fertu DIT, Diaconu M, Pavel LV, Hlihor RM, Drăgoi EN, Curteanu S, Lenz M, Corvini PFX, Gavrilescu M (2016) Rhizobacteria and plant symbiosis in heavy metal uptake and its implications for soil bioremediation. New Biotechnol
Soobhany N, Mohee R, Garg VK (2015) Comparative assessment of heavy metals content during the composting and vermicomposting of municipal solid waste employing Eudrilus eugeniae. Waste Manag 39:130–145
Stewart EJ (2012) Growing unculturable bacteria. J Bacteriol 194:4151–4160
Stürzenbaum SR, Georgiev O, Morgan AJ, Kille P (2004) Cadmium detoxification in earthworms: from genes to cells. Environ Sci Technol 38:6283–6289
Sumi H, Kunito T, Ishikawa Y, Sato T, Park HD, Nagaoka K, Aikawa Y (2015) Plant roots influence microbial activities as well as cadmium and zinc fractions in metal-contaminated soil. Chem Ecol 31:105–110
Sun Z, Chen J, Wang X, Lv C (2016) Heavy metal accumulation in native plants at a metallurgy waste site in rural areas of northern China. Ecol Eng 86:60–68
Talebi AF, Tabatabaei M, Mohtashami SK, Tohidfar M, Moradi F (2013) Comparative salt stress study on intracellular ion concentration in marine and salt-adapted freshwater strains of microalgae. Not Sci Biol 5(3):309–315
Tomanek L (2014) Proteomics to study adaptations in marine organisms to environmental stress. J Proteome 105:92–106. https://doi.org/10.1016/j.jprot.2014.04.009
Tomar P, Suthar S (2011) Urban wastewater treatment using vermi-biofiltration system. Desalination 282:95–103
Tunali S, Cabuk A, Akar T (2006) Removal of lead and copper ions from aqueous solutions by bacterial strain isolated from soil. Chem Eng J 115(3):203–211
Ullah A, Mushtaq H, Ali H, Munis MFH, Javed MT, Chaudhary HJ (2015) Diazotrophs-assisted phytoremediation of heavy metals: a novel approach. Environ Sci Pollut Res 22:2505–2514
van Dijk EL, Auger H, Jaszczyszyn Y, Thermes C (2014) Ten years of next-generation sequencing technology. Trends Genet 30:418–426
van Gestel CAM, Koolhaas JE, Hamers T, van Hoppe M, van Roovert M, Korsman C, Reinecke SA (2009) Effects of metal pollution on earthworm communities in a contaminated floodplain area: linking biomarker, community and functional responses. Environ Pollut 157(3):895–903
Vido K, Spector D, Lagniel G, Lopez S, Toledano MB, Labarre J (2001) A proteome analysis of the cadmium response in Saccharomyces cerevisiae. J Biol Chem 276(11):8469–8474
Visioli G, Marmiroli N (2013) The proteomics of heavy metal hyperaccumulation by plants. J Proteome 79:133–145
Volesky B (2007) Biosorption and me. Water Res 41(18):4017–4029
Volesky B, Holan ZR (1995) Biosorption of heavy metals. Biotechnol Prog 11:235–250. 40
Wang D, Li H (2006) Effect of earthworms on the phytoremediation of zinc-polluted soil by ryegrass and Indian mustard. Biol Fertil Soils 43:120–123
Wang W, Deng Z, Tan H, Cao L (2013) Effects of Cd, Pb, Zn, Cu-resistant endophytic Enterobacter sp. CBSB1 and Rhodotorula sp. CBSB79 on the growth and phytoextraction of Brassica plants in multimetal contaminated soils. Inter J Phytoremed 15:488–497
Wang F, Ji R, Jiang ZW, Chen W (2014) Species-dependent effects of biochar amendment on bioaccumulation of atrazine in earthworms. Environ Pollut 186:241–247
Wang HQ, Zhao Q, Zeng DH, YL H, ZY Y (2015) Remediation of a magnesium -contaminated soil by chemical amendments and leaching. Land Degrad Dev 26(6):613–619. https://doi.org/10.1002/ldr.2362
Watanabe M, Meyer KA, Jackson TM, Schock TB, Johnson WE, Bearden DW (2015) Application of NMR-based metabolomics for environmental assessment in the Great Lakes using zebra mussel (Dreissena polymorpha). Metabolomics. https://doi.org/10.1007/s11306-015-0789-4

Wei SH, Zhou QX, Wang X (2005) Cadmium-hyperaccumulator Solanum nigrum L. and its accumulating characteristics. Environ Sci 26(3):167–171

Wenzel WW (2009) Rhizosphere processes and management in plant-assisted bioremediation (phytoremediation) of soils. Plant Soil 321:385–408

Weyens N, van der Lelie D, Taghavi S, Newman L, Vangronsveld J (2009) Exploiting plantemicrobe partnerships to improve biomass production and remediation. Trends Biotechnol 27:591–598

Wijnhoven S, Leuven RSEW, Van Der Velde G, Jungheim G, Koelemij EI, deVries FT, Eijsackers HJP, Smits AJM (2007) Heavy-metal concentrations in small mammals from a diffusely polluted floodplain: importance of species-and location-specific characteristics. Arch Environ Contam Toxicol 52:603–613. https://doi.org/10.1007/s00244-006-0124-1

Wood JL, Tang C, Franks AE (2016) Microbial associated plant growth and heavy metal accumulation to improve phytoextraction of contaminated soils. Soil Biol Biochem 103:131–137

Xiao X, Luo S, Zeng G, Wei W, Wan Y, Chen L, Guo H, Cao Z, Yang L, Chen J et al (2010) Biosorpiton of cadmium by endophytic fungus (EF) Microsphaeropsis sp. LSE10 isolated from cadmium hyperaccumulator Solanum nigrum L. Bioresour Technol 101:1668–1674

Yu H, Ni SJ, He ZW, Zhang CJ, Nan X, Kong B, Weng ZY (2014) Analysis of the spatial relationship between heavy metals in soil and human activities based on landscape geochemical interpretation. J Geochem Explor 146:136–148. https://doi.org/10.1016/j.gexplo.2014.08.010

Yuan M, He H, Xiao L, Zhong T, Liu H, Li S, Deng P, Ye Z, Jing Y (2014) Enhancement of Cd phytoextraction by two Amaranthus species with endophytic Rahnella sp. JN27. Chemosphere 103:99–104

Zargar SM, Mahajan R, Nazir M, Nagar P, Kim ST, Rai V, Masi A, Ahmad SM, Shah RA, Ganai NA, Agrawal GK (2017) Common bean proteomics: present status and future strategies. J Proteome. Mar 25

Zeraatkar AK, Ahmadzadeh H, Talebi AF, Moheimani NR, McHenry MP (2016) Potential use of algae for heavy metal bioremediation, a critical review. J Environ Manag 181:817–831

Zhai R, Su S, Lu X, Liao R, Ge X, He M, Huang Y, Mai S, Lu X, Christiani D (2005) Proteomic profiling in the sera of workers occupationally exposed to arsenic and lead: identification of potential biomarkers. Biometals 18(6):603–613

Chapter 10
Quantitative Structure-Activity Modelling of Toxic Compounds

Raghunath Satpathy

Abstract Continuous exposure of living organisms to toxic compounds is a major health issue. Studying the effects of toxic compounds is a difficult task because compounds are present at trace levels in complex media with other toxic compounds. Toxicity evaluation by animal testing is long and costly. Therefore, this chapter reviews an alternative method of toxicity evaluation, named quantitative structure-activity relationship (QSAR) modelling, which is used to predict the acute toxicity of substances. The principle is that the molecular structure is correlated with toxicological effects. The characteristics of toxic compounds are computed and correlated using software tools and databases. Biodegradation features and classification methods are discussed. Various computational tools and databases are presented. This review also presents the discipline of bionformatics, to study risk, toxicity and biodegradability.

Keywords Biodegradation • Computational tools • Database • Descriptors • Environmental toxicity • Quantitative structure-activity relationship modeling • Validation methods • Risk assessment • Toxic compounds

Contents

R. Satpathy (✉)
Department of Biotechnology, MITS Engineering College, Rayagada, Odisha, India
e-mail: rnsatpathy@gmail.com

K M Gothandam et al. (eds.), *Nanotechnology, Food Security and Water Treatment*, Environmental Chemistry for a Sustainable World 11,
https://doi.org/10.1007/978-3-319-70166-0_10

10.1 Introduction

The diversity in the chemistry of the several environmental toxic compounds is one of the important consideration to study their toxicity (Raies et al. 2016; Chen et al. 2013; Basant et al. 2016). Some of these toxic substances are extensively used in day to day life as well as many industrial processes as solvents, as precursors for another compound such as dyes, plasticizers, pharmaceuticals, pesticides etc. Every year, millions of tons of chemicals are released into the environment (Settivari et al. 2015; Awaleh and Soubaneh 2014; Perocco et al. 1983). In order to protect human health and the environment, risk assessments should be conducted for all toxic chemicals, especially for those already found in the environment. Several tons of these toxic compounds produced from industry and their accumulation in the environmental habitats represents a global ecological danger. Also, many of these substances have also been reported to be persistent in nature (Halling-Sørensen et al. 1998; Battaglin and Kolpin 2009; Li et al. 2011; Luo et al. 2015; Satpathy et al. 2015a, b). Toxicity tests are usually used to predict the harmful effect of these compounds by considering the animal models (Raies et al. 2016). However, this is a time-consuming process also it is not economically feasible to conduct toxicity tests on each chemical individually and also difficult to study the combinatorial effects of the toxic substances on organisms. Therefore, in the field of environmental toxicology and environmental chemistry, Quantitative Structure-Activity Relationships (QSARs) is currently used to predict toxicity for a number of toxic chemicals (Doke and Dhawale 2015; Raunio 2011; Boobis et al. 2002).

Quantitative structure-activity relationships (QSARs) referred as a correlation of the chemical structure to a wide variety of physical, chemical, biological (including biological, toxicological, ecotoxicological) properties. Once, the suitable correlations are established and validated, the same can be used to predict properties of other similar unknown compounds, for that the properties are not yet measured (Fig. 10.1). The QSAR process depends on the principle that molecular structures are internal factors are responsible for the physicochemical properties, environmental behavior, and ecotoxicology of organic compounds.

One of the assumptions is to be considered in QSAR analysis is that those chemical compounds share the similar structures must have the similar physicochemical properties thereby exhibiting similar toxicological effects on the environment. The relations can be represented as mathematical models, termed quantitative structure-activity relationships models. Traditional Quantitative structure-activity relationships are regression models between the biological activities called as the dependent variable and structure-related *descriptors* (numerical presentation of

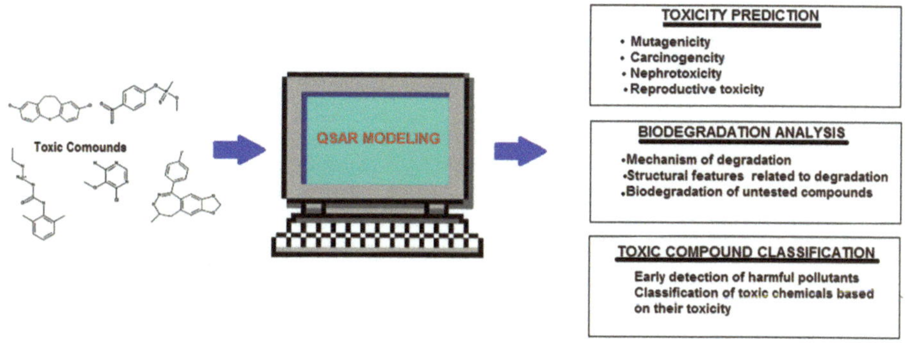

Fig. 10.1 Showing the importance of quantitative structure activity relationship based analysis on toxic compounds. Three major categories of characterization can be made from the analysis

molecular features) called as independent variables. By a proper choice of the variables and an adequate data analysis, QSARs or similar models may account for exposure-related chemical properties as well as for correlations among responses in different test systems. Furthermore, by a proper selection of the training compounds, QSARs can cover large series of similar compounds and thus provide the basis for an efficient and comparative screening assessment of toxic substances. (Nendza et al. 1991; Tunkel et al. 2005; Hulzebos et al. 2005).

One of the important features of QSAR study is the physical chemistry lies at the heart of all biological activity, and it is a reflection of that fact that so many physicochemical parameters have been devised and used in QSAR studies. So, Knowledge about different types of toxicity is essential in *environmental risk assessment*, a study that includes acute toxicity, chronic toxicity and environmental fate and behavior of a chemical substance. The quantitative evaluation of the toxicity of compounds enables to derive potential models in the different environment such as soil, air, water, and sediment.

10.2 Quantitative Structure Activity Relationship Methods

The foremost important thing in QSAR methods is to correlate the molecular property of the toxic compounds with the other biological and environmental property. The molecular properties are coded by the representation of descriptors. The molecular descriptor is computed by a logical and mathematical method that transduces the chemical information of molecule in a numerical form as shown and explained in Tables 10.1 and 10.2.

The QSAR method begins by taking a suitable group of compounds known as training data set. The required molecular features of a suitable training set are data

Table 10.1 Common categories of descriptors derived from the toxic molecules for quantitative activity relationship analysis

Sl No.	Types of descriptors	Explanation	Common examples
1	Constitutional	Based on atomic constitution of the chemical structure (molecule)	Total number of atoms and bonds in the molecule, numbers of atoms of certain chemical identity (Carbon, Hydrogen, Oxygen, Nitrogen, etc.) in the molecule, numbers of single, double, triple, aromatic or other bonds in the molecule, total number of rings
2	Topological descriptors	Numbers that give information about the bonding arrangement in a molecule.	Weiner index, Randic indices, Kier and Hall indices, connectivity index and so on
3	Electrostatic descriptors	Single values that give information about the molecular charge distribution	Polarity indices, multipoles, and polarizability, topological polar surface area
4	Geometrical descriptors	Values that describe the molecule's size and shape	Moments of inertia, molecular volume, molecular surface area, and other parameters that describe length, height, and width
5	Quantum chemical	Give information about the electronic structure of the molecule	Highest occupied molecular orbital and highest occupied molecular orbital energies, refractivity, total energy, ionization potential, electron affinity, and energy of protonation

Compararive analysis among different types of descriptors are usually helpful for better prediction of toxicity

Table 10.2 Types of biological and environmental properties of the toxic compounds

Sl No.	Properties of compounds	Explanation	Common examples
1	Biological properties	Properties related to biological activities	Binding affinity, lethal dose, inhibition concentration, mutagenicity, carcinogenicity, anti-inflammatory activity, antidepressant activity, skin sensitization and so on
2	Environmental properties	Parameters associated with environmental fate of toxic pollutants	Biodegradation, bio-concentration, Biochemical oxygen demand, Chemical oxygen demand, half -life time, mobility, atmospheric persistence

may be taken from literature, various databases and sometimes computed from by the sophisticated software tools. The parameter to be studied and correlated with the molecular features may be either a physicochemical property, such as water solubility, vapor pressure, Henry Law constant or octanol/water distribution coefficient (log P) or a biological activity, for instance, acute toxicity values, bioaccumulation

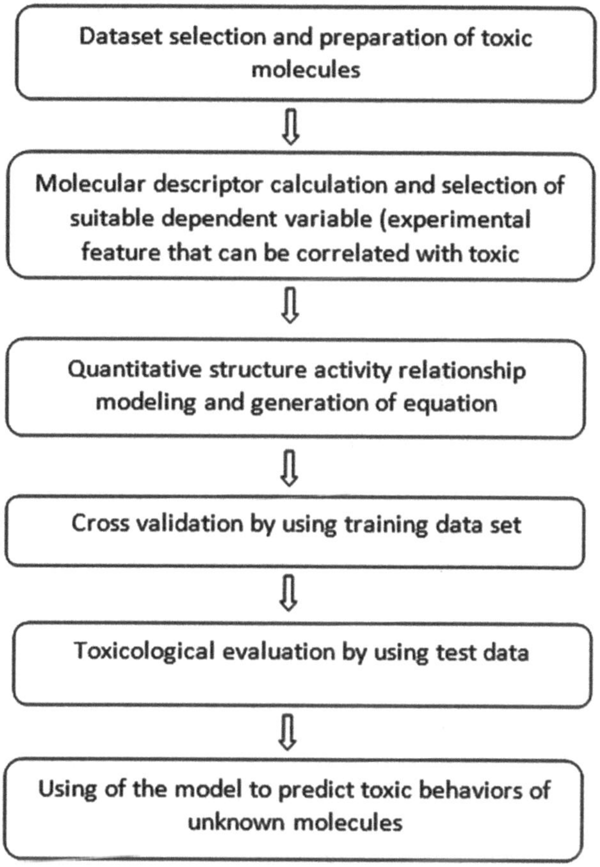

Fig. 10.2 Steps of quantitative structure activity relationship modeling methods for prediction of toxic compounds. After end of the analysis an equation and plot is obtained to represent the relationship between toxicity and molecular feature of the compound

or biodegradation data. After proper selection of training set, the actual QSAR modeling starts. The modeling procedure is to statistically correlate the molecular feature along with the biological/ environmental properties of consideration. (Fig. 10.2).

The correlations among both dependent and independent variables are most commonly expressed by a linear equation. For examples, if the molecular properties (independent variables) are x, y, z. . ., and the desired activity (dependent variable) of study is A.

For compound i, the relationship can be written as

$$A_i = a\,x_i + b\,y_i + c\,z_i + \beta$$

Where a, b, and c are the expression of the correlation of the particular molecular property with the activity of the compound (A_i), and β is a constant. The correlation in the above equation is most frequently calculated using various statistical methods. The most common QSAR models have been developed to predict the environmental toxicity is based on the physicochemical parameters of the toxic compounds. Various common physicochemical parameters and variables considered for toxicity analysis are the size and structure, water solubility, vapor pressure, octanol/water partition coefficient (corresponds to lipophilicity) and adsorption coefficient etc. QSARs have been developed with acceptable results for most industrial chemicals. In the case of different types of toxic compounds different dependent variable, parameters have been analysed in QSAR analysis by many researchers (Table 10.3).

10.3 Quantitative Structure Activity Relationship Validation Methods

To characterize the established QSAR models, the quality of goodness-of-fit (Kolossov and Stanforth 2007; Tropsha et al. 2003). The traditional parameters used for the measure of goodness-of-fit are as follows:

(a) Determination coefficient (R^2) or adjusted determination coefficient (Radj2): R ^{2} is a measure of the quality of fit between model-predicted and experimental values. But due to over-fitting criteria, obtaining of high R^2 values resulting from low degrees of freedom may lead to a predict a weak QSAR model. Therefore, the adjusted R^2 by the freedom degree (Radj2) should be adopted. The higher the Radj2 value, the better is the goodness-of-fit.
(b) Summary square error (SSE): This value reflects the deviation of predicted values from observed values, and depends on the number of data points.
(c) Root mean square error (RMSE), mean absolute residual (MAR), and standard error (SE) or standard deviation (SD) are commonly used to indicate the precision of prediction.
(d) F-test: It is a variance test method of the overall significance level and is only applicable to QSAR models derived basically from multivariate linear regression (MLR).

In addition to this, the overfitting and underfitting of QSAR models should be further accessed by the stability analysis of the model (Eriksson et al. 2003; Baumann et al. 2002; Wehrens et al. 2000; Burden et al. 2000). Usually, model instability is analyzed, by figuring out the wrong set of variables selected in an experiment. Many of the commonly used cross-validation methods are used such as leave-one-out cross-validation (LOO-CV) and bootstrapping. An alternative method is to split the original dataset into a training set, used for establishing QSAR models, a test set, and a set for external validation. The objective of using this method is to ensure that in the case of all these types of data set the evaluation parameter should not be too dissimilar.

Table 10.3 Earlier literature study about QSAR analysis to predict the toxicity of different types of toxic chemicals

Sl No.	Toxic Compounds	Dependent variable parameter	References
1	Phenols and thiophenols	EC50	Ghamali et al. (2017)
2	Nitroaromatics compounds	IGC_{50}	Artemenko et al. (2011)
3	Non-polar narcotic	EC50	Aruoja et al. (2014)
4	1-(3,4-dichlorophenyl)-3-methlyurea (DCPMU), 3-(3-chlorophenyl)-1,1-dimethylurea (MCPDMU), and 1-(3,4-dichlorophenyl)urea (DCPU)	EC50	Neuwoehner et al. (2010)
5	Poly-substituted benzenes	EC50	Netzeva et al. (2004)
6	Nitrobenzene	EC50	Altenburger et al. (2005)
7	Alcohol ethoxylate surfactants	EC50	Wong et al. (1997)
8	Normal and branched alkanes, alkylbenzenes, polyaromatic hydrocarbons, alkanols, polyols, phenylcarbinols, aliphatic primary amines etc	P_{OW}	Chicu et al. (2000)
9	Linear alkyl benzene sulphonates and ester sulphonates	EC50	Hodges et al. (2006)
10	Carboxylic acids	IGC50	Seward and Schultz (1999)
11	Anilines and phenols	Kow	Damborsky and Schultz (1997)
13	Organophosphorus pesticides	LC_{50}	de Bruijn and Hermens (1993)

Major dependent variables were used is EC_{50}
Note: **EC50= effective concentration 50**, refers to the concentration of a toxic substances, that indices 50 % of mortality in cells after a specified exposure time ; **IGC50=50% inhibition growth concentration,** *IGC 50* stands for Inhibited the Growth of Cells by 50% ; **Kow and Pow= n-octanol-water partition coefficients**, is widely used widely used property for assessing the partitioning behaviour of chemicals in the environment to estimate the fate, behaviour and effects of toxic chemicals in the environment; **LC 50= lethal concentration 50,** LC_{50} value is the concentration of a material in air that will kill *50%* of the test subjects (animals, typically mice or rats) when administered as a single exposure (typically 1 or 4 h)

10.4 Computational Tools Used in Quantitative Structure Activity Relationship Study of Toxic Compounds

Currently, more and more studies have been carried out by utilizing computational programs to predict the toxicity of the hazardous chemical compounds. The main challenge is to discover the novel chemical descriptors, new algorithms, and statistical perspectives for classification purpose. Another factor for using computer programs is to calculate toxicity of compounds as for many toxic compounds, the experimental value for many toxic compounds is not available. The basis for performing a quantitative structure-activity relationship (QSAR) model is that a chosen toxic property of chemicals can be described in relation to its features of the chemical compound that is described by using certain parameters. Therefore, implementation of a suitable modeling method is required that include a good mathematical algorithm. However, in using the same algorithm with a chemical descriptor calculated using different programs, it is likely that variation in results may obtain (Benfenati 2007). Another potential use of computational methods, this is used as an alternative to the *in vitro* and *in vivo* toxicity tests, because that require animal testing, are a high-cost and time-consuming process. In addition to this, these *in silico* methods are able to predict the toxicity feature of the molecules even before they are chemically (industrially) synthesized (Madan et al. 2013).

QSAR based toxicology research utilizes a wide variety of computational tools (Pirhadi et al. 2016; Liao et al. 2011), such as databases for storing data about chemicals, their toxicity, and chemical properties, software for generating molecular descriptors and simulation tools to generate the QSAR equation and validation (Tables 10.4, 10.5 and 10.6). However, the good predictive models for toxicity parameters depend crucially on selecting the right mathematical approach, the right molecular descriptors for the particular toxicity endpoint.

10.5 Applications of Quantitative Structure Activity Relationship Analysis of Environmental Toxic Compounds

In the field of environmental toxicology, quantitative structure-activity relationships (QSARs) methods have been used as robust tools for predicting the toxicity of chemicals whenever there is no or little amount of data are available. As per the statistics, there are more than one million toxic chemicals are exposed to the environment throughout the world but among them about few as 1000–5000 compounds toxicity data are available. Also the industrial point of view, some high volume producing compounds having a risk to induce toxicity. However, it is often difficult to determine whether or not a chemical possesses a particular mechanism of the toxic action, can be solved by QSAR analysis. A basic and fundamental understanding of toxicological principles has been considered crucial to the continued acceptance and application of these techniques as biologically relevant (Gopi Mohan et al. 2007; Pölloth and Mangelsdof 1997). As a consequence, many novel QSAR methods have been developed and implemented to deduce the consistent with assumptions regarding modes of toxic action of several toxic compounds. Thus, in this way, the applicability of a QSAR model will help in the understanding of both toxic mechanisms and the critical structural

Table 10.4 Software tools and servers details for calculating variables in toxicity prediction in case of compounds

Sl. No.	Software	Availability	Application
1	ADMET Predictor	http://www.simulations-plus.com/	Quantitative prediction of oestrogen receptor toxicity in rats.
2	ACD ToxSuite (ToxBoxes)	http://www.acdlabs.com/products/admet/tox/	Prediction of Endoplasmic Reticulum (ER) binding affinity prediction.
3	CAESAR	http://www.lhasalimited.org/	Two classification models for developmental toxicity
4	Derek	https://www.lhasalimited.org/	Different levels of classification models (based on developmental toxicity
5	Leadscope	http://www.leadscope.com/	Classification models for developmental toxicity in the rodent fetus
6	MolCode Toolbox	http://molcode.com/	Quantitative prediction of rat ER binding affinity and AhR binding affinity
7	MultiCASE	http://www.multicase.com/	Classification models for developmental toxicity associated with drugs
8	OSIRIS property explorer	http://www.organic-chemistry.org/prog/peo/	Predicts mutagenicity, tumorigenicity, irritating effects and reproductive effects
9	PASS	http://ibmc.p450.ru/PASS//	Classification models giving the probability of reproductive toxic effects.
10	T.E.S.T.: The Toxicity Estimation Software Tool	http://oasis-lmc.org/	Developmental toxicity estimation.
11	TOPKAT	http://www.accelrys.com	Classification model for developmental toxicity of pesticides, industrial chemical
12	Toxboxespharma algorithms	http://pharma-algorithms.com/tox_boxes.htm	A classification model for the prediction of ER binding.
13	VirtualToxLa	http://www.biograf.ch	Classification model for endocrine disrupting potential
14	HAZARD EXPERT	http://www.compudrug.com/hazardexpertpro	Human carcinogenicity and genotoxicity prediction
15	Toxline	https://toxnet.nlm.nih.gov/cgi-bin/sis/htmlgen?TOXLINE	Human neurotoxicity prediction
16	BCABAF	https://www.epa.gov/tsca-screening-tools/epi-suitetm-estimation-program-interface	Prediction of bio- concentration of toxic substances
17	PCKOCWIN	cpas.mtu.edu/cencitt/oppt/tsld019.htm	Prediction of soil sorption with the toxic chemicals

(continued)

Table 10.4 (continued)

Sl. No.	Software	Availability	Application
18	BIOWIN	envirosim.com/products/biowin	Prediction of biodegradability of toxic substances
19	KOWWIN	https://www.epa.gov/tsca-screening-tools/epi-suitetm-estimation-program-interface	Estimates octanol-water partition coefficient of toxic chemicals
20	AMBIT	http://ambit.sourceforge.net/intro.html	Chemical structure search, experimental data and predictive model can be obtained
21	PreADMET	https://preadmet.bmdrc.kr/	A web-based application for predicting ADME data and also toxicity prediction

The dependent variables are to be obtained from the below computational tools are used for evaluation of toxicity of compounds

characteristics of a chemical. The various potential applications of the QSAR analysis with respect to the toxicity of chemical compounds are presented below.

10.5.1 Prediction of Toxicity

As stated above, the biochemical behavior of a molecule is directly associated with its molecular structure as well as the chemical properties. The establishment of a relationship between the structures and chemical properties of molecules and the toxicities, the QSAR models can be used to predict the toxicity of analogous chemicals. Also, it provides much significant information for the design and modification of new molecules. The QSAR analysis has been used to predict a specific chemical class that behaves in a toxicologically similar manner (Pavan et al. 2008; Chen et al. 2004; Li et al. 2006). In the case of human being, the prediction of the acute toxicity of a compound is an important task in order to justify the in the regulatory assessment of particular compounds. However, mostly this information is obtained from the animal studies that is related to animal ethics and cost considerations. Therefore, the alternative method alternatives to animal experiments are preferable (Lapenna et al. 2010; Raies and Bajic 2016). As a powerful technique, the QSAR methods have been widely applied in toxicology by many researchers as described below. Cronin et al. (2003) emphasized about the application of QSAR techniques to predict ecologic effects and environmental fate of chemicals for facilitating the regulatory agencies and authorities will find them to be acceptable alternatives to chemical testing. Roberts, D. W. (1991) studied the acute lethal toxicity data for a range of anionic and non-ionic surfactants by QSAR modeling to predict by calculated log *P* (octanol/water) values. Similarly, the acute aquatic toxicity of reactive inorganic compounds on have been reviewed by (Cronin

Table 10.5 Describes about major 9 types of databases to compute the descriptor calculation in case of toxic compounds

Sl. No.	Databases	Availability	Application
1	Toxicology Data Network (Toxnet) Developmental and Reproductive Toxicology Database (DART)	http://toxnet.nlm.nih.gov/cgibin/sis/htmlgen?DARTETIC	Contains references to the aspects of developmental and reproductive toxicology.
2	Endocrine Disruptor Knowledge Base (EDKB)	http://www.fda.gov/ScienceResearch/BioinformaticsTools/EndocrineDisruptorKnowledgebase	Contains in vitro and in vivo experimental data for more than 3000 chemicals
3	Endocrine Active Substances information system (EASIC)	https://eurl-ecvam.jrc.ec.europa.eu/databases/eas_database	Searchable database giving information chemical structure, toxicity
4	NureXbase	http://www.nursa.org	Information on chemical structure, crystal structure, physical descriptors, nuclear receptors and mechanism of endocrine action
5	OECD (Q)SAR Toolbox	http://www.oecd.org/env/ehs/riskassessment/theoecdqsartoolbox.htm	Contains several databases, including reproductive toxicity data
6	Acute Toxicity Database	https://www.cerc.usgs.gov/data/acute/acute.html	Database of aquatic acute toxicity test results for thousands of chemicals
7	Distributed Structure-Searchable Toxicity (DSSTox) Database	https://www.epa.gov/chemical-research/distributed-structure-searchable-toxicity-dsstox-database	Resource for supporting improved predictive toxicology.
8	TerraBase	http://www.terrabase-inc.com/	Provide for the quick search of compounds with specific biological effects and properties
9	EXTOXNET	http://extoxnet.orst.edu/pips/ghindex.html	Pesticide information including experimental toxicity value

These resources are freely available on web to facilitate the researchers for toxicity study

and Dearden 1995) and suggested about the importance of validity, quality and quantity of toxicological data to fit the model. The toxicity prediction about active ingredients in pharmaceutical products and their importance and mechanism has been reviewed by (Kruhlak et al. 2007). The successful prediction of genotoxicity of the compounds like 2-amino-9H-pyrido[2,3-b]indole (AαC) and

Table 10.6 Standard list of softwatre tools details used for QSAR analysis

Sl. No.	Simulation software	Availability
1	Molecular Operating Environment (MOE)	https://www.chemcomp.com/MOE-Cheminformatics_and_QSAR.htm
2	BIOVIA QSAR Workbench	http://accelrys.com/products/collaborative-science/biovia-qsar-workbench/
3	VEGAHUB	https://www.vegahub.eu/
4	WEKA	http://www.cs.waikato.ac.nz/ml/weka/
5	KNIME	https://www.knime.org
6	BuildQSAR	http://www.profanderson.net/files/buildqsar.php
7	Orange	http://orange.biolab.si/
8	Rapid Miner	https://rapidminer.com/
9	MALLET	http://mallet.cs.umass.edu/
10	R	www.r-project.org

The programs are basically data mining tools involves in classification, clustering, modeling, validation of the model

2-aminoacetophenone (2-AAP) by QSAR have been studied by (Worth et al. 2013). Comparative evaluation and prediction of mammalian acute toxicity, by considering lethal dose (LD_{50}) as a dependent variable has been studied by GonellaDiaza et al. 2015) in a dataset of 7417 toxic chemical compounds. Also, the prediction for *no observed effect level* (NOEL), developmental and reproductive toxicities have been successfully predicted by Hisaki et al. (2015) from total 892 numbers of toxic chemicals.

10.5.2 Biodegradation Analysis

Another important application of QSAR analysis is known as Quantitative Structure Biodegradability Relationship (QSBR) model to the study of mechanisms of degradation of toxic chemicals. This is an interesting approach to understand the chemical structure of toxic pollutant molecules and their biodegradability as there is no direct relationship among them (Philipp et al. 2007; Yin and XI 2007). Also, different classes of chemicals are likely to have different mechanisms of biodegradation, it would be expected that the biodegradability of groups of diverse chemicals would be difficult to model using QSAR techniques (Dearden 2002). A QSBR model for microbial degradation of substances usually performed by considering the similar features toxic molecule same as that of classical QSAR study but the selection of the dependent variable which is the important task (Zhang et al. 2006). Molecular connectivity indices were taken and studied by Okey and Stensel (1993) to obtain the direct relationship between biodegradation of substances in an activated sludge. Eriksson et al. (1995) studied about multivariate quantitative structure-biodegradability relationships (QSBRs) were developed for a series of

20 halogenated aliphatic hydrocarbons investigated for their microbial bio dehalogenation (expressed as half-lives) can be further extrapolated to predict bio dehalogenation properties for yet untested compounds. Lu et al. (2003) studied about the biodegradation of substituted phenols and benzoic acids by QSAR methods. In the study, the quantum mechanical descriptors were found suitable whenever linear regression method was implemented in the modeling method. Yang et al. (2006) predicted about the features (descriptors) about the 46 types of aromatic compounds that biodegraded anaerobically. Similarly, the aerobic biodegradability of chlorinated aromatic compounds has been analyzed by Liu and Feng (2012) and the fundamental structural parameters of the compounds are related to the biodegradation was obtained from their study. Biodegradation rate constants that are correlated better with the quantum chemical descriptors of the halide substituent are able to predict the activity and possibly enzyme binding features of the halogenated compounds have been described by Satpathy et al. (2015a, b).

10.5.3 Classification of Toxic Compounds

Many of the toxic compounds have been identified only after human exposure and create health hazards. Therefore early detection and classification of such chemicals are required to reduce the risk of exposure to developmental hazards (Gomba et al. 1995; Sussman et al. 2003). Another application by using the QSAR analysis it is possible to classify the toxic chemicals based on their mode of action, the study is known as predictive toxicology. But the prediction is constant for a particular variables or organisms considered for the prediction (Nendza and Wenzel 2006). Similarly out of hundreds of compounds, it is also possible to assign a class to a particular class of toxic chemicals, has been studied by many researchers (von der et al. 2005; Lin et al. 2003). Verhaar et al. (1992) presented a scheme that classifies a large number of organic pollutants into one of four classes such as inert chemicals, less inert chemicals, reactive chemicals and specifically acting chemicals by QSAR analysis. Subsequent classification of toxic compounds by Verhaar (1994) estimated the effect of reactive compounds specifically on the toxicity of aquatic biota. Vaal et al. (1997), studied and classified the chemicals as non-polar narcotics, polar narcotics, reactive compounds in terms of acute toxicity data for aquatic species. Similarly, binary classification of toxic compounds into nephrotoxin versus non-nephrotoxin has been classified by QSAR methods have been described by Lee et al. (2013). In a recent study by Ghorbanzadeh et al. (2016), a novel binary QSAR classification models were developed and validated, based on the OECD QSAR validation principles, to discriminate developmental toxic compounds from non-toxic ones in zebrafish. To discriminate between baseline and excess toxicants in the case of fish acute toxicity a scheme based on the physicochemical property proposed by Nendza et al. (2017).

10.6 Advantage of Quantitative Structure Activity Relationship Based Study on Toxic Substances

The QSAR based study in toxic compounds having a lot of advantages such as summarized below:

- Prediction of the environmental fate of the toxic compounds such as bio-concentration, soil sorption and biodegradation and so on.
- Since the prediction methods are computer based, therefore they provide a rapid assessment of toxicity of these compounds.
- Further, they have the capability of reducing, and even replacing, animal tests for toxicological assessment of the pollutant compounds.
- Industrial users can apply these models to screen new compounds and to assist in the process of designing out toxic features of new chemical entities, may be this information can be used by the regulatory agencies use, and helps in impose the regulation of new and existing chemical compounds.
- Prediction of toxicity can be applied to environmental risk assessments for common pollutants.

10.7 Challenges in Quantitative Structure Activity Relationship Based Modeling on Toxic Compounds

Although QSAR based methods having enormous potential for analyzing the toxic profile of compounds, however, certain challenges should be overcome.

- **Problems in Biological dataset**

  The foremost important thing in QSAR analysis is the data. One of the limitation is very little amount biochemical data is available in terms of mechanism of toxic action. Therefore, for validation purpose, it faces a problem thereby causing inconsistency in prediction.

- **Better selection of dependent and independent variable and domain applicability**

  Predictions of toxicity should be made within the domain of applicability of an appropriately validated QSAR. An appropriate choosing of descriptor variables and dependent variable lead to a good prediction of the model. Also, the number of independent variable in case of specific chemicals is important for model generation and prediction.

- **Variability in toxicity action of compounds**

  Usually, it is expected that similar toxic chemicals possess similar mechanism of action in toxicity, but if any of the compounds that do not possess the same mechanism of action will show up as outliers; that is, they will not be well modeled by the QSAR.

10.8 Conclusion

The development of models for quantitative structure-activity relationships (SARs) and its application for toxicological effects are of great importance. Since a million numbers of chemicals from industry and other sources are released and exposed to the environment, but little is known about the toxicity of them. It would be difficult, expensive and time-consuming to test all such chemicals for toxicity. Due to this, QSAR methods have been used to interpret the toxicity also many regulatory agencies are beginning to accept toxicities predicted by QSAR. The basic principles behind QSAR analysis are a prediction of biological activities from chemical structures that are closely related. However while performing the analysis many of the aspects are to be considered such as avoiding the false correlation of the data and perfect experimental design. Therefore validation process for the experiment must be observed carefully.

References

Altenburger R, Schmitt H, Schüürmann G (2005) Algal toxicity of nitrobenzenes: combined effect analysis as a pharmacological probe for similar modes of interaction. Environ Toxicol Chem 24(2):324–333. https://doi.org/10.1897/04-032R.1

Artemenko AG, Muratov EN, Kuz'min VE, Muratov NN, Varlamova EV, Kuz'mina AV et al (2011) QSAR analysis of the toxicity of nitroaromatics in Tetrahymena pyriformis: structural factors and possible modes of action. SAR QSAR Environ Res 22(5–6):575–601. https://doi.org/10.1080/1062936X.2011.569950

Aruoja V, Moosus M, Kahru A, Sihtmäe M, Maran U (2014) Measurement of baseline toxicity and QSAR analysis of 50 non-polar and 58 polar narcotic chemicals for the alga Pseudokirchneriella subcapitata. Chemosphere 96:23–32. https://doi.org/10.1016/j.chemosphere.2013.06.088

Awaleh MO, Soubaneh YD (2014) Waste water treatment in chemical industries: the concept and current technologies. Hydrol Current Res 5(1):1–12. https://doi.org/10.4172/2157-7587.1000164

Basant N, Gupta S, Singh KP (2016) In silico prediction of the developmental toxicity of diverse organic chemicals in rodents for regulatory purposes. Toxicol Res 5(3):773–787. https://doi.org/10.1039/C5TX00493D

Battaglin WA, Kolpin DW (2009) Contaminants of emerging concern: introduction to a featured collection1. JAWRA 45(1):1–3. https://doi.org/10.1111/j.1752-1688.2008.00299.x

Baumann K, Von Korff M, Albert H (2002) A systematic evaluation of the benefits and hazards of variable selection in latent variable regression. Part II practical applications. J Chemom 16 (7):351–360. https://doi.org/10.1002/cem.730

Benfenati E (2007) Predicting toxicity through computers: a changing world. Chem Cent J 1(1):32. https://doi.org/10.1186/1752-153X-1-32

Boobis A, Gundert-Remy U, Kremers P, Macheras P, Pelkonen O (2002) In silico prediction of ADME and pharmacokinetics: report of an expert meeting organised by COST B15. Eur J Pharm Sci 17(4):183–193. https://doi.org/10.1016/S0928-0987(02)00185-9

Burden FR, Ford MG, Whitley DC, Winkler DA (2000) Use of automatic relevance determination in QSAR studies using Bayesian neural networks. J Chem Inf Comput Sci 40(6):1423–1430. https://doi.org/10.1021/ci000450a

Chen J, Harner T, Ding G, Quan X, Schramm KW, Kettrup A (2004) Universal predictive models on octanol-air partition coefficients at different temperatures for persistent organic pollutants. Environ Toxicol Chem 23(10):2309–2317. https://doi.org/10.1897/03-341

Chen L, Lu J, Zhang J, Feng KR, Zheng MY, Cai YD (2013) Predicting chemical toxicity effects based on chemical-chemical interactions. PLoS One 8(2):e56517. https://doi.org/10.1371/journal.pone.0056517

Chicu SA, Herrmann K, Berking S (2000) An approach to calculate the toxicity of simple organic molecules on the basis of QSAR analysis in Hydractinia echinata (Hydrozoa, Cnidaria). Quant Struct-Act Relat 19(3):227–236. https://doi.org/10.1002/1521-3838(200006)19:3<227::AID-QSAR227>3.0

Cronin MT, Dearden JC (1995) QSAR in toxicology. 2. Prediction of acute mammalian toxicity and interspecies correlations. Mol Inf 14(2):117–120. https://doi.org/10.1002/qsar.19950140202

Cronin MT, Jaworska JS, Walker JD, Comber MH, Watts CD, Worth AP (2003) Use of QSARs in international decision-making frameworks to predict health effects of chemical substances. Environ Health Perspect 111(10):1391. https://doi.org/10.1289/ehp.5760

Damborsky J, Schultz TW (1997) Comparison of the QSAR models for toxicity and biodegradability of anilines and phenols. Chemosphere 34(2):429–446. https://doi.org/10.1016/S0895-3988(08)60028-8

de Bruijn J, Hermens J (1993) Inhibition of acetylcholinesterase and acute toxicity of organophosphorous compounds to fish: a preliminary structure-activity analysis. Aquat Toxicol 24 (3–4):257–274. https://doi.org/10.1016/0166-445X(93)90075-C

Dearden JC (2002) Prediction of environmental toxicity and fate using quantitative structure-activity relationships (QSARs). J Braz Chem Soc 13(6):754–762. https://doi.org/10.1590/S0103-50532002000600005

Doke SK, Dhawale SC (2015) Alternatives to animal testing: a review. Saudi Pharm J 23 (3):223–229. https://doi.org/10.1016/j.jsps.2013.11.002

Eriksson L, Hermens JL (1995) A multivariate approach to quantitative structure-activity and structure-property relationships. In: Chemometrics in environmental chemistry-applications. Springer, Berlin/Heidelberg, pp 135–168

Eriksson L, Jonsson J, Tysklind M (1995) Multivariate QSBR modeling of biodehalogenation half-lives of halogenated aliphatic hydrocarbons. Environ Toxicol Chem 14(2):209–217. https://doi.org/10.1002/etc.5620140205

Eriksson L, Jaworska J, Worth AP, Cronin MT, McDowell RM, Gramatica P (2003) Methods for reliability and uncertainty assessment and for applicability evaluations of classification-and regression-based QSARs. Environ Health Perspect 111(10):1361. https://doi.org/10.1289/ehp.5758

Ghamali M, Chtita S, Ousaa A, Elidrissi B, Bouachrine M, Lakhlifi T (2017) QSAR analysis of the toxicity of phenols and thiophenols using MLR and ANN. J Taibah Univ Sci 11(1):1–10. https://doi.org/10.1016/j.jtusci.2016.03.002

Ghorbanzadeh M, Zhang J, Andersson PL (2016) Binary classification model to predict developmental toxicity of industrial chemicals in zebrafish. J Chemom 30(6):298–307. https://doi.org/10.1002/cem.2791

Gomba VK, Enslein K, Blake BW (1995) Assessment of developmental toxicity potential of chemicals by quantitative structure-toxicity relationship models. Chemosphere 31 (1):2499–2510. https://doi.org/10.1016/0045-6535(95)00119-S

GonellaDiaza R, Manganelli S, Esposito A, Roncaglioni A, Manganaro A, Benfenati E (2015) Comparison of in silico tools for evaluating rat oral acute toxicity. SAR QSAR Environ Res 26 (1):1–27. https://doi.org/10.1080/1062936X.2014.977819

Gopi Mohan C, Gandhi T, Garg D, Shinde R (2007) Computer-assisted methods in chemical toxicity prediction. Mini-Rev Med Chem 7(5):499–507. https://doi.org/10.2174/138955707780619554

Halling-Sørensen B, Nielsen SN, Lanzky PF, Ingerslev F, Lützhøft HH, Jørgensen SE (1998) Occurrence, fate and effects of pharmaceutical substances in the environment – a review. Chemosphere 36(2):357–393. https://doi.org/10.1016/S0045-6535(97)00354-8

Hisaki T, Née Kaneko MA, Yamaguchi M, Sasa H, Kouzuki H (2015) Development of QSAR models using artificial neural network analysis for risk assessment of repeated-dose, reproductive, and developmental toxicities of cosmetic ingredients. J Toxicol Sci 40(2):163–180. https://doi.org/10.2131/jts.40.163

Hodges G, Roberts DW, Marshall SJ, Dearden JC (2006) The aquatic toxicity of anionic surfactants to *Daphnia magna* – a comparative QSAR study of linear alkylbenzenesulphonates and ester sulphonates. Chemosphere 63(9):1443–1450. https://doi.org/10.1016/j.chemosphere.2005.10.001

Hulzebos E, Walker JD, Gerner I, Schlegel K (2005) Use of structural alerts to develop rules for identifying chemical substances with skin irritation or skin corrosion potential. QSAR Comb Sci 24(3):332–342. https://doi.org/10.1002/qsar.200430905

Kolossov E, Stanforth R (2007) The quality of QSAR models: problems and solutions. SAR QSAR Environ Res 18(1–2):89–100. https://doi.org/10.1080/10629360601053984

Kruhlak NL, Contrera JF, Benz RD, Matthews EJ (2007) Progress in QSAR toxicity screening of pharmaceutical impurities and other FDA regulated products. Adv Drug Deliv Rev 59 (1):43–55. https://doi.org/10.1016/j.addr.2006.10.008

Lapenna S, Gatnik MF, Worth AP (2010) Review of QSA R models and software tools for predicting acute and chronic systemic toxicity. Publications Office of the European Union, Luxembourg. https://doi.org/10.2788/60766

Lee S, Kang YM, Park H, Dong MS, Shin JM, No KT (2013) Human nephrotoxicity prediction models for three types of kidney injury based on data sets of pharmacological compounds and their metabolites. Chem Res Toxicol 26(11):1652–1659. https://doi.org/10.1021/tx400249t

Li X, Chen J, Zhang L, Qiao X, Huang L (2006) The fragment constant method for predicting octanol–air partition coefficients of persistent organic pollutants at different temperatures. J Phys Chem Ref Data 35(3):1365–1384. https://doi.org/10.1063/1.2203356

Li J, Lu Y, Shi Y, Wang T, Wang G, Luo W, Yan F (2011) Environmental pollution by persistent toxic substances and health risk in an industrial area of China. J Environ Sci 23(8):1359–1367. https://doi.org/10.1016/S1001-0742(10)60554-2

Liao C, Sitzmann M, Pugliese A, Nicklaus MC (2011) Software and resources for computational medicinal chemistry. Future Med Chem 3(8):1057–1085. https://doi.org/10.4155/fmc.11.63

Lin Z, Zhong P, Yin K, Wang L, Yu H (2003) Quantification of joint effect for hydrogen bond and development of QSARs for predicting mixture toxicity. Chemosphere 52(7):1199–1208. https://doi.org/10.1016/S0045-6535(03)00329-1

Liu CN, Feng CJ (2012) Prediction of the aerobic biodegradability of chlorinated aromatic compounds using QSBR. J Petrochem Univ 5:002. https://doi.org/10.3969/j.issn.1006-396X.2012.05.002

Lu GH, Wang C, Yuan X, Zhao YH (2003) QSBR study of substituted phenols and benzoic acids. J Environ Sci 15(1):88–91

Luo J, Hu J, Wei X, Fu L, Li L (2015) Dehalogenation of persistent halogenated organic compounds: a review of computational studies and quantitative structure–property relationships. Chemosphere 131:17–33. https://doi.org/10.1016/j.chemosphere.2015.02.013

Madan AK, Bajaj S, Dureja H (2013) Classification models for safe drug molecules. In: Reisfeld B, Mayeno AN (eds) Computational toxicology, vol 930. Humana Press, New York, pp 99–124

Nendza M, Wenzel A (2006) Discriminating toxicant classes by mode of action-1.(Eco) toxicity profiles. Environ Sci Pollut Res 13(3):192–203. https://doi.org/10.1065/espr2006.01.013

Nendza M, Dittrich B, Wenzel A, Klein W (1991) Predictive QSAR models for estimating ecotoxic hazard of plant-protecting agents: target and non-target toxicity. Sci Total Environ 109:527–535. https://doi.org/10.1016/0048-9697(91)90206-T

Nendza M, Müller M, Wenzel A (2017) Classification of baseline toxicants for QSAR predictions to replace fish acute toxicity studies. Environ Sci Proc Imp 19(3):429–437. https://doi.org/10.1039/c6em00600k

Netzeva TI, Dearden JC, Edwards R, Worgan AD, Cronin MT (2004) QSAR analysis of the toxicity of aromatic Compounds to *Chlorella vulgaris* in a novel short-term assay. J Chem Inf Comput Sci 44(1):258–265. https://doi.org/10.1021/ci034195g

Neuwoehner J, Zilberman T, Fenner K, Escher BI (2010) QSAR-analysis and mixture toxicity as diagnostic tools: influence of degradation on the toxicity and mode of action of diuron in algae and daphnids. Aquat Toxicol 97(1):58–67. https://doi.org/10.1016/j.aquatox.2009.12.005

Okey RW, Stensel HD (1993) A QSBR development procedure for aromatic xenobiotic degradation by unacclimated bacteria. Water Environ Res 65(6):772–780. https://doi.org/10.2175/WER.65.6.10

Pavan M, Netzeva TI, Worth AP (2008) Review of literature-based quantitative structure–activity relationship models for bioconcentration. QSAR Comb Sci 27(1):21–31. https://doi.org/10.1002/qsar.200710102

Perocco P, Bolognesi S, Alberghini W (1983) Toxic activity of seventeen industrial solvents and halogenated compounds on human lymphocytes cultured in vitro. Toxicol Lett 16(1):69–75. https://doi.org/10.1016/0378-4274(83)90012-7

Philipp B, Hoff M, Germa F, Schink B, Beimborn D, Mersch-Sundermann V (2007) Biochemical interpretation of quantitative structure-activity relationships (QSAR) for biodegradation of N-heterocycles: a complementary approach to predict biodegradability. Environ Sci Technol 41(4):1390–1398. https://doi.org/10.1021/es061505d

Pirhadi S, Sunseri J, Koes DR (2016) Open source molecular modeling. J Mol Graph Model 69:127–143. https://doi.org/10.1016/j.jmgm.2016.07.008

Pölloth C, Mangelsdof I (1997) Commentary on the application of (Q) SAR to the toxicological evaluation of existing chemicals. Chemosphere 35(11):2525–2542. https://doi.org/10.1016/S0045-6535(97)00323-8

Raies AB, Bajic VB (2016) In silico toxicology: computational methods for the prediction of chemical toxicity. Wiley Interdiscip Rev: Comput Mol Sci 6(2):147–172.

Raunio H (2011) In silico toxicology–non-testing methods. Front Pharmacol 2:33. https://doi.org/10.3389/fphar.2011.00033

Roberts DW (1991) QSAR issues in aquatic toxicity of surfactants. Sci Total Environ 109:557–568. https://doi.org/10.1016/0048-9697(91)90209-W

Satpathy R, Konkimalla VB, Ratha J (2015a) In silico based prediction and correlation of dehalogenase enzyme activity of some haloalkane compounds by quantitative structure biodegradation relationship (QSBR) analysis. Curr Environ Eng 2(2):122–126. https://doi.org/10.2174/2212717802021601221922808

Satpathy R, Konkimalla VB, Ratha J (2015b) Application of bioinformatics tools and databases in microbial dehalogenation research: a review. Appl Biochem Microbiol 51(1):11–20. https://doi.org/10.1134/S0003683815010147

Settivari RS, Ball N, Murphy L, Rasoulpour R, Boverhof DR, Carney EW (2015) Predicting the future: opportunities and challenges for the chemical industry to apply 21st-century toxicity testing. J Am Assoc Lab Anim Sci 54(2):214–223

Seward JR, Schultz TW (1999) QSAR analyses of the toxicity of aliphatic carboxylic acids and salts to Tetrahymenapyriformis. SAR QSAR Environ Res 10(6):557–567. https://doi.org/10.1080/10629369908033224

Sussman NB, Arena VC, Yu S, Mazumdar S, Thampatty BP (2003) Decision tree SAR models for developmental toxicity based on an FDA/TERIS database. SAR QSAR Environ Res 14 (2):83–96. https://doi.org/10.1080/1062936031000073126

Tropsha A, Gramatica P, Gombar VK (2003) The importance of being earnest: validation is the absolute essential for successful application and interpretation of QSPR models. QSAR Comb Sci 22(1):69–77. https://doi.org/10.1002/qsar.200390007

Tunkel J, Mayo K, Austin C, Hickerson A, Howard P (2005) Practical considerations on the use of predictive models for regulatory purposes. Environ Sci Technol 39(7):2188–2199. https://doi.org/10.1021/es049220t

Vaal M, van der Wal JT, Hoekstra J, Hermens J (1997) Variation in the sensitivity of aquatic species in relation to the classification of environmental pollutants. Chemosphere 35 (6):1311–1327. https://doi.org/10.1016/S0045-6535(97)00167-7

Verhaar HJ, Van Leeuwen CJ, Hermens JL (1992) Classifying environmental pollutants. Chemosphere 25(4):471–491.

von der Ohe PC, Kühne R, Ebert RU, Altenburger R, Liess M, Schüürmann G (2005) Structural alerts a new classification model to discriminate excess toxicity from narcotic effect levels of organic compounds in the acute daphnid assay. Chem Res Toxicol 18(3):536–555. https://doi.org/10.1021/tx0497954

Wehrens R, Putter H, Buydens LMC (2000) The bootstrap: a tutorial. ChemomIntell Lab Syst 54:35–52. https://doi.org/10.1080/02701367.1997.10608865

Wong DC, Dorn PB, Chai EY (1997) Acute toxicity and structure-activity relationships of nine alcohol ethoxylate surfactants to fathead minnow and Daphnia magna. Environ Toxicol Chem 16(9):1970–1976. https://doi.org/10.1002/etc.5620160929

Worth AP, Lapenna S, Serafimova R (2013) QSAR and metabolic assessment tools in the assessment of genotoxicity. Comput Toxicol II:125–162. https://doi.org/10.1007/978-1-62703-059-5_6

Yang H, Jiang Z, Shi S (2006) Aromatic compounds biodegradation under anaerobic conditions and their QSBR models. Sci Total Environ 358(1):265–276. https://doi.org/10.1016/j.scitotenv.2005.04.004

Yin LI, XI DL (2007) Quantitative structure-activity relationship study on the biodegradation of acid dyestuffs. J Environ Sci 19(7):800–804. https://doi.org/10.1016/S1001-0742(07)60134-X

Zhang S, Golbraikh A, Tropsha A (2006) Development of quantitative structure-binding affinity relationship models based on novel geometrical chemical descriptors of the protein-ligand interfaces. J Med Chem 49(9):2713–2724. https://doi.org/10.1021/jm050260x

Index

K M Gothandam et al. (eds.), *Nanotechnology, Food Security and Water Treatment*, Environmental Chemistry for a Sustainable World 11,
https://doi.org/10.1007/978-3-319-70166-0

FSC
www.fsc.org
MIX
Papier | Fördert gute Waldnutzung
FSC® C083411

Kolibri 360 GmbH
Ferdinand-Jühlke-Straße 7
99095 Erfurt, Deutschland
produktsicherheit@kolibri360.de